HUMIC SUBSTANCES

Molecular Details and Applications in Land and Water Conservation

HUMIC SUBSTANCES

Molecular Details and Applications in Land and Water Conservation

Edited by

Elham A. Ghabbour

Northeastern University, Boston, USA and
Soil, Water and Environmental Research Institute, Alexandria, Egypt

Geoffrey Davies

Northeastern University, Boston, USA

Based on the proceedings of the 7th Humic Substances Seminar held on 17-19 March 2004 at Northeastern University, Boston, Massachusetts, USA

Taylor and Francis, Inc.
New York

The front cover illustration is taken from the contribution by Michael U. Kumke and Sascha Eidner, Scheme 9.1, p. 132.

Denise T. Schanck, Vice President
Robert L. Rogers, Senior Editor
Summers Scholl, Assistant Editor
Savita Poornam, Marketing Manager

Published in 2005 by
Taylor & Francis
270 Madison Avenue
New York, NY 10016
www.taylorandfrancis.com

Published in Great Britain by
Taylor & Francis
2 Park Square
Milton Park, Abingdon
Oxon, OX14 4RN
www.taylorandfrancis.co.uk

Printed in the United States of America on acid-free paper.

10 9 8 7 6 5 4 3 2 1

Library of Congress Cataloging-in-Publication Data

Humic substances : molecular details and applications in land and water conservation / edited by Elham A. Ghabbour, Geoffrey Davies.-- 1st ed.
p. cm.
Includes index.
ISBN 1-59169-031-5
1. Humus. 2. Soils--Humic acid content 3. Soil chemistry. 4. Water chemistry. I. Ghabbour, Elham A. II. Davies, Geoffrey, 1942-
S592.8.H83 2005
631.4'17--dc22

2004027414

Contents

Preface

Humic substances are major regulators of plants, sediments, soils and water, and, as such, they are the lynchpins of conserving our soil and water resources. This is why natural organic matter (NOM) and its humic components (the relatively long-lived humic and fulvic acids, and humin) continue to attract well-justified, worldwide attention.

This book is a companion to "Humic Substances: Nature's Most Versatile Materials," which contains selected papers from Humic Substances Seminar VI and the Twentieth Anniversary Biennial Conference of the International Humic Substances Society held at Northeastern University, Boston, USA in July 2002. Both books demonstrate the impressive progress being made in understanding and utilizing humic substances' remarkable properties.

Humic Substances Seminar VII was convened at Northeastern in March 2004. It was dedicated with thanks for the work of Professor Fritz Hartmann Frimmel of the University of Karlsruhe, Germany, and Professor Russell Christman of the University of North Carolina at Chapel Hill was the Honorary Chair. The papers presented at Seminar VII and the chapters of this book represent the best current research on humic substances from around the world. Each paper adds significantly to our understanding of HSs sources, structures and properties. The chapters follow a natural progression of humic substances separation, characterization, chemical reactivity, metal binding and transport, as well as their role in stabilizing soils against erosion. The results direct us to fight drought and pollution with the aid of humic substances.

SOURCES AND CHARACTERIZATION

In Chapter 1, Fritz Frimmel and colleagues address a fundamental question about the huge interaction zone between human beings and Nature: Is the organic matter in wastewater effluent (WWE) the same as the NOM of a lake or river into which the effluent is being discharged? With a combination of multistage ultrafiltration, size-exclusion chromatography and NMR (including dipolar dephasing), Frimmel et al. show that 1) HAs and FAs from a protected brown water lake are more aromatic and much less biodegradable than HAs and FAs isolated from WWE; 2) The brown water lake humics are mainly derived from terrestrial lignocelluloses and tannins, whereas those from WWE are mainly derived from amino acids and sugars that originate in bacteria; 3) the differences can only be recognized if the NOM from each source is fractionated. The work from Karlsruhe demonstrates that discharging WWE into natural waters can change the energy balance and the whole character of an ecosystem.

Size-exclusion chromatography (SEC) is popular for humic substances fractionation. Mass spectrometry is still in its infancy for analysis of whole soils or unfractionated humic samples. To serve as an alternative approach, much effort has been made to calibrate SEC columns for molar mass determinations. Mass standards have included globular proteins and polystyrene sulfonates (PSS), but the solid phase affects the properties of the analytes and standards differently. Remove the solid phase of chromatography and you have field flow fractionation (FFF). In Chapter 2, Manfred Wolf and colleagues show with asymmetrical FFF that the size to mass ratios of polyacrylic acids, other polycarboxylic acids and fulvic acids are similar and larger than for PSS. The polystyrene sulfonates are available in a wide range of molar masses, often with narrow dispersion. However, Chapter 2 concludes that PAAs and similar standards give more accurate molar mass data for humics than obtained with PSS, as confirmed by comparison of molar masses measured by SEC calibrated with PAA with those from mass spectrometry.

Black carbon is widespread hydrogen-depleted organic material resulting from incomplete fuel combustion and natural fires. Its influence on the sorptive properties of soils and sediments is being increasingly recognized, but quantitative analytical methods for black carbon measurements are lacking, as also is the case for humins, the humic fraction that is insoluble at all pH values. In Chapter 3, Guangwei Ding and James Rice use long range recoupled C-H dipolar dephasing direct polarization (DP) with total sideband suppression (CP/T_1-TOSS) nuclear magnetic resonance spectroscopy to measure the black carbon content of the humin isolated from the International Humic Substances Society leonardite reference material. This technique, coupled with direct polarization (DP), is well suited for identifying black carbon in natural organic matter. The apparent black carbon content of the IHSS leonardite reference humin and an IHSS reference peat (including a lipid-depleted sample) is about 20% of the total aromatic carbon content. In NMR experiments, the aromatic carbon in humin decays more slowly than for lignin, indicating that it is less protonated than the aromatic carbon in lignin.

Nuclear waste storage is a major problem in nuclear power generation. Clay-kerogen composites may release dissolved humic and fulvic acids and increase metal mobility by forming soluble metal complexes. For this reason, humic release from metal repository clays is a serious concern, especially for radionuclides. In Chapter 4, Thorsten Schäfer and colleagues have used synchrotron based C-, K-, Ca-, O- and Fe-edge X-ray absorption near-edge spectroscopy (XANES), scanning transmission X-ray microscopy and FT-IR microscopy to identify the spatial distribution of clay organic matter with different functional groups. The results show that Oxfordian argillite and Opalinus clay are sources of HSs that originate in the smectic-rich and illite/mixed layer mineral regions of the repository and probably are released from surface layers that differ in hydrophobicity and other features. The nature of HSs monolayers and bilayers on mineral surfaces certainly is of high current interest. Binding of curium(III) by the released materials resembles its binding by HAs and FAs, as indicated by time-resolved laser fluorescence spectroscopy. The methods described in this paper can be used to identify the sources of HAs and FAs in any undisturbed mineral sample with nanoscale resolution. Radionuclide leakage from nuclear waste sites is a long-term hazard that requires constant attention.

SORPTION AND REACTIVITY

The debate over macromolecular or supramolecular structures for humic and fulvic acids continues. Increasing acceptance of the supramolecular concept can be found elsewhere in this volume and in recent literature. In Chapter 5, Cooper Langford and Joseph Melton explore the view that humics *are* supramolecules and that, as such, they form a *combinatorial library.* In a *dynamic* combinatorial library, the connections between building blocks (H-bonds, hydrophobic interactions, charge transfer) are reversible and in constant flux. Any perturbation (for example, metal binding by one of the components that lowers its "free" concentration) causes the system to adjust in much the same way as a buffer adjusts on addition of a strong acid or base. In this sense, the behavior of the system is *co-operative.* New data from light scattering and water proton NMR relaxation times in the presence of Laurentian humic acid and copper(II) are interpreted in terms of such equilibria. The factors include the proportions of "free" and copper(II)-bound water, as addressed in Chapter 12 by Avanadula and Ryan. Effects of aggregation and conformation on metal binding are natural consequences of a dynamic combinatorial library of humic constituent molecules.

Chapter 6 by Lourdes Pancheco and Josef Havel has two main thrusts. The first discounts MALDI-TOF mass spectrometry as an efficient way to measure molar masses of humics. Instead, the solid humic acid forms its own matrix in laser desorption ionization. The results support the proof that ionizable fractions of humic samples are molecular families, as indicated in Chapter 6 and recent work by Stenson, Marshall and Cooper (*Anal. Chem.*, 2003; 75:1275). The fact is that mass spectra of humic samples of different origin have common features. The molar masses in both studies are all less than 1 kDa, and Chapter 6 indicates that the detected ions are monocations. Molecular formulas for the principal ions are derived. Chapter 6 also demonstrates by capillary zone electrophoresis the interactions of a coal-derived humic acid with ions, including nitrate, $Fe(CN)_6^{3-}$ and $Fe(CN)_6^{4-}$, with $Fe(CN)_6^{4-}$ produced from $Fe(CN)_6^{3-}$ on incubation with the HA. Complexation of Na^+ ions with 18-crown-6 ether results in an unexpectedly stronger interaction with Chemapex HA.

The photochemistry and photophysics of humic acids in surface waters is an important subject of continuing interest. In Chapter 7, Andrea Paul and co-workers in Berlin, Germany have investigated the quenching of photogenerated singlet oxygen, $^1O_2$ (a strong oxidant), by humic substances from a wide variety of sources. Singlet oxygen luminescence at 1269 nm is a measure of the photo-reactivity of humic substances. Natural organic matter in surface waters is molecularly transformed in the summer sun, and the result is evident in the different quantum yields of $^1O_2$ in the presence of NOM from the same site but at different seasons. Under the same conditions, humic acids of aquatic and terrestrial origin react more rapidly ($k_Q \approx 35$ - 40 L g^{-1} s^{-1}) with $^1O_2$ than do NOM and fulvic acids (with rate constants up to 12 L g^{-1} s^{-1}). The quenching rate constants are correlated with HSs aromaticity. Importantly, it is predicted that the photo-reactivity of receiving waters increases on introduction of HSs from terrestrial sources, for example by soil erosion, but for how long this increased activity persists is still unknown.

Treatment of water containing NOM with chlorine produces a mixture of halogenated and other disinfection byproducts (DBP). Of particular concern is trichloromethane (THM), which causes cancer and other health problems in humans. Chapter 8, by David Reckhow and colleagues, reports investigations of the fraction of dissolved organic carbon (DOC) that reacts with chlorine to form THM. The data are compared with the overall organic carbon levels (TOC) and examined for trends that suggest THM precursor origins and processes that affect their concentrations. It is apparent from an extensive data analysis that the THM precursor content of natural waters can vary with the source type, location and time of year. For surface waters there is a consistent trend of increasing Sp-THMFP with decreasing TOC. Here, THMFP is the THM formation potential of the source water and Sp-THMFP is THMFP divided by TOC for each data point. Sp-THMFP is an intensive property of NOM. Evidence from laboratory leaching and degradation studies suggests that increasing Sp-THMFP with decreasing TOC is due to selective biodegradation of freshly solubilized NOM. Biodegradation actually results in a substantial increase in Sp-THMFP, which is attributed to the relative buildup of highly reactive tannins.

Michael Kumke and Sascha Eidner end this second section with Chapter 9, which describes fluorescence properties and fluorescence quenching of substituted benzoic acids by Tb^{3+}. The properties of these models resemble the fluorescence behavior of HSs. However, while Tb^{3+} is a weak quencher of substituted benzoic acids fluorescence, it very effectively quenches HSs fluorescence, perhaps because of the need for conformational reorientation in HSs fluorescence. The luminescence of bound Tb^{3+} is sensitized on excitation of the model compounds and HSs, but it depends on the HSs origin. This phenomenon may characterize structural differences of HSs in solution and the relative abundance of chromophores with triplet state energies suitable for energy transfer. Inter-lanthanide energy transfer between HSs-bound Tb^{3+} and Nd^{3+} is used to determine the distance between the bound metal sites. This distance R depends on the HSs origin and reflects different HSs structures. The distance R decreases with increasing Nd^{3+} concentration to values in the range 0.85 – 1.2 nm. At $[Nd^{3+}] = 20$ μM with Suwannee River fulvic acid, the distance R increases with decreasing pH. It seems that inter-lanthanide energy transfer measurements can be used to monitor conformational changes in HSs as a function of pH, ionic strength, metal binding and solute sorption. This breakthrough in HSs research promises the same molecular detail as achieved for intra-protein electron transfer and protein folding.

METAL BINDING AND MOBILITY: DATA AND THEORY

This section covers HSs-metal binding and mobilization, which are important attributes of HSs in natural systems. In Chapter 10, Gunnar Buckau covers a lot of ground in addressing the fundamental problem of describing HSs-metal binding: what is the nature of the humic acid ligand and how can we express its concentraction? Attention is focused on Am(III), Cm(III) and U(VI) binding by humic acids. Analysis of a general system model shows that the effective humate ligand concentration can be described by the number concentration of ionized humic acid molecules. This implies that individual HA molecules are inactivated for metal binding by protonation of their proton exchanging groups, and that an ionized humic acid molecule becomes deactivated by the binding of one metal ion. The study provides a physico-chemical explanation for the frequently used simple approach of metal ion-humic acid complexation described by an experimentally determined effective humate ligand concentration (or binding or complexation capacity) and one binding constant. The unit slope of the plot for all the Am(III)/Cm(III) data in Figure 10.6 indicates a single binding constant at pH 6.

Interactions of HA with metal ions can proceed in the liquid phase or at the liquid/solid interface. Chapter 11 by Martina Klučáková and Miloslav Pekař describes copper(II) interactions with humic acid colloids, sols and water-swollen gels. Carboxylic acid and phenol groups bind metals and H^+ ions are liberated by metal complexation. The acidity and conductivity should increase if the system attains equilibrium, but the kinetic pH-curves for colloids show minima (diagnostic of a reaction intermediate) and the conductivity also changes with time at constant pH. However, pH curves obtained for humic acid sols and gels correspond with conductivity measurements. Because of the high affinity of copper(II) for HAs, it presumably occupies most of the HA binding sites and the appropriate amount of H^+ ions is liberated. The binding sites that cannot connect with Cu^{2+} ion are protonated. The interaction mechanism is strongly influenced by the physical form of a humic system. In sols, the important factor is the mobility of humic particles, whereas metal ion diffusion plays a significant role in HA solids and gels.

Urine and other human samples reveal hundreds of ^1H NMR peaks once the overwhelming water peak has been suppressed with a presaturation technique such as WATERGATE. In Chapter 12, Raja Avanadula and David Ryan describe the use of this technique to measure uncomplexed metals in an HSs-metal equilibrium. Copper(II) binds up to 6 water molecules. Protons of partially dissociated $Cu(H_2O)_5OH^+$ have NMR spectra that differ from H-O-H protons. Uncomplexed copper(II) concentrations were measured indirectly by analyzing the protons of water molecules attached to the metal. Results were obtained at metal ion concentrations ranging from 0 to 1 mM. The amount of metal ion was measured in the presence of a known complexing agent, such as EDTA, at pH ranging from 4-7. The results were used to quantitate the free metal in the presence of HSs in order to fully describe the HSs complexation equilibria.

The paper by Norman Lee and colleagues reports progress in using quantitative ^{27}Al NMR spectroscopy to study Al binding by a soil-derived fulvic acid. Their study covers aluminum concentrations ranging from 27 to 135 mg/L and soil fulvic acid concentrations from 1000 to 7500 mg/L at pH 3.0, 4.0 and 5.0. Aluminum iodide dissolved in anhydrous acetonitrile in an NMR insert is used as a standard. The stability constants (log K) obtained range from 2.5 to 3.6 for K_1 and 2.4 to 2.8 for K_2, which are similar to results obtained with Scatchard plots (2.4 to 3.5 for K_1 and 2.7 to 3.6 for K_2) and Langmuir regression (2.6 to 3.7 for K_1 and 3.5 to 4.0 for K_2). Oxalic acid was used as a model compound and the experimental NMR results were found to agree within 10 % of the theoretical aluminum-oxalate speciation predicted using the geochemical modeling program MINTEQ. An interesting conclusion of this paper is that the metal-binding affinity of one site is affected by the affinity of other sites. In other words, Al binding by the soil fulvic acid used appears to be *co-operative.*

Chapter 14 by Miloslav Pekař and co-workers is aimed at formulating organic-mineral preparations for soil remediation using lignite-derived humic acids (HAs). The affinities of soil and lignitic HAs for copper(II) and cadmium(II) were compared at room temperature. Sequential metal desorption with ammonium nitrate and EDTA was studied. The results are similar, and lignite HA has the higher affinity for copper(II). Regarding desorption, at low cadmium concentrations HAs prepared from fluvisol and chernozem showed the best affinity, and lignite and luvisol materials were the best at high initial metal concentrations. Superior affinity of lignite HAs for metals was observed. Although the name "humic acids" denotes some special class of substances, there are differences in their properties. HAs bind heavy metals in general but the nature and extent of such interactions depend on the HA origin.

Soil-derived humic acids often bind trace metals by ion exchange and ligand binding with oxygen, nitrogen, and sulfur donor atoms. The paper by Anahita Dua and co-workers uses capillary electrophoresis-inductively coupled plasma-mass spectrometry (CE-ICP-MS) to determine trace metals bound to soil HAs that were successively extracted from north and south slope soils of Mt. Toby in western Massachusetts, USA. The O/R ratios of DRIFTS spectra indicate decreased presence of -COOH, phenol and alcohol groups with successive extractions. Addition of tricine to the CE buffer reduced HA-capillary wall interactions without increasing Joule heating. Reproducible ion electropherograms with well-resolved elemental peaks were obtained for HA samples preconcentrated by flow field-flow fractionation (FlFFF). Tightly bound trace metals (Pb, Cu, Zn, Co, Fe, Al) were observed by CE-ICP-MS in all extracted humic acids. As anticipated, the HA fractions obtained by successive extraction showed different metal-HA complexing properties. The extent of trace metal binding by these humic acids declined with diminishing oxygen-rich moieties in the HA structure.

SOIL AGGREGATION

Soil organic matter (SOM) is important to soil function and productivity. Increased tillage intensity promotes the loss of SOM and the susceptibility of a soil to erosion. The final paper demonstrates that long-term field experiments on adjacent land provide a great opportunity to compare the results of different soil management strategies. Joseph Pikul and colleagues have determined the effect of tillage on components of SOM and the stability of soil aggregates. Measurements were made on adjacent farms in eastern South Dakota. One farm used no tillage (NT) and the other used chisel tillage (CT). The soil is a clay loam. A rotary sieve was used to separate 10 kg soil samples into 6 size groups. The water stability of aggregates from groups 3, 4 and 5 was measured by wet-sieving. Humic acid (HA) was extracted from each soil with 0.5 M NaOH. Immunoreactive total glomalin (IRTG) and soil aggregating basidiomycete fungi (BF) were measured with an enzyme-linked immunosorbent assay (ELISA). Soil C was greater under NT compared with CT in all aggregate groups 1-6. The basidiomycete population and IRTG also were greater under NT. There was a difference in the C:N ratio of HA between NT and CT. Fine particulate organic matter (POM) and the water stability of aggregates were greater under NT compared with CT in aggregate groups 3-5. Multiple regression reveals that fine POM, the C:N of HA, and BF account for 63% of the variability in water stability of dry soil aggregates. Organic cementing agents (humic materials or microbial exudates) are not uniformly distributed among aggregate groups. The important conclusion of this study is that improved soil aggregation under no tillage management is due to increased proportions of organic matter.

The contributions in this book are representative of the high standard of work presented by authors from many countries at the Humic Substances Seminars. We are confident that Nature's most versatile materials will one day be fully understood and appreciated for their ability to make an immeasurable contribution to conservation of our precious land and water.

ACKNOWLEDGEMENTS

We thank the authors for their enthusiasm and co-operation, the contributors to Humic Substances Seminar VII, the staff of Northeastern University who give so much to their success and the sponsors who make all this progress possible. Robert Rogers and his staff at Taylor & Francis ensure that our work as Editors is straightforward, pleasant and enjoyable. We sincerely thank them all.

Elham A. Ghabbour
Geoffrey Davies
Editors
Boston, Massachusetts
August, 2004

Contributors

Dula Amarasiriwardena	Professor, School of Natural Science, Hampshire College, Amherst, MA 01002, USA
Raja M Avanadula	Graduate Student, Department of Chemistry, University of Massachusetts Lowell, Lowell, MA 01854, USA
Gabriela Barancikova	Researcher, Soil Science and Conservation Research Institute, Regional laboratory, Reimanova 2, 080 01 Presov, Slovakia
Ramon Barnes	Director, University Research Institute for Analytical Chemistry, 85 North Whitney St., Amherst, MA 01002-1869, and Professor Emeritus, Department of Chemistry, University of Massachusetts, Amherst, MA 01003-9336, USA
Andreas Bauer	Doctor, Institut für Nukleare Entsorgung (INE), Forschungszentrum Karlsruhe, 76021 Karlsruhe, Germany
Darleen Bryan	Graduate Student, Department of Civil and Environmental Engineering, University of Massachusetts, Amherst, MA 01003, USA
Gunnar Buckau	Senior Scientist, Institut für Nukleare Entsorgung, Forschungszentrum Karlsruhe, P.O. Box 3640, 76021 Karlsruhe, Germany
TheCan Caesar	Microbiologist, USDA-ARS, Northern Plains Agricultural Research Laboratory, Agricultural Systems Research Unit, 1500 North Central Ave., Sidney, MT 59270, USA
Véronique Chanel	Engineer, Institute of Groundwater Ecology, GSF-National Research Center for Environment and Health, D-85764 Neuherberg, Germany
Francis Claret	Doctor, Marie Curie Fellowship, Institut für Nukleare Entsorgung (INE), Forschungszentrum Karlsruhe, 76021 Karlsruhe, Germany
John Danku	Doctoral Student, Department of Chemistry, University of Massachusetts, Amherst, MA 01003-9336, and University Research Institute for Analytical Chemistry, 85 North Whitney St., Amherst, MA 01002-1869, USA

Guangwei Ding Postdoctoral Research Associate, Department of Chemistry and Biochemistry, South Dakota State University, Brookings, SD 57007, USA

Anahita Dua Undergraduate Student, School of Natural Science, Hampshire College, Amherst, MA 01002, USA

Sascha Eidner Doctoral Student, Institute of Chemistry, University of Potsdam, Karl-Liebknecht-Str. 24-25, 14476 Potsdam-Golm, Germany

Michael Ellsbury Research Entomologist, USDA-Agricultural Research Service, Northern Grain Insect Research Laboratory, 2923 Medary Ave., Brookings, SD 57006, USA

Fritz H. Frimmel Professor, Universität Karlsruhe (TH), Engler-Bunte-Institut, Bereich Wasserchemie, Engler-Bunte-Ring 1, 76131 Karlsruhe, Germany

Steffen Hackbarth Postdoctoral Fellow, Institute of Physics, Humboldt University Berlin, Newtonstr. 5, D-12489 Berlin, Germany

Ma. de Lourdes Pacheco Hernandez PhD, Analytical Chemistry Department, Autonomous University of Hidalgo State, Carretera Pachuca-Tulancingo Km 4.5, CP 42076, Pachuca, Hidalgo, Mexico

Josef Havel Professor and Head, Department of Analytical Chemistry, Faculty of Science, Masaryk University, Kotlarska 2, 611 37 Brno, Czech Republic

Chris Jacobsen Professor, Department of Physics and Astronomy, State University of New York, SUNY at Stony Brook, New York 11794, USA

Jane M. F. Johnson Soil Scientist, USDA-ARS, North Central Soil Conservation Research Laboratory, 803 Iowa Ave., Morris, MN 56267, USA

Martina Klučáková Assistant Professor, Institute of Physical and Applied Chemistry, Faculty of Chemistry, Brno University of Technology, Purkyňova 118, 612 00 Brno, Czech Republic

Michael U. Kumke Senior Scientist, Institute of Chemistry, University of Potsdam, Karl-Liebknecht-Str. 24-25, 14476 Potsdam-Golm, Germany

Cooper H Langford Professor of Chemistry and Coordinator of the Science, Technology, and Society program, University of Calgary, Calgary, AB, T2N 1N4 Canada

Ulrich Lankes Universität Karlsruhe (TH), Engler-Bunte-Institut, Bereich Wasserchemie, Engler-Bunte-Ring 1, 76131 Karlsruhe, Germany

Norman C.Y. Lee Doctoral Student, Department of Chemistry, University of Massachusetts Lowell, Lowell, MA 01854, USA

Mirna Lerotic Doctoral Student, Department of Physics and Astronomy, State University of New York, SUNY at Stony Brook, New York 11973, USA

Hans-Dietrich Lüdemann Karl-Anselm-Strasse 6, 93051 Regensburg, Germany

Mikulas Madaras Researcher, Soil Science and Conservation Research Institute, Regional laboratory, Reimanova 2, 080 01 Presov, Slovakia

Gladys Makdissy Post-Doctoral Research Associate, Department of Civil and Environmental Engineering, University of Massachusetts, Amherst, MA 01003, USA

Jarmila Makovnikova Researcher, Soil Science and Conservation Research Institute, Regional laboratory, Mladeznicka 3, 974 04 Banska Bystrica, Slovakia

Joseph R. Melton Doctoral Student, University of Victoria, Victoria, BC V8W 2Y2, Canada

Margit B. Müller Universität Karlsruhe (TH), Engler-Bunte-Institut, Bereich Wasserchemie, Engler-Bunte-Ring 1, 76131 Karlsruhe, Germany

Andrea Paul Postdoctoral Fellow, Leibniz Institute of Freshwater Ecology and Inland Fisheries, Müggelseedamm 301, D-12561 Berlin, Germany

Miloslav Pekař Associate Professor, Institute of Physical and Applied Chemistry, Faculty of Chemistry, Brno University of Technology, Purkyňova 118, 612 00 Brno, Czech Republic

Joseph L. Pikul Jr. Soil Scientist, USDA-Agricultural Research Service, Northern Grain Insect Research Laboratory, 2923 Medary Ave., Brookings, SD 57006, USA

Thomas Rabung Doctor, Institut für Nukleare Entsorgung (INE), Forschungszentrum Karlsruhe, 76021 Karlsruhe, Germany

Geetha Rajesh Doctoral Student, Department of Chemistry, University of Massachusetts Lowell, Lowell, MA 01854, USA

David A. Reckhow Professor, Department of Civil and Environmental Engineering, University of Massachusetts, Amherst, MA 01003, USA

Paula L. Rees Assistant Professor, Department of Civil and Environmental Engineering, University of Massachusetts, Amherst, MA 01003, USA

James A. Rice Professor and Head, Department of Chemistry and Biochemistry, South Dakota State University, Brookings, SD 57007-0896, USA

Beate Röder Professor of Physics and Head of the Division of Photobiophysics, Institute of Physics, Humboldt University Berlin, Newtonstr. 5, D-12489 Berlin, Germany

David K. Ryan Professor, Department of Chemistry, University of Massachusetts Lowell, Lowell, MA 01854, USA

Thorsten Schäfer Doctor, Institut für Nukleare Entsorgung (INE), Forschungszentrum Karlsruhe, 76021 Karlsruhe, Germany

Christian E.W. Steinberg Professor of Freshwater Ecology, Humboldt University at Berlin, and Leibniz Institute of Freshwater Ecology and Inland Fisheries, Müggelseedamm 310, D-12587 Berlin, Germany

Manfred Wolf Senior Scientist, Institute of Groundwater Ecology, GSF National Research Center for Environment and Health, 85764 Neuherberg, Germany

Sara E. Wright Soil Scientist, USDA-ARS, Sustainable Agricultural Systems Laboratory, Building 001, 10300 Baltimore Ave., Beltsville, MD 20705, USA

Baoshan Xing Professor, Department of Plant and Soil Sciences, University of Massachusetts, Amherst, MA 01003, USA

Elke Zwirnmann Senior Scientist, Leibniz Institute of Freshwater Ecology and Inland Fisheries, Müggelseedamm 301, D-12561 Berlin, Germany

Part 1

SOURCES AND CHARACTERIZATION

Chapter 1

STRUCTURAL CHARACTERIZATION OF HUMIC SUBSTANCES FROM WASTE WATER

Fritz H. Frimmel,[1] Ulrich Lankes,[1] Hans-Dietrich Lüdemann[2] and Margit B. Müller[1]

[1]Universität Karlsruhe (TH), Engler-Bunte-Institut, Bereich Wasserchemie, Engler-Bunte-Ring 1, 76131 Karlsruhe, Germany
[2]Karl-Anselm-Strasse 6, 93051 Regensburg, Germany

1.1. INTRODUCTION

Water usage in industrialized countries leads to the production of large amounts of waste water (WW), for example, in Germany about 100 m^3/person·annum. Depending on its origin, WW contains organic waste from industry and smaller production units (16%), households (44%), and precipitation runoff (40%) [1]. Both biogenic and synthetic compounds contribute to the organic load of WW. Some of the organic matter (OM) can be used as a growth substrate by various microorganisms, whereas other organic compounds are of refractory nature [2]. After collection, WW normally is treated mechanically and microbiologically in large technical plants. Through the biological treatment, the water loses most of its readily biodegradable load. Other water constituents and microorganisms can adsorb onto the biologically active sludge, and are eliminated from the water phase with the solids formed. As a consequence, the waste water effluent (WWE) is relatively enriched in inorganic material and soluble organic compounds that are refractory from the biotechnical point of view.

The biodegradability of organic compounds is mainly a function of their chemical structure. Humic substances (HSs) are the refractory leftovers of natural organic compounds [3,4]. Besides this biochemical definition, HSs are operationally defined as the fraction of the OM that can be isolated using a sophisticated isolation procedure with adsorption/desorption on synthetic polymers (XAD-8 isolation [5-7]). With the XAD-8 procedure, weak hydrophobic acids, neutrals, and bases ("humic substances") are isolated from the OM. The HSs are further divided into humic acids (HAs) and fulvic acids (FAs) based on their solubility at different pH values. The fraction that does not adsorb onto the XAD adsorbent is termed non-humic substances (NHS). In natural waters, HSs make up the major part of the OM. In particular, this holds true for brown water lakes [3]. HSs from Lake Hohloh, a brown water lake in the Black Forest, Germany, are refractory. For waste water effluents it has not been investigated in detail whether the HSs fraction of the OM indeed is refractory and which chemical structures are present in the FA and HA fractions.

In this paper, emphasis is given to a comparative elucidation of the structural details of the OM of a WWE and a reference natural brown water (BW). Size exclusion chromato-graphy (SEC) was used to study the outer structure ("size, shape, and weight") of the molecules. Nuclear magnetic resonance (NMR) spectroscopy was used to study the inner ("chemical") structure of the molecules. These analytical methods are supplemented by multistage ultrafiltration (mst-UF) fractionation experiments to get information on both the outer and the inner structures.

SEC with ultraviolet light (UV) and dissolved organic carbon (DOC) detection has become a powerful tool for the characterization of dissolved organic matter (DOM) [8-10]. SEC separation of natural organic matter (NOM) is based on the differences in molecular size, shape, and molar mass of the sample components as well as on intermolecular interactions, and interactions of sample components with the stationary gel phase. Both polar (electrostatic, H-bridges) and non-polar (e. g., van der Waals) interactions can contribute to the separation mechanism [11-13]. Swift and Posner [14] assumed for OM eluting between the exclusion volume V_0 and the permeation volume V_p a separation that is mainly based on differences in the hydrodynamic radii of the molecules. Recently, fractionation of DOM by SEC and mst-UF has shown that special care must be taken when interpreting the results merely based on molecular size considerations [15].

^{13}C high-resolution solid state NMR is an excellent tool for structural assignment of HSs [16,17]. The ^{13}C NMR spectra are divided into chemical shift regions, which are assigned to structural groups. The relative amount of carbon that contributes to the signal intensity of the spectral parts is determined by integration. This NMR analysis of FAs and HAs of various BW samples has already been carried out in another study [18]. The integration yielded an average amount of nearly 40% of aromatic carbon, about 25% O-alkyl carbon, nearly 20% remaining aliphatics, around 15% of carboxyl functions and 3% of carbonyl groups. The values for FAs are slightly different from those for HAs.

Besides of this straightforward quantitative evaluation, the intensity patterns of the spectra contain additional structural information. Closer inspection of the spectra (chemical shifts and intensities of signal maxima, signal shapes) reveals detailed information about substitution patterns and the degree of branching of the organic carbon structures. With the help of these specific data, the precursors of the OM can be identified.

The aim of this work was to

(1) isolate, quantify, and fractionate the humic matter from the effluent of a waste water treatment plant,
(2) characterize the OM by size exclusion chromatography and NMR spectroscopy,
(3) compare the results with those obtained for brown water OM, and
(4) deduce the influence of the waste water effluents on the receiving waters.

1.2. MATERIALS AND METHODS

1.2.1. Samples

The municipal waste water treatment plant of Karlsruhe, a city in the south-west of Germany, was used for sampling. Karlsruhe has a population around 270 000, with about one quarter of employees working in production and three quarters in services [19]. Hence, the WW is mainly of domestic origin. Samples were taken from the effluent of the waste water treatment plant (WWE) and from the waste water (WW) at the treatment plant inflow. Sampling dates of WWE samples were in March 1995 and November 2001.

For comparison, samples from Lake Hohloh, a humic-rich brown water lake in the Northern Black Forest, Germany, were taken. The lake is situated in a protected pine forest area and is naturally preserved. The brown water (BW) has served as a source of standardized humic matter. The OM of this lake is of natural origin and has been investigated in previous studies [20]. Sampling dates of BW samples were in July 1998 and July 2001.

1.2.2. Sample Fractionation

The BW (sampling date July 1998) and WWE (sampling date March 1995) were fractionated according to the XAD-8 method [7] to isolate the fulvic acids (FAs), humic acids (HAs) and non-humic substances (NHS). A schematic overview of the isolation procedure is given in Figure 1.1.

The BW (sampling date July 2001) and WWE (sampling date November 2001) were also fractionated by multistage ultrafiltration (mst-UF) using regenerated cellulose membranes with nominal molar mass cut offs (MWCO) of 30, 10, 3, and 1 kDa. A detailed description of the fractionation procedure is given in Müller and Frimmel [15] and Müller et al. [21]. A fractionation scheme is shown in Figure 1.2.

1.2.3. Size Exclusion Chromatography with DOC Detection (SEC-DOC)

The molecular size distribution of the samples was investigated by size exclusion chromatography. The analytical SEC system was used as described earlier [8]. The column size was 250 mm x 20 mm, and the stationary phase was TSK HW 50 (S) (Tosoh Bioscience, Japan). A phosphate buffer with an ionic strength of 40 mmol/L and a pH value of 6.8 was used as the mobile phase. The flow rate was 1 mL/min. The injected sample volume was 2 mL and when by-passing the SEC column it was 0.5 mL.

1.2.4. ^{13}C High-Resolution Solid State NMR

^{13}C MAS NMR measurements were carried out on a Bruker Avance 400 MHz-spectrometer at a ^{13}C resonance frequency of 100.6 MHz. The ^{13}C CPMAS spectra were recorded on a Bruker MSL-100 system at a ^{13}C resonance frequency of 25.1 MHz. Magic angle spinning (MAS) was performed at 15 kHz (5 kHz for cross polarization (CP) mode). A single pulse experiment (MAS) was carried out for acquisition (90° ^{13}C pulse of 5 μs duration). For CP experiments, a contact time of 1 ms was used (90° ^{1}H pulse of 5 μs duration). Dipolar dephasing (DD) spectra were recorded after a DD time of 60 μs. During the DD time, a 180° ^{13}C pulse was inserted to minimize phase shifts. High power proton decoupling was applied during acquisition. To ensure full relaxation of all ^{13}C nuclei, the repetition time of the MAS experiments was set at 5 s. For the CP experiments, the repetition time was set at 0.5 s. The number of recorded scans ranged from about 10^4 to 10^5 scans per sample. Chemical shifts were calibrated externally to the carboxyl resonance of glycine (176.03 ppm) with respect to the tetramethylsilane (TMS) scale. The total integral (220 to 0 ppm) of each spectrum was set to 100% (calibration) unless otherwise indicated. Structural assignments of chemical shift regions according to Lambert and Lankes [22] are compiled in Table 1.1. Relative carbon contents of the different structural groups were calculated by determining the signal areas of individual chemical shift regions.

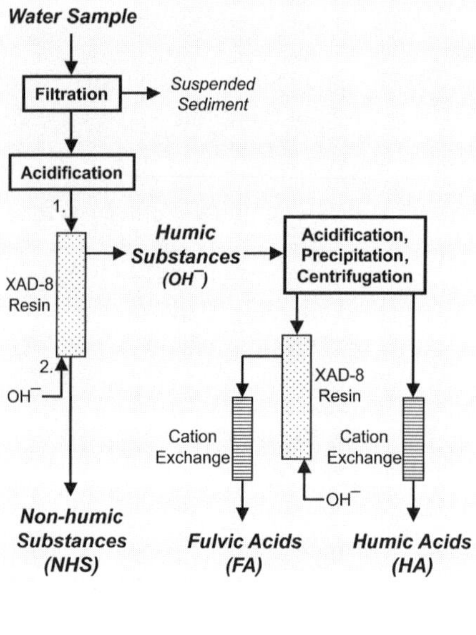

Figure 1.1 Scheme of the XAD-8 isolation procedure

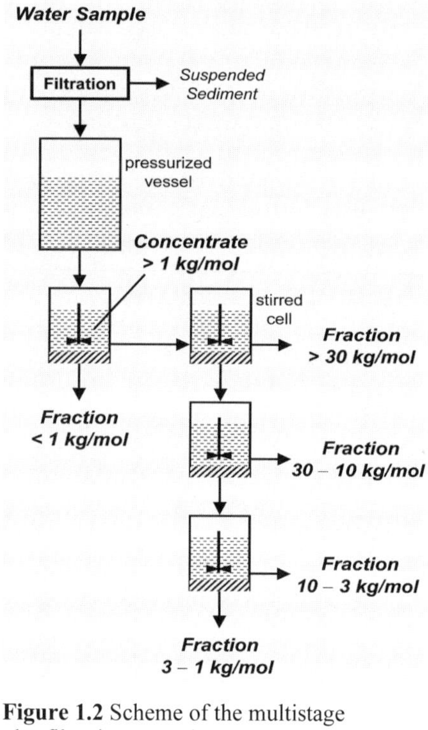

Figure 1.2 Scheme of the multistage ultrafiltration procedure

Table 1.1 Assignment of ^{13}C chemical shifts (TMS scale) to structural groups

Chemical Shift, ppm	Structural assignment
220 – 190	carbonyl carbon
190 – 160	carboxyl carbon: COOH, ester, amide
160 – 140	O-/N-substituted aromatic carbon
140 – 100	C-/unsubstituted aromatic carbon
100 – 50	O-/N-alkyl carbon, e. g., carbohydrates
50 – 0	aliphatic carbon

1.3. RESULTS

1.3.1. Basic Characterization and Mass Balances of Organic Matter

Basic Parameters. Representative values for the basic parameters of the municipal waste water and the effluent of the treatment plant are given in Table 1.2. The data for water from the brown water lake are shown for comparison. Elimination of the WW organic matter caused by physicochemical removal and biological degradation results in a decrease of the DOC concentration by 84%. The degree of elimination mainly depends on the composition of the biocenosis and physicochemical conditions (e. g., residence time, temperature, biomass concentration) of the activated sludge treatment step of the Karlsruhe WW treatment plant. The content of inorganic matter (salts), which is roughly reflected by the electrical conductivity values, in the WW and WWE is fairly high compared to the salt content in the BW. The specific UV absorbance (SUVA) as well as the specific visible light absorbance (SVisA) show an increase after the treatment procedure. The SUVA is a fast and simple measure of the unsaturated character of the samples [23,24]. The biodegradability of the WW OM is quite high compared to a lower one of the WWE OM. The NOM of the BW sample is rather refractory. The biodegradability is indicated by the BOD_5 value relative to the DOC concentration. BOD_5 is biochemical oxygen demand during 5 days of incubation [25]. A low ratio (BOD/DOC) indicates a highly refractory character.

Table 1.2 Representative characteristics of the original waters. WW: waste water, WWE: waste water effluent, BW: brown water

Parameter	WW	WWE	BW
DOC, mg/L	50	8	22
Electrical conductivity, μS/cm	1100	988	32
pH value	8.3	8.0	4.5
$SUVA_{254\ nm}{}^{a}$, L/mg·m	0.9	2.2	4.8
$SVisA_{436\ nm}{}^{b}$, L/mg·m	0.1	0.2	0.4
$BOD_5{}^{c}$, mg/L	300	10	3

[a]$SUVA_{254\ nm}$: specific UV absorbance at 254 nm; [b]$SVisA_{436\ nm}$: specific light absorbance at 436 nm; [c]BOD_5: biochemical oxygen demand during 5 days of incubation

The isolated humic material was mainly organic. Atomic ratios from elemental analysis data of the freeze dried samples are given in Table 1.3. The HSs of the WWE sample have significantly higher values for the H/C and N/C ratios.

Mass Balance of the DOC Fractionations. Table 1.4 shows the relative DOC mass amounts (DOC balance) of FA, HA, and NHS, which were obtained after XAD-8 fractionation. Only 18% of the WWE organic carbon could be isolated as humic material, whereas much more organic carbon (64%) was found in FA and HA from BW.

The DOC mass balance of the mst-UF is given in Table 1.5. There are great differences for the WWE sample and the BW sample. A high amount (46%) of the WWE DOC penetrates the membrane of a nominal MWCO of 1 kDa, whereas most of the BW DOC (91%) is retained. The partitioning into molar mass fractions is also rather different for both samples. The remaining organic carbon (> 1 kDa) of the WWE is equally distributed across the three higher molar mass fractions (each about 12%). A high proportion (63%) of the BW organic carbon is found in the highest molar mass fraction (> 30 kDa). In the low molar mass fraction (3 – 1 kDa), the relatively lowest amounts are collected (5% for WWE, 2% for BW).

Table 1.3 Atomic ratios (H/C, N/C) of different samples

Sample	H/C	N/C
BW	0.97	0.03
BW FA	0.93	0.01
BW HA	0.97	0.03
BW > 30 kDa	0.93	0.02
BW 10 – 3 kDa	0.87	0.02
WWE FA	1.19	0.05
WWE HA	1.50	0.14

Table 1.4 DOC mass balance for the fractions of the XAD-8 isolation procedure of organic matter from the waste water effluent (WWE) and the brown water (BW)

Sample	WWE DOC,%	BW DOC,%
Original	100	100
FA	11	45
HA	7	19
NHS	73	33
loss	9	3

Table 1.5 DOC mass balance for the fractions of the multistage ultrafiltration removal of organic matter from the waste water effluent (WWE) and the brown water (BW)

Sample	WWE DOC,%	BW DOC,%
Original	100	100
Concentrate (>1 kDa)	54	91
> 30 kDa	12	63
30 kDa – 10 kDa	13	10
10 kDa – 3 kDa	12	11
3 kDa – 1 kDa	5	2
remaining part[a]	58	14

[a]remaining part: loss + fraction < 1 kDa

1.3.2. Information about the Outer Structure from Size Exclusion Chromatography

Molecular Size Distribution of Organic Matter from BW and WWE. Figure 1.3 shows the size exclusion chromatograms of the original WWE and BW samples. The OM of both samples differs in its elution behavior from the SEC column. The maximum of the main peak of the WWE sample lies at a greater elution volume (V_e) than that of the BW sample (about 32 mL for WWE as opposed 37 mL for BW). Furthermore, a significant fraction of the WWE sample (about 30%) elutes after the permeation volume (V_p). Both the WWE and the BW sample contain OM, which elutes at the exclusion volume of the column ($V_0 = 22$ mL). The proportion of this large molecular size OM in the BW sample is higher than in the WWE sample. For the BW, the relative area of the peak at V_0 is about 10% of the total area, whereas it is only 4% for the WWE.

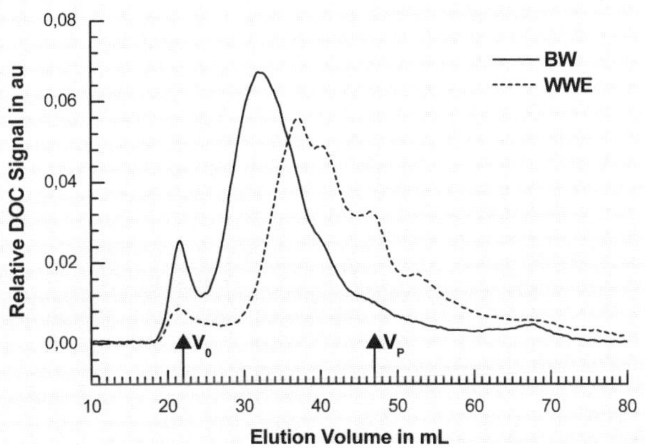

Figure 1.3 Size exclusion chromatograms of the brown water (BW) and the waste water effluent (WWE). TSK HW 50-(S) column, DOC detection. V_0: exclusion volume, V_p: permeation volume. The total areas of the chromatograms were equalized

SEC-DOC Analysis of Molecular Size Fractions from mst-Ultrafiltration. The size exclusion chromatograms of the concentrate of the BW sample and its mst-UF fractions (> 30 kDa; 10 – 3 kDa) are depicted in Figure 1.4. The chromatogram of fraction > 30 kDa, which contains about 60% of the DOC of the original BW sample, is similar to that of the concentrate except that the maximum of the main peak is found at a smaller elution volume compared to the chromatogram of the concentrate (shift from about 32 to 30 mL). The chromatogram of fraction 10 – 3 kDa clearly differs from both the concentrate and fraction > 30 kDa. No peak is detected at the exclusion volume, and the maximum of the main peak is at 34 mL as opposed to 32 mL in the concentrate. The chromatograms of the ultrafiltration fractions of the BW sample show comparatively clear cuts, whereas the separation was less efficient for the WWE sample [15].

1.3.3. Inner Structure Data from ^{13}C NMR Spectroscopy

Total Dissolved Organic Matter from Different Origin. The ^{13}C CPMAS NMR spectra of the freeze-dried BW and WWE sample are given in Figure 1.5. At first glance, the intensity distribution over all chemical shift regions is similar for both samples. The highest signal intensity is found at 70 ppm in both spectra. This signal is mainly attributed to carbohydrate material. Due to the high carbonate content of the WWE, a detailed quantitative comparison could not be carried out. Carbonates give an intense and narrow signal at 167 ppm, which overlaps strongly with the signals of organic carboxyl groups (around 172 ppm). The carbonate signal was qualitatively removed by subtracting a fitted gaussian function. For extended investigations, the inorganic material has to be removed. However, desalination normally results in an undefined loss of OM.

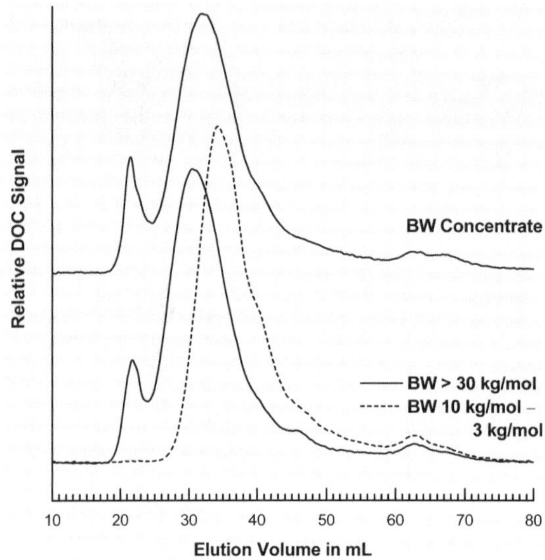

Figure 1.4 Size exclusion chromatograms of the brown water (BW Concentrate) and the mst-UF fractions with nominal size > 30 kDa and 10 kDa – 3 kDa. TSK HW 50-(S) column, DOC detection. The total areas of the chromatograms were equalized. Data according to Müller et al. [21]

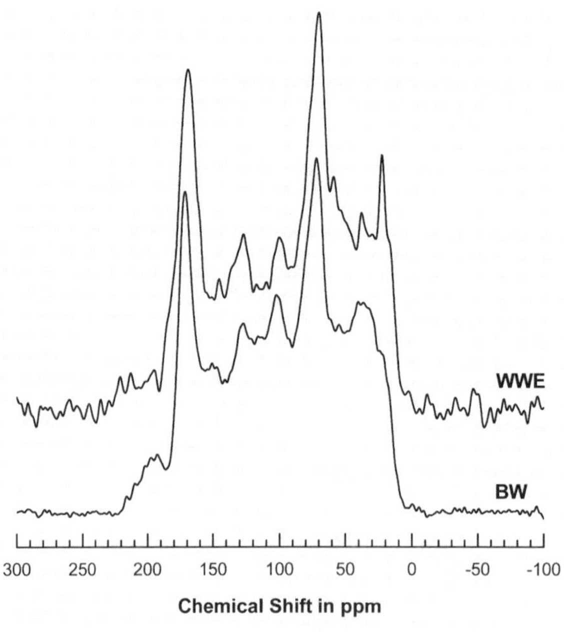

Figure 1.5 ^{13}C CPMAS NMR spectra of the freeze-dried water samples WWE and BW. The strong signal of carbonates has been removed from the WWE spectrum (see text for details). The total areas of the two spectra between 220 – 0 ppm were equalized

The spectrum of the BW sample shows a fairly high signal-to-noise ratio because of the low content of inorganic material. This water sample can be used without further treatment (e. g., dialysis) for NMR investigations of fractions from mst-UF to obtain structural information for the molecular size fractions. The ^{13}C MAS NMR spectra of two molecular size fractions are shown in Figure 1.6. The spectrum of the high molar mass fraction (> 30 kDa) shows a higher signal intensity at 70 ppm, which is attributed to a higher content of carbohydrate material. The spectrum of the low molar mass fraction (10 – 3 kDa) shows a higher signal intensity at 40 ppm and 172 ppm. In this fraction, branched aliphatic structures and structures with higher carboxyl group content are relatively enriched.

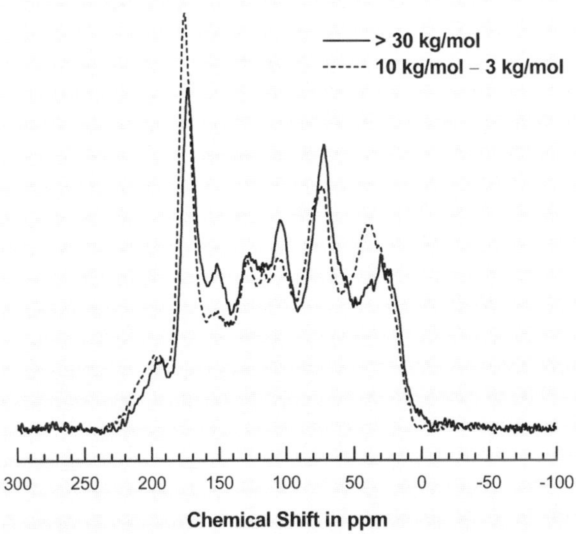

Figure 1.6 ^{13}C MAS NMR spectra of mst-UF fractions > 30 kDa and 10 – 3 kDa. Data according to Müller et al. [21]

Non-Humic Substances. The XAD-8 enrichment of OM was carried out for both the BW and the WWE sample. The non-adsorbed part of the OM (NHS fraction) was also collected and analyzed. A ^{13}C CPMAS NMR spectrum of this freeze-dried WWE fraction is depicted in Figure 1.7. For the BW NHS fraction, the occurrence of hygroscopic salts in the sample as well as hydrolysis reactions, which took place during sample treatment, prevented the acquisition of a spectrum that was suitable for quantitative comparison with the WWE NHS spectrum. Instead, the WWE NHS spectrum is compared to the spectrum of the corresponding part of the BW OM, which was generated by subtracting the spectra of the FA and HA sample according to their relative amount of DOC from the spectrum of the original sample. Carbohydrate signals (70 ppm) are dominant in both of the spectra shown in Figure 1.7. Integration data for the two spectra are given in Tables 1.6 and 1.7. They show about 46% O-alkyl carbon in this BW fraction (NHS + loss), and about 42% O-alkyl carbon in the WWE NHS sample. This O-alkyl carbon can mainly be assigned to carbohydrates.

Balance of Functional Groups Occurring in XAD-8 Fractions. Recalculation of the sum spectrum from the spectra of the three WWE fractions FA, HA and NHS with subsequent integration yields quantitative information on the relative contents of the different functional groups (Table 1.7). Comparison with the data from the original sample (not shown) show good agreement except for carboxylic functions, which are overestimated in the original sample because of its high carbonate content. The data show a higher amount of sp^3-carbon (64%) compared to the original sample of BW (52%). The NMR data of the sample "sum", which represents 91% of the DOC of the original sample, indicate a higher content of carbohydrates and other aliphatic structures in the original WWE sample compared to the original BW sample.

Table 1.6 Integrals (in %) of ^{13}C CPMAS NMR spectra of the original sample and XAD-8 fractions of the BW sample

Structural groups, %	Original	FA	HA	NHS + loss[a]
Carbonyl groups	3.3	3.4	2.6	3.3
Carboxyl groups	15.0	16.6	12.7	14.2
O-/N-subst. arom. C	8.3	10.8	10.7	3.9
C-/unsubst. arom. C	21.0	25.0	26.6	12.7
O-/N-alkyl C	33.5	25.4	29.5	45.6
Aliphatic C	18.9	18.8	17.9	20.3

[a]NHS + loss: calculated values from DOC mass balance (Table 1.4) and the integrals of the spectra from the samples Original, FA, and HA; see text for details

Table 1.7 Integrals (in %) of ^{13}C CPMAS NMR spectra of XAD-8 fractions of the WWE sample

Structural groups, %	sum[a]	FA	HA	NHS
Carbonyl groups	2	1.9	0.1	2
Carboxyl groups	11	13.7	12.9	11
O-/N-subst. arom. C	6	7.5	4.0	7
C-/unsubst. arom. C	17	19.4	13.1	17
O-/N-alkyl C	40	25.3	40.7	42
Aliphatic C	24	32.2	29.2	21

[a]sum: calculated values from DOC mass balance (Table 1.4) and the integrals of the spectra from the samples FA, HA, and NHS

Based on the NMR data that have been presented so far, it can be deduced that the relative amounts of carbohydrates must be reduced in the humic fractions of both samples BW and WWE. Closer inspection of the integral data for the humic fractions (FA and HA) shows that aromatic structures are relatively enriched in the BW humics whereas aliphatic structures are relatively enriched in the WWE humics. This leads to an average value of the aromatic-to-aliphatic carbon ratio of 2.0 for BW humics and 0.7 for WWE humics. Therefore, BW humics have a much more aromatic character than WWE humics. Additionally, the integration data show that O-alkyl carbon is more abundant in the HA fractions (30% in BW, 41% in WWE) than in the FA fractions (25% in BW and WWE). For WWE, an enrichment of aromatic carbon in the FA fraction (27%) compared to the HA fraction (17%) is observed.

Aromatic Structures in Humic Substances. The chemical shift regions of sp^2-carbon of the ^{13}C CPMAS NMR spectra for FA and HA are depicted in Figure 1.8. The areas of the spectra between 160 ppm and 100 ppm were equalized to focus on the comparison of intensity patterns in the aromatic chemical shift region.

No difference in the intensity distribution of aromatic carbon signals is seen between the FA and HA of BW. O-substituted aromatic carbon is observed as a signal group at around 150 ppm. Three other intensity maxima are found in the region between 140 ppm and 100 ppm, where C- and unsubstituted unsaturated carbon signals overlap.

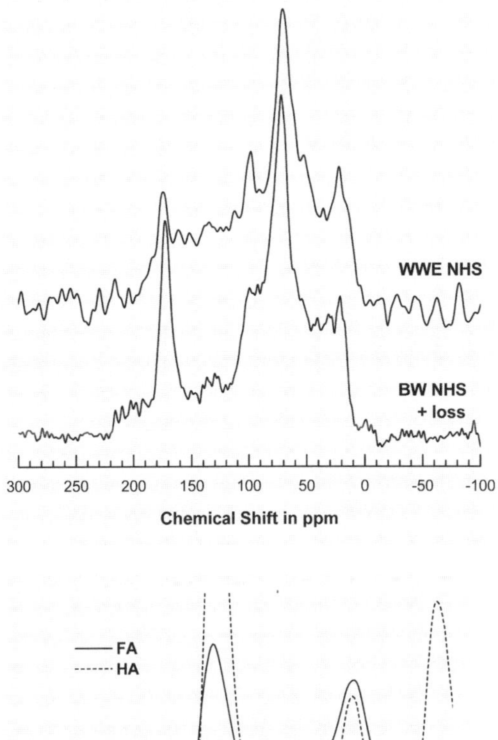

Figure 1.7 ^{13}C CPMAS NMR spectrum of the freeze-dried sample WWE NHS. The ^{13}C NMR spectrum of BW NHS + loss was produced by subtracting the CPMAS spectra of FA and HA from the spectrum of the total OM (freeze-dried original sample BW; the total areas were referred to the DOC mass balance)

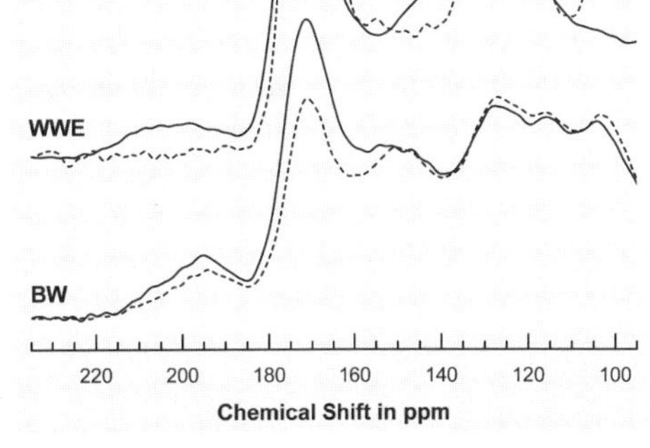

Figure 1.8 "Unsaturated" section (signals of sp^2-carbon) of the ^{13}C CPMAS NMR spectra of FA and HA of WWE and BW. The areas of the aromatic spectral region (160 – 100 ppm) were equalized

A better resolution of this aromatic intensity pattern can be obtained when dipolar dephasing is applied. ^{13}C DD CPMAS NMR spectra of the HA fractions are depicted in Figure 1.9. Here, only carbon that has no covalently bound H-atom is detected. Additionally, carbon atoms in highly mobile groups, e. g., rotating methyl groups, yield signals. The signal groups at 128 ppm and 105 ppm in the DD spectrum of the BW sample are therefore assigned to C-substituted aromatic carbon. The latter is a key signal of polyphenolics/tannin derivatives [26]. The presence of a signal of methoxy groups at 56 ppm indicates that lignins are another major precursor of the aromatic structures of the BW HA. The signal group at 116 ppm in the CPMAS spectrum is assigned to unsubstituted aromatic carbon.

In the spectra of the two WWE isolates, only one aromatic peak at 128 ppm is dominant (Figure 1.8). There is no distinct signal group in the 160 – 140 ppm region. A small peak at 140 ppm is found in the FA spectrum. This sharp signal is prominent in the DD spectrum (not shown) and reflects substituted unsaturated carbon. Key signals of tannins (105 ppm) and lignins (56 ppm) are completely absent. The NMR spectra in Figures 1.8 and 1.9 reveal fundamental differences in the aromatic structures between the WWE and the BW humics.

Finally, evidence concerning the different degrees of aromatic substitution is compiled from the quantitative data of Lankes and Lüdemann [18]. They reported a ratio of unsubstituted to substituted aromatic carbon of 1.0 for the FA of the WWE sample and of 0.6 for the HA, as opposed to 0.3 for the BW humics. The amount of unsubstituted aromatic carbon is therefore significantly higher in the WWE humics than in the BW humics.

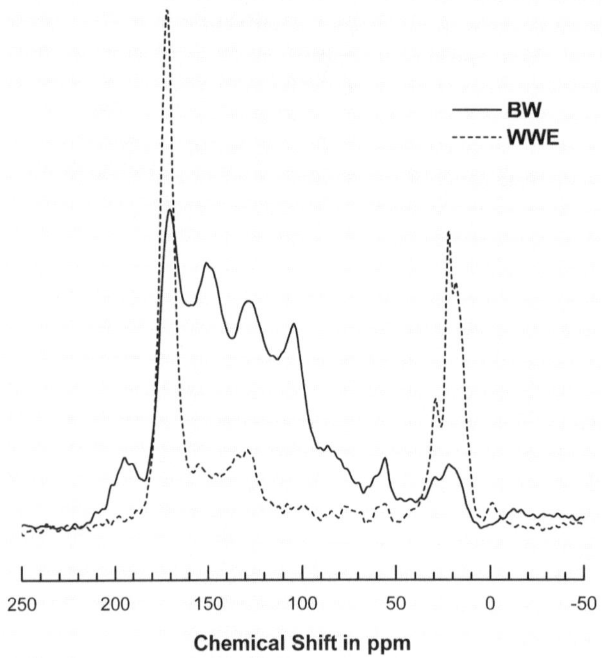

Figure 1.9 ^{13}C DD CPMAS NMR spectra of HAs of WWE and BW. The areas of the chemical shift region of carboxyl groups (190 – 160 ppm) were equalized

Aliphatic Structures in Humic Substances. The intensity distribution in the chemical shift range of sp^3-carbon is poorly structured. In general, there is only one intensity maximum in the range from 50 to 0 ppm. However, structural differences can be determined from the shape of the broad aliphatic signal group. The chemical shift region of sp^3-carbon of the ^{13}C CPMAS NMR spectra of FA and HA is depicted in Figure 1.10 (spectral areas between 50 and 0 ppm were equalized).

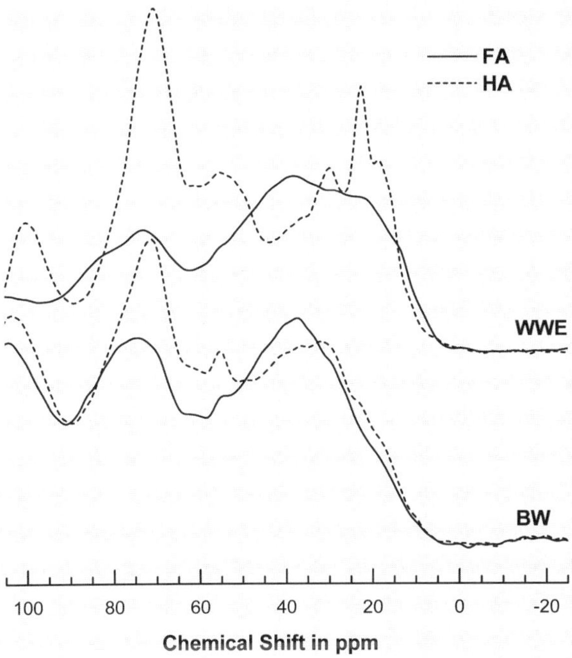

Figure 1.10 "Saturated" section (signals of sp^3-carbon) of the ^{13}C CPMAS NMR spectra of FA and HA of WWE and BW. The areas of the aliphatic spectral region (50 – 0 ppm) were equalized

All aliphatic carbon signals overlap strongly in the ^{13}C CPMAS NMR spectra so that a quantitative evaluation of different signal groups cannot be performed. Separation of methyl groups can be achieved by the DD technique. Methyl groups have a signal intensity maximum at about 22 ppm. Signals mainly of branched aliphatic carbon appear at 40 ppm. Non-branched aliphatic carbon, i. e. methylene groups, have an intensity maximum around 30 ppm.

In the spectra of the brown water FA and HA, the aliphatic intensity maxima appear at 38 and 33 ppm, respectively. For the FA of WWE, the maximum is also detected at 38 ppm. For the spectrum of the HA of WWE the result is rather different. A clearly structured intensity pattern with relatively sharp signals can be observed. The intensity maximum is found at 22 ppm (methyl groups) and another maximum is seen at 30 ppm (methylene groups). A third signal group with a small maximum at 18 ppm is also assigned to methyl groups. Those three signal groups are clearly seen in the DD spectrum as well (Figure 1.9). The signal of methylene groups (maximum at 29 ppm in the DD spectrum) corresponds to the end groups of highly mobile long alkyl chains.

The ratio of the intensity at 30 ppm to the intensity of 40 ppm gives information about the degree of branching of the aliphatic material. The average values for various BW and WWE samples are compiled in column 1 of Table 1.8. The FA of BW show an average value of 0.8, the HA a ratio of 1.0, and the FA of WWE a ratio of 0.9. For the WWE HA, the ratio is significantly higher (1.5). This means that the FA contains more branched structures than the HA. In other words, the WWE HA has a highly non-branched character. Lankes and Lüdemann [18] estimated the content of methyl groups, which indicate that the structural differences in the aliphatic region are more pronounced. The relative content of methyl groups was estimated to be about 15% for the WWE humics and only 5% and 3% for BW FA and HA, respectively. The ratio of the amount of methyl groups to the amount of remaining aliphatics was about 0.3 for the BW humics and about 0.6 for the WWE FA. For the HA of WWE the ratio was 1.1. Such a high ratio is only possible if a significant amount of methyl groups occur in structures other than long-chain aliphatics, e. g., in N-acetyl groups, which are part of certain carbohydrate structures.

O-alkyl Structures and Other Functional Groups. A similar data evaluation can be made for the O-alkyl carbon. The signals at 70 ppm are typical of the main signal of carbohydrates. More branched O-alkyl structures that are found, for example, in side chains of lignins yield signals at about 80 ppm. The ratio of the intensity at 70 ppm to the intensity at 80 ppm is given in column 2 of Table 1.8. It is higher for the HA (1.3) than for the FA (1.0) of the BW. Thus, HA contains a relatively higher amount of carbohydrate material. The ratio in the WWE FA is comparatively low (1.1). The origin of the branched forms of O-alkyls in WWE FA has not yet been fully elucidated. The HA sample shows a fairly high value (2.6), which means that nearly all O-alkyl carbon must be present in carbohydrate forms.

Additional signals in the spectrum of the WWE HA (Figures 1.8 and 1.10) are assigned to α-carbon that is bound to amide nitrogen (55 ppm), to anomeric carbon (100 ppm), and to amide carbon (172 ppm). The signal at around 172 ppm is mainly assigned to carboxyl groups and esters in the spectra of all other samples.

Table 1.8 Ratios of intensity data from ^{13}C CPMAS NMR spectra of FA and HA of various BW and WWE samples

Column	1	2
Intensity ratios[a]	$I_{30\ ppm}/I_{40\ ppm}$	$I_{70\ ppm}/I_{80\ ppm}$
BW HA[b]	1.0 ± 0.1	1.3 ± 0.2
BW FA[b]	0.8 ± 0.0	1.0 ± 0.0
WWE HA	1.5	2.6
WWE FA[c]	0.9	1.1

[a]intensity $I_{30\ ppm}$ = signal intensity of non-branched aliphatic carbon at 30 ppm, intensity $I_{40\ ppm}$ = signal intensity of branched aliphatic carbon at 40 ppm; intensity $I_{70\ ppm}$ = signal intensity of main carbohydrate signal maximum at 70 ppm; intensity $I_{80\ ppm}$ = signal intensity of branched O-alkyl carbon forms at 80 ppm; [b]mean value of 4 samples with standard deviation; [c]mean value of 2 samples.

Carbonyl groups as a minor structural component contribute only up to 5% to the structural composition of the BW HSs [18]. The HA of the WWE shows no significant amount of carbonyl functions. The carbonyl functions of the WWE FA (about 2%) show a signal maximum at higher chemical shift values (198 ppm) compared to the BW HSs (195 ppm) (Figure 1.8). This is attributed to the predominantly aliphatic character of WWE FA.

1.4. DISCUSSION

1.4.1. Non-humic Carbohydrates

The predominant structural components in the NHS fractions of both samples are carbohydrates. Carbohydrates are ubiquitous in nature and they are polymorphous. Carbohydrates occur in structure-forming polysaccharides (e. g., celluloses, chitins, mureins), in storage products like glycogen and starch, and as a huge number of small sugar units from monomers to oligomers. Therefore, a differentiation in complex carbohydrate mixtures is difficult. ^{13}C NMR yields no exact identification of the predominant structures. Wet chemical methods suffer from incomplete recoveries of analytes as well as from degradation reactions (hydrolysis) during the isolation procedure (HCl treatment). However, through a combined analysis of the size distributions and chemical compositions that is explained below, it becomes obvious that different types of carbohydrates occur in both samples. In the BW, high molar mass polysaccharides are the predominant carbohydrate form. By contrast, in the WWE a significant amount of mono- to oligosaccharides is present.

Two thirds of the BW OM are found in the large molar mass fraction (> 30 kDa). Less than one third of the BW OM is found in smaller molar mass fractions (< 30 kDa). It has been shown that carbohydrates are enriched in the large molar mass fraction (> 30 kDa), and depleted in the low molar mass fraction (10 – 3 kDa). Support comes from the SEC-DOC data, which confirm that size separation could be achieved through mst-UF. The OM of the BW sample and its mst-UF fractions eluted mainly between V_0 and V_p. The elution volume of the peak maximum in the large molar mass fraction (> 30 kDa) is smaller than in the concentrated sample (Figure 1.4). This indicates that large molecules have become relatively enriched in this fraction. Smaller molecules, in turn, have become relatively enriched in the low molar mass fraction (10 – 3 kDa). The peak eluting at the exclusion volume of the TSK HW 50-(S) column appeared in the high molar mass mst-UF fraction only. It has been suggested by Huber and Frimmel [27] that the OM of this peak mainly consists of carbohydrates. Müller et al. [28] have verified this hypothesis for the BW sample through preparative SEC fractionation and ^{13}C NMR analysis of the fractions. The structural characterization together with information on molecular size from SEC and mst-UF show that the BW OM contains carbohydrates that are large in molecular size, i. e. polysaccharides.

In the WWE sample, the NHS fraction makes up the major part of the OM (nearly three quarters of the DOC). Data from mst-UF fractionation and SEC analysis both indicate that the WWE contains a greater amount of small organic molecules than the BW OM. Up to half of the DOC is found in the low molar mass fraction (< 1 kDa). About one third is distributed fairly equally across the three larger molar mass fractions (> 30 kDa; 30 – 10 kDa; and 10 – 3 kDa). From the position of the peak maxima of the BW and the WWE sample, it can also be concluded that the OM of the WWE, which elutes between V_0 and V_p, consists of molecules that are smaller than the majority of the BW OM. Based on those results as well as on the fact that the NHS fraction, which is by far the dominant fraction of the WWE OM, mainly consists of carbohydrates, it can be deduced that a significant amount of the carbohydrates in the WWE must be of smaller size than the polysaccharides in the BW OM. Monomeric to oligomeric forms likely have been relatively enriched as a consequence of the 0.45 μm-filtration of the sample prior to XAD-8 isolation. In this filtration step, bacterial cell walls, and thus carbohydrates of large molecular size, are predominantly retained by the membrane and thereby are removed from the sample. Small cytoplasmic components, in turn, pass the membrane as well as the XAD-8 column in the subsequent isolation step and contribute to the NHS fraction.

1.4.2. The Building Blocks of Natural Humic Matter

Two thirds of the BW OM are defined as humic matter. Vascular plant material is the major source of the BW HSs. The NMR data indicate a highly aromatic character. This is also confirmed by the SUVA value of the original BW sample. The key signals of methoxyl groups (56 ppm) and of C-substituted carbon of polyphenolics (105 ppm), which both are especially pronounced in the DD spectra, point to the origin of the aromatic structures. Because of their typical aromatic substitution pattern, phenylpropanols, which occur in lignins, and polyphenolics, which occur in tannins, show a typical intensity distribution pattern of aromatic carbon signals. The signal groups at 150 ppm, 128 ppm, 116 ppm, and 105 ppm are assigned to lignin- and tannin-derived structures. Lignins have been investigated in detail by Lüdemann and Nimz [29], and tannins by Wilson and Hatcher [26] as well as by Lorenz et al. [30]. Such structures occur in higher plants. Derivatives of lignocelluloses are commonly found in forest soils where they become soluble through oxidation processes. Derivatives of tannins occur mainly in forest litter materials and they already are highly soluble. Both products are washed out of the forest soil into the aquatic environment of brown water lakes.

Besides the aromatic carbon of this degraded plant material, alkyl and O-alkyl carbon are found in side chains of lignins, and in hemicellulose and cellulose fragments. Furthermore, some branched alkyl and O-alkyl carbon structures may originate from plant isoprenoids. Isoprenoids are supposed to be the predominant aliphatic forms in the easily soluble FA fractions as well as in the small molar mass fraction of the ultrafiltration separation, as indicated by their highly branched character [31,32]. Non-branched aliphatic forms are more abundant in HA and in the higher molar mass fraction. Alkyl chains probably arise from waxes. Both aliphatic precursors have relatively low methyl group content.

1.4.3. The Building Blocks of Humic Matter in the Waste Water Effluent

Only about 20% of the WWE OM is defined as HSs. Here, microbial material is the predominant precursor. In contrast to the BW humics, the WWE humics have a more aliphatic character and the structural differences between the WWE components in HA and FA are much more pronounced. Carbohydrates are enriched in HA whereas highly branched aliphatic structures are enriched in FA.

It is attractive to assume that the carbohydrates in HA originate from bacterial cell wall fragments. Sharp and intense signals of methyl groups suggest a significant amount of N-acetyl containing sugar units. The N/C ratio of this fraction is very high. The ^{15}N CPMAS NMR spectrum shows that up to 90% of the nitrogen is bound in amide form [18]. Hence, nearly all of the signal intensity in the chemical shift region of carboxyl functions has to be assigned to amide carbon, which appears in peptides and other amides. Additionally, the α-carbon of amino acids and amino sugars yields a strong signal at 55 ppm. N-acetylic sugar units and peptides are known to be present in mureins, a major component of bacterial cell walls.

This is in good agreement with the results of Jahnel et al. [33], who showed that the ratio of the content of fucose + rhamnose to the content of arabinose + xylose is 2.6 for the HA of WWE and thus is significantly higher than in the studied BW samples (ratio of 0.7). Fucose and rhamnose are sugar units in lipopolysaccharides that are typical of bacteria. The presence of lipopolysaccharides also results in signals of long-chained alkyl structures (30 ppm) in the NMR spectra. By contrast, arabinose and xylose are typical sugar units of hemicelluloses and their presence is thus related to the occurrence of lignocelluloses in the OM.

Other key molecules for the identification of OM sources are amino acids. Leucine plays a representative role as an indicator of bacterial biosynthesis proteins [33]. About 10% of the identified amino acids in the HA is leucine. In the FA, leucine accounts for about 20% of the identified amino acids. In the BW humics, the leucine content of 6% is lower and in a normal range. Leucine in particular contributes to a fairly high ratio of the content of methyl groups to the content of remaining aliphatics, as it is found in WWE humics.

The substitution pattern of unsaturated carbon in WWE samples is quite different from that of the BW samples. Aromatic amino acids and unsaturated fatty acids contribute to the highly unsubstituted character. The combination of spectroscopic and hydrolysis data provides evidence for the conclusion that bacteria are by far the most important precursors of the WWE OM.

Additionally, it is likely that aromatic and heterocyclic (O, N, S) compounds of anthropogenic origin contribute with an unknown amount to the FA aromatics. The increase in the SUVA value after waste water treatment suggests that those compounds are rather refractory and thus become enriched during the treatment.

1.5. CONCLUSIONS

Comparison of the NMR spectra of the BW and WWE total organic matter does not reveal major differences between the two samples. However, after closer inspection it becomes obvious that substantial structural differences exist that are obscured by the predominant carbohydrate and broad signals in the spectra. The types of carbohydrates cannot be characterized in detail by NMR. Neverthless, results from the size fractionation via multistage ultrafiltration as well as from SEC-DOC analysis indicate that the molecular size distribution of the carbohydrate material of the BW and WWE samples is rather different. Investigation of the occurrence of characteristic sugar units (hydrolyzable carbohydrates) gives information on their sources. In BW OM, high molar mass polysaccharides derived from plant material (celluloses, hemicelluloses) are the predominant carbohydrate forms. The average molar mass of carbohydrates in WWE OM is lower. Here, carbohydrates mainly originate from microbial sources.

After enrichment of OM via XAD-8 isolation and NMR spectroscopic characterization, the fundamental structural differences between the BW and WWE humic substances can be recognized clearly. These differences result from different precursor materials. The signals of unsaturated aromatic carbon atoms in BW are characteristic of degradation products of lignins and tannins. Consequently, the HSs of the BW mainly originate from vascular plant material.

The HSs of the WWE are predominantly composed of saturated carbon structures. Structure-forming peptidoglycans serve as a major precursor. Peptidoglycans seem to be the only source in the HA fraction of WWE. For FA it is assumed that smaller cell wall fragments as well as cytoplasmic molecules and anthropogenic leftovers are prominent.

The results clearly show that the characterization of bulk samples does not lead to a useful interpretation of their humic properties. Separation methods in combination with advanced analytical methods are most promising to elucidate the complicated structure activity relations.

The OM of the BW has a refractory character. Highly persistent vascular plant materials are not the favourite energy source of microbial communities. In the WWE, a large amount of highly biodegradable compounds is available. From a structural point of view the major part of the humic fractions of the WWE definitely are not so refractory. Those structures originate from microorganisms, and it is well known that simple degradation units such as sugars and amino acids serve as important substrates for microbiological metabolic activity.

Biodegradability is an important factor for the persistence of HSs. As both biocenosis and physicochemical parameters in the receiving waters are likely to be rather different from the ones in the WW treatment plant, no predictions regarding the biodegradability of the WWE OM in natural aquatic systems can be made at this stage. However, it is obvious that refractory microbial organic substances and anthropogenic leftovers become enriched over time in natural environments. The composition of the OM pool therefore mainly depends on the relative proportions of the various sources of OM and the refractory character of the different OM constituents.

Based on the structural characterization of the WWE OM it can be concluded that in the WWE a high proportion of easily biodegradable compounds is discharged to surface waters. In highly populated regions where a large amount of WW is produced, this can cause long- term changes in the OM composition of the receiving waters. Such changes have both ecological and technological implications. The input of bioavailable OM can bring about changes in biocenosis as well as the oxygen content of the surface waters. The transfer of bioavailable substances to existing poorly degradable matter can thus affect the ecological balance of the aquatic system. Furthermore, changes in the composition of the OM pool can also cause problems during water treatment processes if the surface water is used as input for the production of drinking water. From both points of view, sustainable water use is essential for the preservation of this precious resource for the generations to come.

ACKNOWLEDGEMENTS

This work was supported by grants from the German Research Foundation (DFG), the Ministry for Education and Research (BMBF) of Germany and the Priority Research Program of the State of Baden-Württemberg.

REFERENCES

1. Esch B, Ewens H-P. Stand der öffentlichen Abwasserbeseitigung (1991). *Korrespondenz Abwasser*, 8/1994; 41: 1375-1376.
2. Tchobanoglous G, Burton FL. *Wastewater engineering: Treatment, disposal, and reuse*. 3rd Edn. New York: McGraw Hill, 1991.

3. Frimmel, FH. Aquatic humic substances. In: Steinbüchel A, Hofrichter M eds. *Biopolymers*. Weinheim: Wiley-VCH, 2001: 301-324.
4. Leenheer JA, Croué J-P. Characterizing aquatic dissolved organic matter. *Environ. Sci. Technol.*, 2003; 37: 18A-26A.
5. Mantoura RFC, Riley JP. Analytical concentration of humic substances from natural waters. *Anal. Chim. Acta*, 1975; 76: 97-106.
6. Thurman EM, Malcolm RL. Preparative isolation of aquatic humic substances. *Environ. Sci. Technol.*, 1981; 15: 463-466.
7. Abbt-Braun G, Frimmel FH. The relevance of reference materials – isolation and general characterization. In: Frimmel FH, Abbt-Braun G, Heumann KG, Hock B, Lüdemann H-D, Spiteller M eds. *Refractory organic substances in the environment*. Weinheim: Wiley-VCH, 2002: 1-38.
8. Huber S, Frimmel FH. A liquid chromatographic system with multi-detection for the direct determination of hydrophilic organic compounds in natural waters. *Fresenius' J. Anal. Chem.*, 1992; 342: 198-200.
9. Specht CH, Kumke MU, Frimmel FH. Characterization of NOM adsorption to clay minerals by size exclusion chromatography. *Water Res.*, 2000; 34: 4063-4069.
10. Her N, Amy G, Foss D, Cho J, Yoon Y, Kosenka P. Optimization of method for detecting and characterizing NOM by HPLC-size exclusion chromatography with UV and on-line DOC detection. *Environ. Sci. Technol.*, 2002; 36: 1069-1076.
11. Posner AM. Importance of electrolyte in the determination of molecular weights by Sephadex gel filtration, with special reference to humic acid. *Nature*, 1963; 198: 1161-1163.
12. Lindqvist I. Adsorption effects in gel filtration of humic acid. *Acta Chimica Scandinavica*, 1967; 21: 2564-2566.
13. Specht CH, Frimmel FH. Specific interactions of organic substances in size-exclusion chromatography. *Environ. Sci. Technol.*, 2000, 34: 2361-2366.
14. Swift RS, Posner AM. Gel chromatography of humic acid. *J. Soil Sci.*, 1971; 22: 237-249.
15. Müller MB, Frimmel FH. A new concept for the fractionation of DOM as a basis for its combined chemical and biological characterization. *Water Res.*, 2002; 36: 2643-2655.
16. Hatcher PG, Breger IA, Dennis LW, Maciel GE. Solid-state ^{13}C NMR of sedimentary humic substances: New revelations on their chemical composition. In: Gjessing ET, Christman RF eds. *Aquatic, and terrestrial humic materials*. Ann Arbor: Ann Arbor Science Publishers, 1983: 37-81.
17. Wilson MA. *NMR techniques and applications in geochemistry and soil chemistry*. Oxford: Pergamon Press, 1987.
18. Lankes U, Lüdemann H-D. Structural characterization of refractory organic substances by solid-state high-resolution ^{13}C and ^{15}N nuclear magnetic resonance. In: Frimmel FH, Abbt-Braun G, Heumann KG, Hock B, Lüdemann H-D, Spiteller M eds. *Refractory organic substances in the environment*. Weinheim: Wiley-VCH, 2002: 96-114.
19. Statistisches Landesamt Baden-Württemberg. Struktur- und Regionaldatenbank, 2002: http://www.statistik.baden-wuerttemberg.de
20. Frimmel FH, Abbt-Braun G, Heumann KG, Hock B, Lüdemann H-D, Spiteller M eds. *Refractory organic substances in the environment*. Weinheim: Wiley-VCH, 2002.

21. Müller MB, Fritz W, Lankes U, Frimmel FH. Ultrafiltration of nonionic surfactants and dissolved organic matter. *Environ. Sci. Technol.*, 2004; 38: 1124-1132.
22. Lambert J, Lankes U. Application of nuclear magnetic resonance spectroscopy to structural investigations of refractory organic substances – principles and definitions. In: Frimmel FH, Abbt-Braun G, Heumann KG, Hock B, Lüdemann H-D, Spiteller M eds. *Refractory organic substances in the environment*. Weinheim: Wiley-VCH, 2002: 89-95.
23. Abbt-Braun G, Lankes U, Jahnel JB, Lambert J, Lüdemann H-D, Frimmel FH. Chemical and spectroscopic data of the reference samples – comparison and evaluation. In: Frimmel FH, Abbt-Braun G, Heumann KG, Hock B, Lüdemann H-D, Spiteller M eds. *Refractory organic substances in the environment.* Weinheim: Wiley-VCH, 2002: 302-309.
24. Weishaar JL, Aiken GR, Bergamaschi BA, Fram MS, Fujii R, Mopper K. Evaluation of specific ultraviolet absorbance as an indicator of the chemical composition and reactivity of dissolved organic carbon. *Environ. Sci. Technol.*, 2003; 37: 4702-4708.
25. *German standard methods for water, wastewater and sludge analysis*. Evaluation in an aqueous medium of the ultimate aerobic biodegradability of organic compounds – method of determining the oxygen demand in a closed respirometer. *DIN EN 29408*. Berlin: Beuth Verlag, 1993.
26. Wilson MA, Hatcher PG. Detection of tannins in modern and fossil barks and in plant residues by high-resolution solid state ^{13}C nuclear magnetic resonance. *Org. Geochem.*, 1988; 12: 539-546.
27. Huber SA, Frimmel FH. Gelchromatographie mit Kohlenstoffdetektion (LC-OCD): Ein rasches und aussagekräftiges Verfahren zur Charakterisierung hydrophiler organischer Wasserinhaltsstoffe. *Vom Wasser*, 1996; 86: 277-290.
28. Müller MB, Lankes U, Abbt-Braun G, Frimmel FH. Structural characterization of brown water SEC fractions by ^{13}C MAS NMR spectroscopy. To be submitted.
29. Lüdemann H-D, Nimz, H. Carbon-13 nuclear magnetic resonance spectra of lignins. *Biochem. Biophys. Res. Commun.*, 1973; 52: 1162-1169.
30. Lorenz K, Preston CM, Raspe S, Morrison IK, Feger KH. Litter decomposition and humus characteristics in Canadian and German spruce ecosystems: Information from tannin analysis and ^{13}C CPMAS NMR. *Soil Biol. Biochem.*, 2000; 32: 779-792.
31. Lankes U. Festkörper-NMR-Spektroskopie zur Charakterisierung von refraktären organischen Säuren in Gewässern. *Ph. D. thesis*, Universität Regensburg, 2001.
32. Leenheer JA, Nanny MA, McIntyre C. Terpenoids as major precursors of dissolved organic matter in landfill leachates, surface water, and groundwater. *Environ. Sci. Technol.,* 2003; 37: 2323-2331.
33. Jahnel JB, Brinkmann T, Abbt-Braun G, Frimmel FH. Occurrence of amino acids, carbohydrates, and low-molecular-weight organic acids in refractory organic substances. In: Frimmel FH, Abbt-Braun G, Heumann KG, Hock B, Lüdemann H-D, Spiteller M eds. *Refractory organic substances in the environment*. Weinheim: Wiley-VCH, 2002: 264-281.

Chapter 2

ASYMMETRICAL FLOW FIELD-FLOW FRACTIONATION OF HUMIC SUBSTANCES: COMPARISON OF POLYACRYLIC ACIDS AND POLYSTYRENE SULFONATES AS MOLAR MASS STANDARDS

Manfred Wolf,[1] Gunnar Buckau[2] and Véronique Chanel[1]

[1] GSF-National Research Center for Environment and Health, Institute for Groundwater Ecology, 85764 Neuherberg, Germany
[2] Research Center Karlsruhe, Institute for Nuclear Waste Management, 76021 Karlsruhe, Germany

2.1. INTRODUCTION

Humic substances (HSs) are very complex mixtures of organic acids (mainly polycarboxylic acids with various functionality) with only partly known structures [1-3]. HSs can be subdivided in fulvic acid (FA), humic acid (HA) and humin, and they are ubiquitous in the terrestrial and aquatic environment [4,5]. The ability of HSs to form relatively stable complexes with heavy metals and radionuclides is important with respect to the migration of these pollutants in groundwater [6,7]. To understand the physical/chemical interactions of HSs with these pollutants, it is essential to elucidate not only structural information for HSs but also their molecular size/mass distribution. The measurement of molecular size/mass distribution of HSs can be made by various methods [8,9]. For example, the methods of gel-permeation chromatography (GPC) [10] or size-exclusion chromatography (SEC) [11], high-performance size-exclusion chromatography (HPSEC) [12-14], flow field-flow fractionation (FFFF) [15-20] and asymmetric flow field-flow fractionation (AFFFF) [21-24] are based on the different diffusion coefficients of molecules with different hydrodynamic sizes in solution. By contrast, special mass spectrometric methods, for example electrospray-ionization or laser desorption ionization mass spectrometry (ESI or LDI MS) [1-3,25-29] or time-of-flight secondary ion mass spectrometry (TOF-SIMS) [9,13,24,30], directly measure the molar mass distribution of HSs.

Indirect deduction of molar mass distributions requires directly comparable size to mass ratios of standards used and the investigated HSs. It was found in recent HPSEC studies that, in comparison to HSs, the lower size to mass ratios of globular proteins, dextrans and polystyrene sulfonate (PSS) standards results in overestimation of the molar masses of HSs, but that by using simple polycarboxylic acids (PCAs) and polyacrylic acids (PAAs) as molar mass standards similar molar mass distributions and consequently similar size to mass ratios result [9,13,14]. Therefore, the aim of this paper was to find out if calibration of AFFFF with HSs and similar PAA standards together with other PCAs also gives more reliable molar mass distributions of HSs than the often-used PSS standards. For this purpose the AFFFF system was calibrated with these two different types of standards and the derived molecular size distributions of HSs were compared and discussed with molar mass distributions or data from recent literature.

2.2. MATERIALS AND METHODS

2.2.1. Materials

Humic Substances. Natural aquatic humic and fulvic acids (Gohy-573 HA and FA) were isolated from groundwater of the deep borehole Gohy-573 in the Gorleben aquifer (Lower Saxony, Germany), purified and characterized [31]. The borehole Gohy-573 is located above the Gorleben salt dome with an inflow of groundwater from Präelster sediments in the depth range 134-137 m. The groundwater of this borehole is dark brown and has an elevated DOC (dissolved organic carbon) content of 97.2 mg C L^{-1} with the main DOC components being HA (60 mg C L^{-1}) and FA (17 mg C L^{-1}) [32,33]. Additionally, Suwannee River fulvic acid (SRFA) standard was used as delivered from the International Humic Substances Society (IHSS).

Other Materials. PSS and PAA standards were obtained from American Polymer Standards Corp., Mentor, Ohio. The polycarboxylic acids benzene-1,3-dicarboxylic acid and glycyrrhizic acid (as its ammonium salt) were obtained from Fluka; benzene-1,2,4,5-tetracarboxylic acid was obtained from Aldrich and 2-(4-(2,2-dicarboxy-ethyl)-2,5-dimethoxy-benzyl)-malonic acid (DMA) was obtained from Sigma-Aldrich. The chemicals sodium azide and 4-ethylbenzene sulfonic acid were obtained from Aldrich; tris-(hydroxymethyl)-aminomethane hydrochloride (buffer substance with pH 9.0 at 25°C, Tris-buffer) was obtained from Fluka. All substances were purchased in the highest purity available and used as delivered. All aqueous solutions were prepared with high purity water (Milli-Q_{PLUS}, Millipore, Bedford, Massachusetts).

2.2.2. Methods

Asymmetric Flow Field-Flow Fractionation (AFFFF). Field-flow fractionation is a technique for size fractionation of macromolecules and colloids. It was first proposed in 1966 by Giddings [15] and later also applied for humic substances [16-20]. In this method, the analyte dissolved in a carrier medium is pumped with constant velocity through a thin ribbon-like channel equipped with a membrane permeable for the carrier. The flow-field is established by a cross-flow perpendicular to the channel flow and the size fractionation is a function of the diffusion coefficient. Two types of equipment are available: a) the first developed was a symmetrical flow field-flow fractionation (FFFF) using a symmetrical channel with constant width and two permeable walls and b) the more recently developed asymmetrical flow field-flow fractionation (AFFFF), which uses an asymmetrical channel with continuously reduced channel width and one permeable wall [21-24].

The AFFFF measurements were carried out at room temperature with a system delivered from Wyatt Technology using an asymmetrical fractionation channel from ConSenxus (Germany) with a length of 286 mm and a spacer thickness of 0.54 mm. The regenerated cellulose membrane has a cut-off of 1 kDa (related to globular proteins). As eluent, 5 mM Tris-buffer [19] with pH ~9.1 was used at a channel flow of 0.6 mL min^{-1} and a cross-flow of 3.0 mL min^{-1}. The absorbance of the effluent was recorded with an UV/Vis detector (K-2500, Knauer, Germany) at 210 nm. The fractionated sample volume was 20 μL. Analyte concentrations in the range of 3 – 300 mg L^{-1} dissolved in the eluent were analyzed.

Calibration Standards and Data Evaluation of AFFFF. The calibration of the AFFFF system was carried out with two sets of standards, namely a) PAA standards with molar masses at the elution peak (Mp values) of 1.25, 4.1 and 16.0 kDa, various polycarboxylic acids with molar masses (M_r) between 166 and 823 Da and sodium azide and b) PSS standards with Mp values of 1.37, 4.6, 6.2 and 16.9 kDa, 4-ethylbenzene sulfonic acid with M_r 186 Da and sodium azide. The AFFFF fractograms were recorded simultaneously during the measurements with the software ASTRA 4.50 (Wyatt Technology). Before and after the standard or HSs measurements, blank samples (only eluent) were measured. The derived fractogram of the blank was subtracted from the original standard or HSs fractogram using the Corona 1.40 software (Wyatt Technology).

2.3. RESULTS AND DISCUSSION

2.3.1. Calibration of the AFFFF System

The results of the AFFFF measurements of the PAA and PSS standards, various PCAs and sodium azide at concentrations in the range 3 – 300 mg L^{-1} are plotted in Figures 2.1 and 2.2. We found the concentration dependence of the elution volume in accordance with Thang et al. [19]. In the studied concentration range (3 – 300 mg L^{-1}), the individual elution volumes seem to be linearly correlated with the logarithm of the concentration. For this reason, extrapolation of the elution volumes measured at higher concentrations to lower concentrations (e.g., 3 mg L^{-1}) is possible. Furthermore, the extent of this concentration dependence is correlated to the molar mass and type of standard used and therefore probably to the molecular size.

The results of the calibration with the different molar mass standards at low concentrations (3 and 30 mg L^{-1}) are plotted in Figures 2.3 and 2.4. Due to the stronger concentration dependence of the elution volume with higher molar masses of the standards, the calibration curves obtained show strong differences in the higher molar mass range. Comparison of the calibration curves obtained at measured concentrations of 30 mg L^{-1} shows that those PAA standards and other PCAs and PSS standards fall on different calibration curves (Figure 2.5). Higher size to mass ratios in comparison to PSS standards were found for all PCAs and PAA standards in the lower molar mass range. However, similar size to mass ratios were found for PAA and PSS standards with the highest molar masses (about 16 and 17 kDa). All calibration curves can be fitted to exponential functions given in Figures 2.3-2.5.

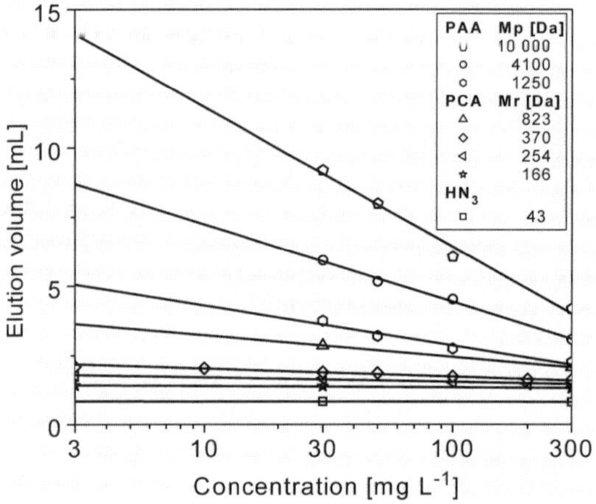

Figure 2.1 Concentration dependence of the elution volume of PAA standards, other PCAs and sodium azide

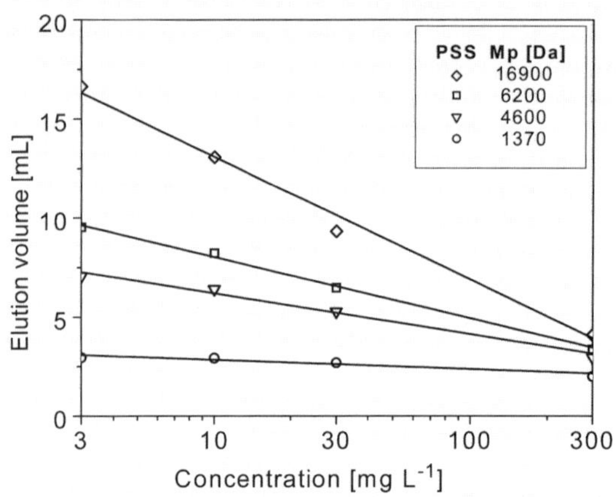

Figure 2.2 Concentration dependence of the elution volume of PSS standards

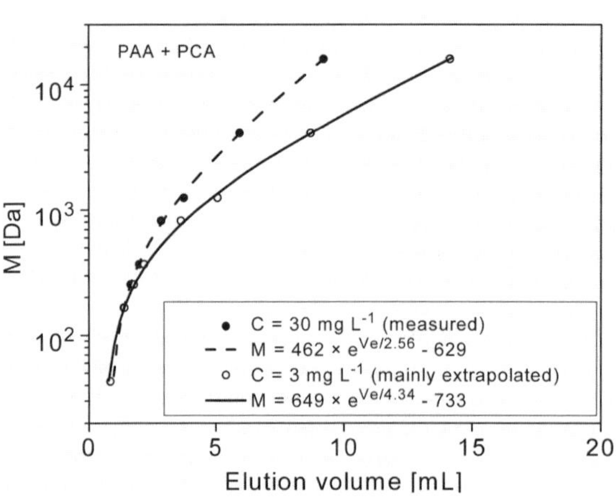

Figure 2.3 AFFFF calibration curves of PAA standards and other PCAs at concentrations of 30 and 3 mg L^{-1}

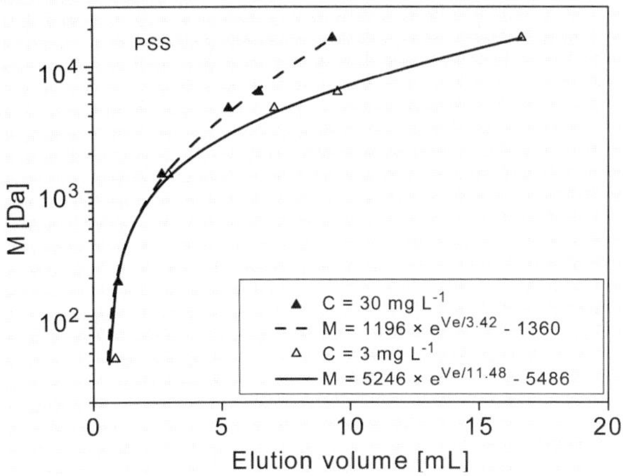

Figure 2.4 AFFFF calibration curves of PSS standards at concentrations of 30 and 3 mg L^{-1}

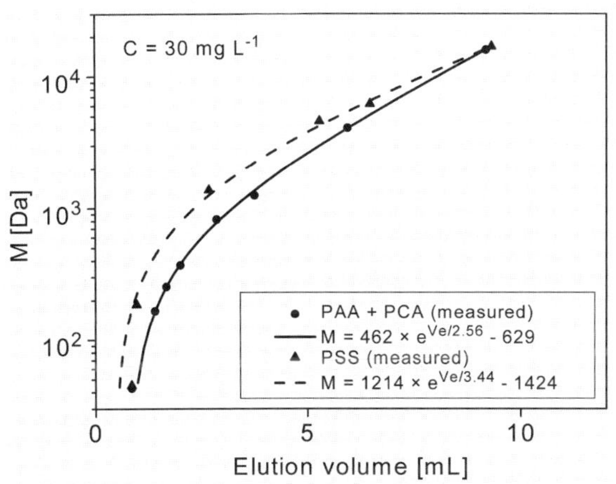

Figure 2.5 Comparison of AFFFF calibration curves of PAA standards + other PCAs and PSS standards at concentrations of 30 mg L^{-1}

To construct the calibration curve for very low concentrations (1-5 mgL^{-1} [19] necessary to get more realistic molecular size distributions), we measure or extrapolate the concentration dependence of the different standards down to the lowest possible measurable concentration (in our case 3 mgL^{-1}). The final calibration curves for PAA standards and other PCAs, and PSS standards used for calculation of the molecular size distribution (Figure 2.6), also show in the molar mass range up to ~ 10 kDa higher size to mass ratios for PAA standards and other PCAs than for PSS standards. Both calibration curves can be fitted to exponential functions given in Figure 2.6.

Loss of analyte with lower molecular sizes/masses through the membrane may shift the molecular size distribution to higher values and therefore the recovery of various analytes in the AFFFF system was estimated. The recovery was related to the integrated UV signal of the respective analyte measured under the same conditions without the AFFFF channel. It was found that, under our conditions and measurements at low concentrations (3 and 30 mg L^{-1}), about 80-90 % recovery for all analytes with molar masses of 254 Da and higher was achieved. Significantly lower recoveries were found for benzene-1,3-dicarboxylic acid (M_r 166) and very low recoveries were found for sodium azide, as expected.

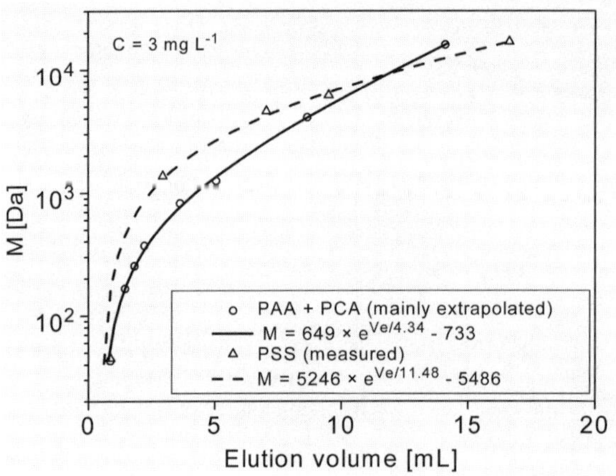

Figure 2.6 Comparison of AFFFF calibration curves of PAA standards + other PCAs and PSS standards at concentrations of 3 mg L^{-1}

2.3.2. Humic Substances

The fractograms obtained (after blank subtraction) for the analyzed HSs Gohy-573 HA, Gohy-573 FA and SRFA are shown in Figure 2.7. It is interesting to note that both FAs show differences in the UV absorption but have a similar range of elution volumes. Therefore, similar molecular size distributions for both FAs are expected. However, the HA shows a much larger range of elution volume. Therefore, a much broader molecular size distribution in comparison to both FAs is expected for this HA.

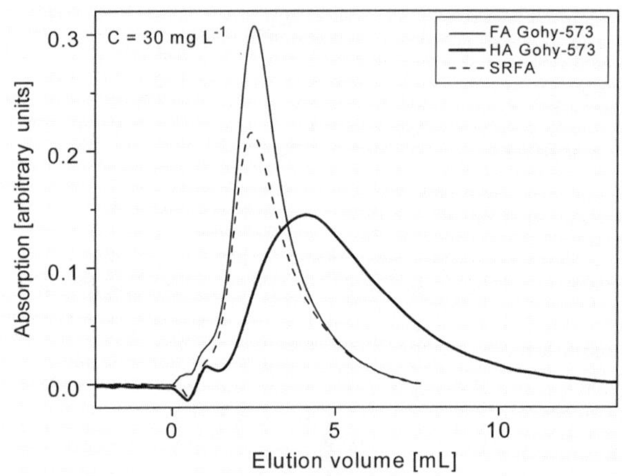

Figure 2.7 AFFFF fractograms of the HSs Gohy-573 HA, Gohy-573FA and SRFA

The transformation of the AFFFF fractograms into size distribution (relative mass concentration related to the mass of the standards used) plotted on a logarithmic mass scale was done by a) smoothing the fractogram obtained (UV signal versus elution volume Ve), b) calculating from Ve the corresponding molar mass with the calibration curves shown in Figure 2.6, c) correction of the UV signal (proportional to mass concentration) due to the nonlinearity of the calibration curve by the equation $H_{corr} = H_{meas}/\Delta(\log M)$, and d) transformation of the corrected UV signal to relative UV signal = relative mass concentration.

The derived molecular size distributions of the analyzed HSs are plotted in

Figure 2.8 for PAA + PCA calibration and in Figure 2.9 for PSS calibration. Compared to PSS calibration, all HSs show significantly lower values in the case of PAA + PCA calibration. Consequently, all molecular size distribution parameters derived from the molecular size distributions with PAA + PCA calibration are significantly lower than with PSS calibration (Table 2.1). Evaluated from PAA + PCA calibration, both FAs show the smallest but relatively similar size distributions with molecular sizes up to ~ 3 kDa, whereas the HA shows larger size distributions up to ~ 14 kDa. Also, the distribution parameters M_p' (peak value in the molecular size distribution plotted on a logarithmic molecular size scale), M_p (peak value in the molecular size distribution plotted on a linear molecular size scale), M_n (number average molecular size) or M_w (weight average molecular size) are relatively similar for both FAs and fall in the range 390 Da (M_p) – 650 Da (M_w). However, the HAs show in all cases significantly higher values with a range of 700 Da (M_p) – 1760 Da (M_w). As expected, the polydispersity of both FAs have lower values (1.3 – 1.4) than those of the HA (1.8). With PSS calibration all distribution parameters are much higher but the polydispersity is slightly lower than with PAA + PCA calibration.

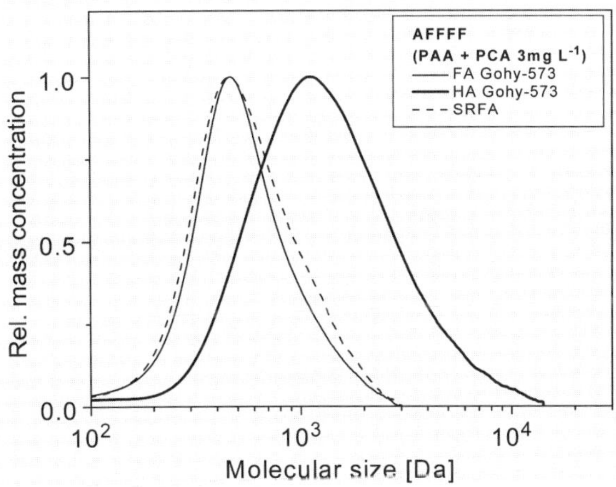

Figure 2.8 Molecular size distributions of Gohy-573 FA, Gohy-573 HA and SRFA evaluated with the PAA + PCA calibration curve (3 mg L^{-1})

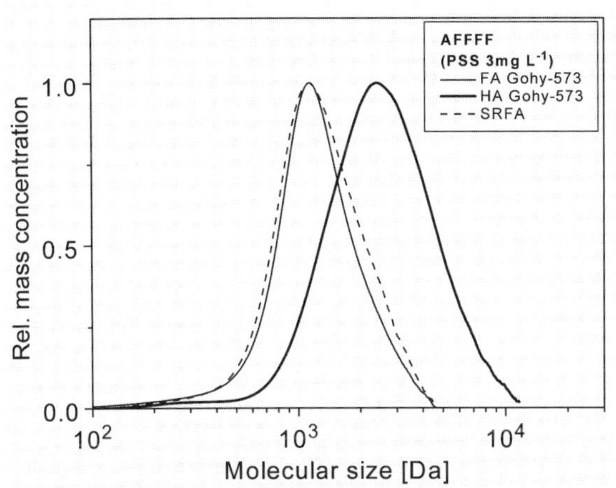

Figure 2.9 Molecular size distributions of Gohy-573 FA, Gohy-573 HA and SRFA evaluated with the PSS calibration curve (3 mg L^{-1})

2.3.3. Comparison with Other Methods

Indirect deduction of molar mass distributions requires directly comparable size to mass ratios of the standards used and the investigated substances. Comparison of the molar mass distribution of fulvic acid from application of time-of-flight secondary ion mass spectrometry (TOF-SIMS) [9,13,30] and electrospray quadrupole time-of-flight mass spectrometry [1,29] shows that the size to mass ratios of PAAs and various polycarboxylic acids are similar but size to mass ratios of PSSs are lower than that of fulvic acid. Therefore, with PSS standards corrections for the deviation in mass to size ratio under given experimental conditions are required. These findings are supported by some recently published molar mass data of SRFA (Table 2.2) that are very similar to our data derived with PAA + PCA calibration.

Table 2.1 Molecular size distribution parameters of Gohy-573 FA, Gohy-573 HA and SRFA obtained by AFFFF measurements and related to PAA + PCA or PSS calibration

HSs	Calibration	M_p' [Da]	M_p [Da]	M_n [Da]	M_w [Da]	M_w/M_n
Gohy-573 FA	PAA	460	410	470	620	1.3
Gohy-573 HA	PAA	1120	700	990	1760	1.8
SRFA	PAA	460	390	480	650	1.4
Gohy-573 FA	PSS	1140	1040	1140	1380	1.2
Gohy-573 HA	PSS	2410	1860	2000	2910	1.5
SRFA	PSS	1140	990	1080	1410	1.3

Table 2.2 Comparison of molecular mass/size data of SRFA from recent literature with our results obtained by AFFFF measurements and related to PAA + PCA calibration

Method	Type of distribution parameter	Data from literature [Da]	This work [Da]
ESI/QMS [2]	M_n	591	480
TOF-SIMS [30]	M_n	423/447	480
Solution equilibria [34]	M_n	551	480
HPSEC [14]	M_p	483	390

2.4. CONCLUSIONS

Calibration of AFFFF using PAA standards and other PCAs shows higher size to mass ratios than PSS standards in the molar mass range up to ~ 10 kDa; in the range ~ 10-17 kDa both types of standards show similar size to mass ratios. Molecular size/mass distribution data of FA measured with AFFFF and calibration with PAA and other PCAs are significantly lower than with PSS calibration but are supported by results from other methods. Therefore the size to mass ratios of PAA standards and other PCAs are more similar to FA than PSS standards and give, in accordance with HPSEC studies, more reliable molar mass distributions of FAs and HAs in the molar mass range below ~ 10 kDa.

ACKNOWLEDGEMENTS

The authors wish to thank Dr. S. Geyer (UFZ-Umweltforschungszentrum Leipzig-Halle, Department of Hydrogeology) for support in sampling of Gorleben groundwater Gohy-573 in the field, D. Jurrat for the very time consuming preparation of large batches of Gohy-573 FA and HA, and G. Teichmann for technical support in the laboratory. We thank the European Commission for partially funding this study under the project HUPA (Humic Substances in Performance Assessment of Nuclear Waste Disposal: Actinide and Iodine Migration in the Far-Field; contract No. FIKW-CT-2001-00128).

REFERENCES

1. Plancque G, Amekraz B, Moulin V, Toulhoat P, Moulin Ch. Molecular structure of fulvic acids by electrospray with quadrupole time-of-flight mass spectrometry. *Rapid Commun. Mass Spectrom.*, 2001; 15: 827-835.
2. Leenheer JA, Rostad CE, Gates PM, Furlong ET, Ferrer I. Molecular resolution and fragmentation of fulvic acid by electrospray ionization/multistage tandem mass spectrometry. *Anal. Chem.*, 2001; 73: 1461-1471.
3. Stenson AC, Marshall AG, Cooper WT. Exact masses and chemical formulas of individual Suwannee River fulvic acids from ultrahigh resolution electrospray ionization Fourier transform ion cyclotron resonance mass spectra. *Anal. Chem.*, 2003; 75: 1275-1284.
4. Stevenson FJ. *Humus chemistry: Genesis, composition, reactions.* 2nd Edn., New York: Wiley, 1994.
5. MacCarthy P. The principles of humic substances: An introduction to the first principle. In: Ghabbour EA, Davies G eds. *Humic substances: Structures, models and functions.* Cambridge: Royal Society of Chemistry, 2001: 19-30.
6. Moulin V, Amekraz B, Barre N, Plancque G, Mercier F, Reiller P, Moulin Ch. The role of humic substances in trace element mobility in natural environments and applications to radionuclides. In: Ghabbour EA, Davies G eds. *Humic substances: Nature's most versatile materials.* New York: Taylor & Francis, 2003: 275-286.
7. Buckau G, Wolf M, Geyer S, Artinger R, Kim JI. Influence of humic substances on the migration of actinides in groundwater. In: Ghabbour EA, Davies G eds. *Humic substances: Nature's most versatile materials.* New York: Taylor & Francis, 2003: 287-296.
8. Janoš P. Separation methods in the chemistry of humic substances. *J. Chromatogr. A,* 2003; 983: 1-18.
9. Wolf M, Buckau G, Geckeis H, Thang NM, Hoque E, Szymczak W, Kim JI. Aspects of measurement of the hydrodynamic size and molecular mass distribution of humic and fulvic acids. In: Ghabbour EA, Davies G eds. *Humic*

substances: Structures, models and functions. Cambridge: Royal Society of Chemistry, 2001: 51-61.

10. Artinger R, Buckau G, Kim JI, Geyer S. Characterization of groundwater humic and fulvic acids of different origin by GPC with UV/Vis and fluorescence detection. *Fresenius J. Anal. Chem.,* 1999; 364: 737-745.
11. Perminova IV, Frimmel FH, Kovalevskii DV, Abbt-Braun G, Kudryavtsev AV, Hesse S. Development of predictive model for calculation of molecular weight of humic substances. *Water Res.,* 1998; 32: 872-881.
12. Chin Yu-P, Aiken G, O'Loughlin E. Molecular weight, polydispersity, and spectroscopic properties of aquatic humic substances. *Environ. Sci. Technol.,* 1994; 28: 1853-1858.
13. Wolf M, Teichmann G, Hoque E, Szymczak W, Schimmack W. Copper speciation in aqueous solutions of fulvic acid and related molecular weight distributions. *Fresenius J. Anal. Chem.*, 1999; 363: 596-599.
14. Hoque E, Wolf M, Teichmann G, Peller E, Schimmack W, Buckau G. Influence of ionic strength and organic modifier concentrations on characterization of aquatic fulvic and humic acids by high-performance size-exclusion chromatography. *J. Chromatogr. A,* 2003; 1017: 97-105.
15. Giddings JC. A new separation concept based on a coupling of concentration and flow uniformities. *Sep. Sci.*, 1966; 1: 123-125.
16. Beckett R, Jue Z, Giddings JC. Determination of molecular weight distributions of fulvic and humic acids using flow field-flow fractionation. *Environ. Sci. Technol.*, 1987; 21: 289-295.
17. Schimpf ME, Petteys MP. Characterization of humic materials by flow field-flow fractionation. *Colloids Surf. A,* 1997; 120: 87-100.
18. Beckett R, Schimpf ME. Characterization of humic substances. In: Schimpf ME, Caldwell K, Giddings JC eds. *Field-flow fractionation handbook.* New York: Wiley-Interscience, 2000: 497-505.
19. Thang NM, Geckeis H, Kim JI., Beck HP. Application of the flow field flow fractionation (FFFF) to the characterization of aquatic humic colloids: evaluation and optimization of the method. *Colloids Surf. A*, 2001; 181: 289-301.
20. Bell J, Amarasiriwardena D, Siripinyanond A, Xing B, Barnes RM. Investigation of colloidal properties and trace metal complexation characteristics of soil-derived fulvic acids by flow field-flow fractionation-inductively coupled plasma-mass spectrometry (Flow FFF-ICP-MS). In: Ghabbour EA, Davies G eds. *Humic substances: Nature's most versatile materials.* New York: Taylor & Francis, 2003: 229-242.
21. Wahlund K-G, Giddings JC. Properties of an asymmetrical flow field-flow fractionation channel having one permeable wall. *Anal. Chem.*, 1987; 59: 1332-1339.
22. Litzén A, Wahlund K-G. Effects of temperature, carrier composition and sample load in asymmetrical flow field-flow fractionation. *J. Chromatogr.,* 1991; 548: 393-406.
23. Wahlund K-G. Asymmetrical flow field-flow fractionation. In: Schimpf ME, Caldwell K, Giddings JC eds. *Field-flow fractionation handbook.* New York: Wiley-Interscience, 2000: 279-294.
24. Wolf M, Szymczak W, Chanel V, Buckau G. Molecular size and mass distributions of humic substances measured by AFFFF and TOF-SIMS. In: Buckau G ed. *Humic substances in performance assessment of nuclear waste disposal: Actinide and iodine migration in the far-field.* Second Technical Progress Report, FZK/INE, Karlsruhe, 2004 (in print).

25. Fievre A, Solouki T, Marshall AG, Cooper WT. High-resolution Fourier transform ion cyclotron resonance mass spectrometry of humic and fulvic acids by laser desorption/ionization and electrospray ionization. *Energy Fuels,* 1997; 11: 554-560.
26. Brown TL, Rice JA. Effect of experimental parameters on the ESI FT-ICR mass spectrum of fulvic acid. *Anal. Chem.,* 2000; 72: 384-390.
27. Reemtsma T, These A. On-line coupling of size exclusion chromatography with electrospray ionization-tandem mass spectrometry for the analysis of aquatic fulvic and humic acids. *Anal. Chem.,* 2003; 75: 1500-1507.
28. These A, Reemtsma T. Limitations of electrospray ionization of fulvic and humic acids as visible from size exclusion chromatography with organic carbon and mass spectrometric detection. *Anal. Chem.,* 2003; 75: 6275-6281.
29. Plancque G. Characterization of fulvic acids by electrospray with quadrupole/time-of-flight mass spectrometry. In: Buckau G ed. *Humic substances in performance assessment of nuclear waste disposal: Actinide and iodine migration in the far-field.* Second Technical Progress Report, FZK/INE, Karlsruhe, 2004 (in print).
30. Szymczak W, Wolf M, Wittmaack K. Characterization of fulvic acids and glycyrrhizic acid by time-of-flight secondary ion mass spectrometry. *Acta hydrochim. hydrobiol.*, 2000; 28: 350-358.
31. Wolf M, Buckau G, Geyer S. Isolation and characterization of new batches of Gohy-573 humic and fulvic acids. In: Buckau G ed. *Humic substances in performance assessment of nuclear waste disposal: Actinide and iodine migration in the far-field.* Second Technical Progress Report, FZK/INE, Karlsruhe, 2004 (in print).
32. Artinger R, Buckau G, Geyer S, Fritz P, Wolf M, Kim JI. Characterization of groundwater humic substances: Influence of sedimentary organic carbon. *Appl. Geochem.*, 2000; 15: 97-116.
33. Buckau G, Artinger R, Geyer S, Wolf M, Fritz P, Kim JI. Groundwater in-situ generation of aquatic humic and fulvic acids and the mineralization of sedimentary organic carbon. *Appl. Geochem.*, 2000; 15: 819-832.
34. Fattahi A, Solouki T. Using solution equilibria to determine average molecular weight of the Suwannee River fulvic acids. *Anal. Chim. Acta,* 2003; 496: 325-337.

Chapter 3

ESTIMATION OF THE BLACK CARBON CONTENT OF THE IHSS LEONARDITE-HUMIN

Guangwei Ding and James A. Rice

Department of Chemistry and Biochemistry,
South Dakota State University, Brookings, SD 57007, USA

3.1. INTRODUCTION

Black carbon encompassing soot, charcoal and products of incomplete combustion contributes significantly to the aromatic components of natural organic matter in soils and sediments [1]. Black carbon is ubiquitous and can be found in air, soils, sediments, waters, and ice [1-6]. Its influence on the transport and bioavailability of organic contaminants is well known [7-9]. However, due to its complexity and its low solubility, analytical techniques for its quantitative determination are limited [10-16].

Golchin et al. [14] reported that the ^{13}C nuclear magnetic resonance (NMR) spectrum of charcoal has a distinctive single resonance centered at approximately 130 ppm. However, C- and H-substituted aromatic carbons in structures such as lignin also occur at similar chemical shifts. It is difficult to discriminate the aromatic signal of charcoal from resonances of other associated components. Skjemstad et al. [12] used a combination of photo-oxidation and NMR spectroscopy to obtain a rough measure of black carbon in several Australian soils. They have demonstrated that up to 30% of the soil organic carbon was charcoal. Their approach suffered from many uncertainties caused mainly by the non-quantitative nature of ^{13}C CP (cross-polarization) MAS (magic-angle-spinning) NMR [1]. The more reliable Bloch decay NMR method is not applicable to large sample sets because of the long measurement times required [12,17].

Dipolar dephasing has been used to probe ^{1}H-^{13}C dipolar interactions [18,19]. It is a simple matter to turn off the high power decoupling for a short time so that the linewidths of the carbons that are close to protons become so broad relative to the linewidths of the other carbons that they are not observed [20-23]. The rate of loss of the ^{13}C NMR signal intensity as a function of dipolar dephasing time is a complex exponential decay influenced mostly by (i) dipolar interaction between carbons and adjacent protons and (ii) rotational motion. Carbons directly bonded to protons decay rapidly unless rotational motion reduces the dipolar interactions.

In this study, recoupled long-range C-H dipolar dephasing solid-state NMR [24,25] was used for distinguishing various types of aromatic residues in natural organic matter fractions. It is achieved by recoupling of long-range C-H dipolar interactions using two ^1H 180° pulses per rotation period. This increases dephasing of unprotonated carbon signals approximately fourfold compared to standard dipolar dephasing without recoupling. It also reduces the effects of spinning-speed dependent effective proton-proton dipolar couplings on the heteronuclear dephasing. This technique, coupled with DP (direct-polarization), is particularly useful for identifying black carbon in natural organic matter.

Humin is defined as the fraction of humic substances that is insoluble in an aqueous solution at any pH value. It has generated relatively little research interest compared with the other humic fractions [26]. However, humin typically represents more than 50% of the organic carbon in a soil [26,27] and more than 70% of the organic carbon in unlithified sediments [28,29]. Furthermore, a significant fraction of organic contaminants introduced into a natural system binds rapidly to humin. Humin's structural aspects and its role in the carbon cycle need further investigation. In this paper, we investigate the black carbon content of an IHSS (International Humic Substance Society) leonardite humin. The IHSS reference peat before and after lipid extraction was used for comparison.

3.2. MATERIALS AND METHODS

3. 2.1. Sample Preparation

The leonardite sample was obtained from IHSS and contains 46.5% C and 13% ash. The IHSS reference peat, referred to as the Pahokee peat, contains 45.7% C and 15% ash [30].

The lipid extracted IHSS peat was prepared by the following procedure. The air-dried sample was placed in a cellulose thimble and extracted with a benzene:methanol azeotrope (3:1, v:v) for a 3-day period in a Soxhlet apparatus.

The IHSS leonardite humin was prepared using methods similar to those described previously [31-33] with some modifications. Air-dried leonardite was extracted with 0.1 M NaOH (three times) with shaking at room temperature for 24 hours. After removal of the extract, the insoluble residue (crude humin) was washed with deionized water until the pH was near 7, dried and ground. 100 mL of 1.0 M HF was added to 10 g of the crude humin in a plastic centrifuge bottle containing a magnetic stir bar. The bottle was shaken for 4 hours and the stir bar was taken out to remove any magnetic particles clinging to it. The bottle was centrifuged at 3000 rpm, the HF was replaced with fresh solution and the stir bar was replaced. The HF treatment was applied a total of 6 times.

3.2.2. Solid-state NMR Experiments

Before setting up the recoupled long-range C-H dipolar dephasing direct-polarization (DP) NMR experiments, the T_1^C (^{13}C spin-lattice relaxation time) was determined by CP/T_1-TOSS (total sideband suppression) experiments to find the shortest recycle delay [24]. Samples were packed in a 4-mm diameter zirconia rotor with a Kel-F cap. Acquisition was at 75 MHz on a Bruker DSX-300 spectrometer at a spinning speed of 8.0 kHz. The ^{1}H 90° pulse length was approximately 3.4 μs and the ^{13}C 180° pulse was approximately 8.0 μs. The contact time was 1 ms. The details of the experiments are reported elsewhere [25].

In all the recoupled long-range C-H dipolar dephasing experiments, TPPM (two pulse phase modulation) decoupling was used during detection. A sweepwidth of 100 kHz was used. The free-induction decay data were processed with a 50-Hz Lorentzian line broading and a 0.005s Gaussian broadening. For the IHSS peat and its lipid-extracted fraction, 1200 scans were acquired with a recycle delay of 12 s. For the leonardite humin, 1500 scans were acquired with a recycle delay of 12 s. The reference spectrum for each sample was run using a reference pulse sequence (DP/TOSS) so that the dipolar dephasing percentage can be compared directly.

Regular transverse (T_2) relaxation can occur in the presence of heteronuclear decoupling. To perform this calibration, the reference pulse sequence (DP/TOSS) was used to run at zero dephasing time and at similar dephasing times for the same sample.

3.3. RESULTS AND DISCUSSION

The series of DP/TOSS solid-state NMR spectra after recoupled dipolar dephasing of IHSS leonardite humin, and the IHSS peat and lipid-extracted IHSS peat are shown in Figures 3.1 and 3.2, respectively. In Figure 3.1, the curve with a 0 ms dipolar dephasing time is a typical humin spectrum indicating the carbon type distribution pattern. Humin is enriched in aliphatic structural units. The band centered at approximately 130 ppm is attributed to charcoal-like aromatic carbons. It is clear that no aliphatic carbon signals are observed with 0.25 ms dipolar dephasing.

Distinct differences were observed between the IHSS peat and the lipid-extracted peat (Figure 3.2). After lipid extraction, the alkyl carbon band (0-50 ppm) (Figure 3.2B) is less intense than in the whole material (Figure 3.2A, 0 ms). Little difference in the aromatic region (108-165 ppm) was observed between the two materials. As the dipolar dephasing time increases, the aromatic carbon signal intensity (107-142 ppm) significantly decreases for leonardite humin and both IHSS peat materials (Figures 3.1 and 3.2).

The relative changes in the aromatic carbon (107-142 ppm) content after recoupled dipolar dephasing decay are listed in Table 3.1. At a dipolar dephasing time of 0.25 ms, almost half of the aromatic carbon signal intensity of the leonardite humin remains while approximately 2/3 of the signal intensity for the IHSS peat samples were dephased (Table 3.1). If we continue to increase the dephasing time to 1 ms, approximately 1/5 of the aromatic carbon intensity is left for each sample. Mao and Schmidt-Rohr [24] reported that at a recoupled dipolar dephasing time of 0.86 ms, almost all the signals of softwood lignin have been dephased while more than half of the charcoal aromatic signals remain in a wood char sample. The decrease in the rate of signal dimensions at longer dephasing times can be attributed to the presence of a slowly-dephasing component whose dephasing appears to resemble that reported for black carbon [24]. Thus, we can estimate the amount of black carbon content in the humin and in IHSS peat.

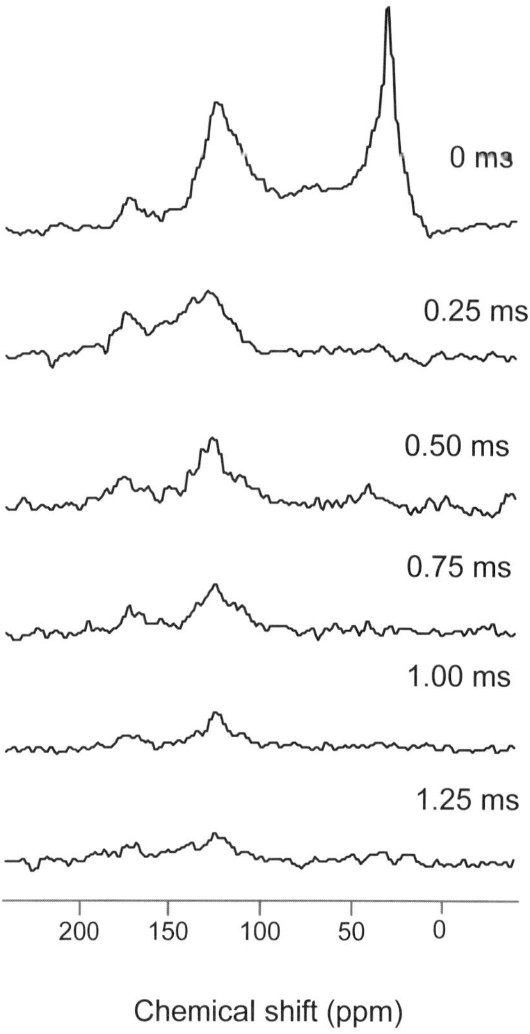

Figure 3.1 DP/TOSS spectra of IHSS leonardite humin after recoupled dipolar dephasing of the indicated duration

The data show that the black carbon-like component of the humin samples is approximately 20% of the total aromatic carbon content. When the humin spectra are compared to the lignin spectra published in Mao and Schmidt-Rohr's paper [24], the aromatic carbon decays more slowly in humin. This indicates that the aromatic carbons in humin are less protonated than the aromatic carbon in lignin.

Figure 3.3 shows the T_2 relaxation curve for leonardite humin at dipolar dephasing times of 0 ms and 1.75 ms. The calibration was conducted based on the signal intensity integration at dephasing time 0 ms (reference) and different dephasing time (e.g., 0.25 ms, 0.50 ms or 1 ms). A similar approach was used for the other two samples. The data points reported here have been corrected for the regular T_2 relaxation. Although there is still no general consensus on the fundamental nature of humin [25], a large black-carbon like aromatic carbon component was observed in our experiment (Figure 3.1).

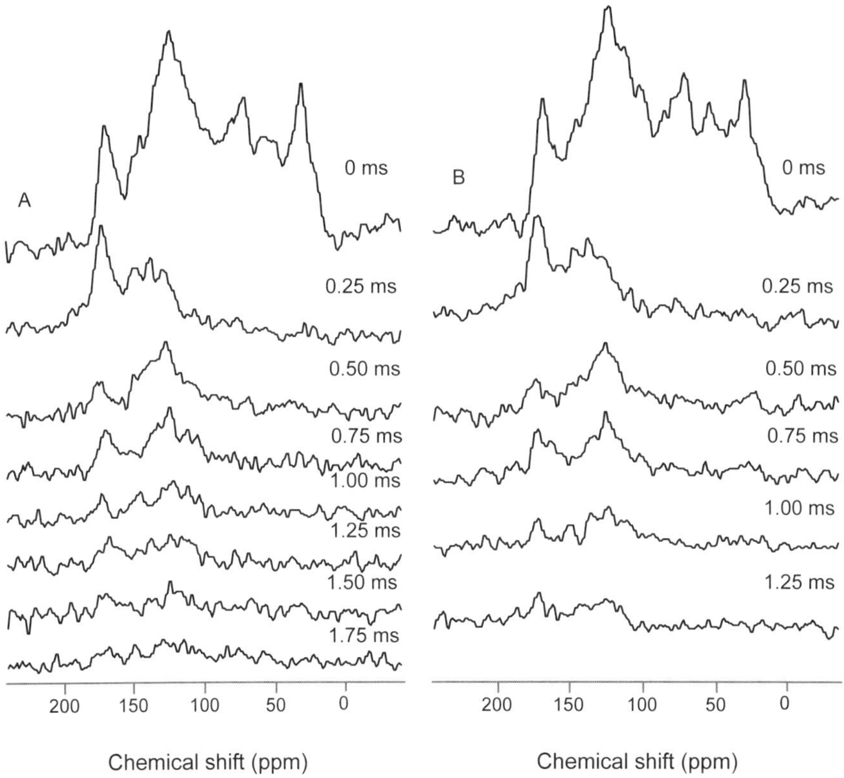

Figure 3.2 DP/TOSS spectra of IHSS peat after recoupled dipolar dephasing of the indicated duration. A: IHSS peat; B: IHSS peat lipid extracted fraction

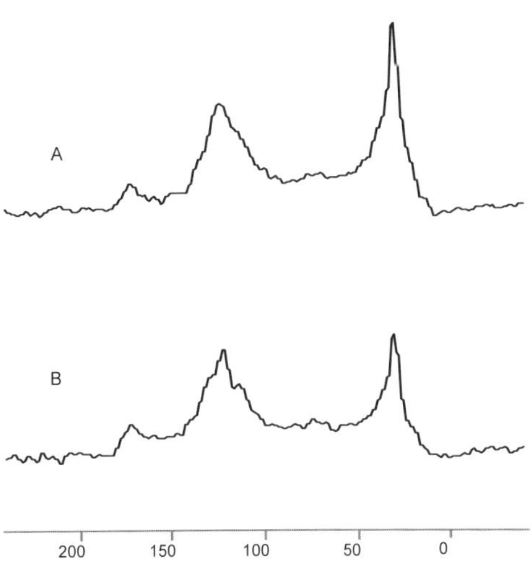

Figure 3.3 T_2 relaxation curve of IHSS leonardite humin: A. with dipolar dephasing time 0 ms. B. with dipolar dephasing time 1.75 ms

Table 3.1 Aromatic-C (107-142 ppm) signal intensity percentage after different recoupled dipolar dephasing decay times

	IHSS peat	IHSS peat lipid-extracted	IHSS leonardite humin
0.25 ms	33.6%	34.9%	47.2%
0.50 ms	29.3%	31.4%	29.7%
1.00 ms	19.8%	20.9%	18.4%
1.75 ms	10.1%		

3.4. CONCLUSIONS

Recoupled long-range C-H dipolar dephasing coupled with CP/T_1-TOSS solid state NMR was used to investigate the black carbon content of a humin. Approximately 20% of the aromatic carbon in the IHSS leonardite humin appears to be similar to black carbon.

ACKNOWLEDGEMENTS

This material is based upon work supported by the National Science Foundation/EPSCoR Grant #EPS-0091948 and by the State of South Dakota. We appreciate the assistance of Drs. Klaus Schmidt-Rohr and Jingdong Mao in implementing the pulse sequence.

REFERENCES

1. Glaser B, Haumaier L, Guggenberger G, Zech W. Black carbon in soils: the use of benzenecarboxylic acids as specific markers. *Org. Geochem.,* 1998; 29: 811-819.
2. Kuhlbusch TAJ. Black carbon and the carbon cycle. *Science*, 1998; 280: 1903-1904.
3. Scholes RJ, Noble IR. Perspectives: climate change: Storing carbon on land. *Science*, 2001; 294: 1012-1013.
4. Skjemstad JO, Taylor JA, Smernik RJ. Estimation of charcoal (char) in soils. *Commun. Soil Sci. Plant Anal.*, 1999; 30: 2283-2298.
5. Novakov T. The role of soot and primary oxidants in atmospheric chemistry. *Sci. Total Environ.*, 1984; 36: 1-10.
6. Masiello CA, Druffel ERM. Black carbon in deep-sea sediments. *Science*, 1998; 280, 1911-1913.
7. Braida WJ, Pignatello JJ, Lu Y, Ravikovitch PI, Neimark AV, Xing B. Sorption hysteresis of benzene in charcoal particles. *Environ. Sci. Technol.*, 2003; 37: 409-417.
8. Jonker MTO, Koelmans AA. Sorption of polycyclic aromatic hydrocarbons and polychlorinated biphenyls to soot and soot-like materials in the aqueous environment: mechanistic considerations. *Environ. Sci. Technol.*, 2002; 36: 3725-3734.

9. Karapanagioti HK, Sabatini DA. Impacts of heterogeneous organic matter on phenanthrene sorption: different aquifer depths. *Environ. Sci. Technol.*, 2000; 34: 2453-2460.
10. Schmidt MWI, Noack AG. Black carbon in soils and sediments: Analysis, distribution, implications, and current challenges. *Global Biogeochem. Cycles*, 2000; 14: 777-793.
11. Schmidt MWI, Skjemstad JO, Czimczik CI, Glaser B, Prentice KM, Gelinas Y, Kuhlbusch TAJ. Comparative analysis of black carbon in soils. *Global Biogeochem. Cycles*, 2001; 15: 163-167.
12. Skjemstad JO, Clarke P, Taylor JA, Oades JM, McClure SG. The chemistry and nature of protected carbon in soil. *Austral. J. Soil Res.*, 1996; 34: 251-271.
13. Kuhlbusch TAJ, Andreae MO, Cachier H, Goldammer JG, Lacaux JP, Shea R, Crutzen PJ. Black carbon formation by savanna fires: Measurements and implications for the global carbon cycle. *J. Geophys. Res.-Atmospheres*, 1996; 101: 23651-23665.
14. Golchin A, Clark P, Oades JM. The heterogeneous nature of microbial products as shown by solid-state ^{13}C CP/MAS NMR spectroscopy. *Biogeochem.*, 1996; 34: 71-97.
15. Kuhlbusch TAJ. Methods for determining black carbon in residues of vegetation fires. *Environ. Sci. Technol.*, 1995; 29: 1695-2702.
16. Gelinas Y, Prentice KM, Baldock JA, Hedges JI. An improved thermal oxidation method for the quantification of soot/graphitic black carbon in sediments and soils. *Environ. Sci. Technol.*, 2001; 35: 3519-3525.
17. Smernik RJ, Oades JM. Solid-state ^{13}C NMR dipolar dephasing experiments for quantifying protonated and non-protonated carbon in soil organic matter and model systems. *Europ. J. Soil Sci.*, 2001; 52: 103-120.
18. Hatcher PG. Chemical structural studies of natural lignin by dipolar dephasing solid-state carbon-13 NMR. *Org. Geochem.*, 1987; 11: 31-39.
19. Smernik RJ, Skjemstad JO, Oades JM. Virtual fractionation of charcoal from soil organic matter using solid-state ^{13}C NMR spectral editing. *Austral. J. Soil Res.*, 2000; 38: 665-683.
20. Wilson MA. *NMR Techniques and applications in geochemistry and soil chemistry*. Oxford: Pergamon Press, 1987.
21. Newman RH. Chemical shift anisotropies can dominate dephasing in "dipolar dephasing" CP/MAS NMR experiments. *J. Magnet. Resonance*, 1992; 96: 370-375.
22. Smernik RJ, Oades JM. The use of spin counting for determining quantitation in solid-state ^{13}C NMR spectra of natural organic matter. 1. Model systems and the effects of paramagnetic impurities. *Geoderma*, 2000; 96: 101-129.
23. Smernik RJ, Oades JM. The use of spin counting for determining quantitation in solid-state ^{13}C NMR spectra of natural organic matter. 2. HF-treated soil fractions. *Geoderma*, 2000; 96: 159-171.
24. Mao JD, Schmidt-Rohr K. Recoupled long-range C-H dipolar dephasing in solid-state NMR, and its use for spectral selection of fused aromatic rings. *J. Magnet. Resonance*, 2003; 162: 217-227.
25. Ding G, Rice JA. Determination of black carbon in natural organic matter samples using recoupled long-range dipolar dephasing solid-state ^{13}C NMR (submitted).
26. Rice JA. Humin. *Soil Sci.*, 2001; 166: 848-857.
27. Stevenson FJ. *Humus chemistry: Genesis, composition, reactions*. 2nd ed. New York: Wiley, 1994.

28. Mayer LM. Relationships between mineral surfaces and organic carbon accumulation in continental shelf sediments. *Geochim. Cosmochim. Acta.*, 1994; 58: 1271-1284.
29. Hatcher PG, Breger I, Maciel G, Szeverenyi N. Geochemistry of humin. In: Aiken GR, McKnight DM, Wershaw RL, MarCarthy P eds. *Humic substances in soil, sediment and water*. New York: Wiley, 1985: 275-302.
30. Kohl SD, Toscano PJ, Hou W, Rice JA. Solid-state ^{19}F NMR investigation of hexafluorobenzene sorption to soil organic matter. *Environ. Sci. Technol.*, 2000; 34: 204-210.
31. Ding G, Mao J, Herbert S, Amarasiriwardena D, Xing B. Spectroscopic evaluation of humin changes in response to soil managements. In Ghabbour EA, Davies, G eds. *Humic substances: Structures, models and functions.*. Cambridge: Royal Society of Chemistry, 2001: 261-271.
32. Preston CM, Newman RH. A long-term effect of N fertilization on the ^{13}C CPMAS NMR of de-ashed soil humin in a second-growth Douglas-fir stand of coastal British Columbia. *Geoderma*, 1995; 68: 229-241.
33. Preston CM, Schnitzer M, Ripmeester JA. A spectroscopic and chemical investigation on the de-ashing of a humin. *Soil Sci. Soc. Amer. J.*, 1989; 53: 1442-1447.

Chapter 4

SOURCE IDENTIFICATION AND CHARACTERIZATION OF HUMIC AND FULVIC ACIDS ISOLATED FROM OXFORDIAN ARGILLITE AND OPALINUS CLAY

Thorsten Schäfer,[1] Francis Claret,[1] Mirna Lerotic,[2] Gunnar Buckau,[1] Thomas Rabung,[1] Andreas Bauer[1] and Chris Jacobsen[2]

[1] Institute für Nukleare Entsorgung (INE), Forschungszentrum Karlsruhe, 76021 Karlsruhe, Germany
[2] Department of Physics & Astronomy, State University of New York, Stony Brook, New York 11794, USA

4.1. INTRODUCTION

Clay-organic matter studies focus predominantly on the process of biogeochemical carbon cycling and the understanding of carbon degradation/preservation pathways in aquatic systems. About 90 % of the preserved organic material (OM) in the global carbon cycle resides in amorphous, kerogen-type macromolecules disseminated in shale or other sedimentary rocks [1]. Over ninety percent of this total sedimentary organic matter cannot be physically separated from its mineral matrix. The observed mineral surface area loading for continental shelf sediments characteristically approaches a mono-layer mineral surface coverage [2]. Mechanisms governing sedimentary organic matter preservation in marine environments (< 0.5 % efficiency [3,4]) have remained unclear. Several theories have been proposed.

Besides the classical degradation-recondensation [5], selective preservation [6] and the natural sulphurization [7] pathways, a new mechanism recently has been proposed [8]. The preservation theory via an sorptive protection pathway [8] is based on the partition coefficient (K_D), which has to be in the order of 10^2-10^4 mL/g or higher to prevent OM degradation. Clay mineral natural biopolymer K_D values (illite 10^2 mL/g; montmorillonite 10^4mL/g) can explain protein preservation via this model [9]. Model refinement includes the combination of the mineral sorption and aggregate formation process to describe the OM preservation mechanism [10,11]. In this case, the OM distribution should be patchy, rather than a monolayer mineral surface coverage. OM distribution in Cenomanian black shale showing an alteration of organic and clay layers of approx. 100 nm thickness rather than molecular monolayers is in accordance with this theory [12]. Pyrolysis-GC/MS analysis of clay-organic associates in recent soils have demonstrated an enrichment of polysaccharides in kaolinite-dominated soils, whereas smectite-dominated soils contained larger amounts of aromatic compounds [13]. However, in ref 13 the total amount of organic carbon in the clay fraction was independent of the clay mineralogy. Combined transmission electron microscopy (TEM) and electron energy-loss spectroscopy (EELS) studies showed a preferential association of organic matter with Ca-containing smectite single platelets rather than with kaolinite and illite [14].

Residual organic matter in diagenetic overprinted shales or argillites of marine or terrestrial origin still show some extractable humic and/or fulvic acid [15-18]. Clay-rich low organic carbon sediments are considered as possible host rock formations for deep geological disposal of nuclear waste. This poses the question, if mineral preserved mostly insoluble organic matter (kerogen) can generate soluble compounds (humic and fulvic acids) having complexation properties with respect to radionuclides, then that might change the radionuclide transport mechanism (for example, colloidal transport) [19-22]. Another aspect (not discussed in detail in this paper) that has to be considered with respect to the long-term evolution of different barriers in the multi-barrier concept of a repository is the change in chemical stability of backfill clay and/or host rock clay due to clay-organic association [17,23,24].

The application of carbon x-ray absorption near edge structure (XANES) in environmental studies includes work on coal, aquatic humic and fulvic acids [23,25-27] and the reactivity of natural organic matter (NOM) inorganic mineral associates [23,24]. The objective of our investigation is to compare XANES spectra measured at the carbon K-edge, potassium L-edge, calcium L-edge, oxygen K-edge and iron L-edge with synchrotron based mid-infrared microspectroscopy and time-resolved laser fluorescence spectroscopy (TRLFS) measurements in order to determine

(a) the nanoscale spatial distribution and functional group composition of organic matter in two different clay-rich, low-carbon sedimentary rocks, namely the Callovo-Oxfordian argillite and the Opalinus shale considered as possible host rock formations for storage of high level nuclear waste (HLW),
(b) to localize possible sources for humic and/or fulvic acid release in these undisturbed sediments, and
(c) to investigate spectroscopically the trivalent actinide complexation behavior of these terrestrial sediment derived HA/FA like materials.

The analytical approach used in this study to localize and characterize HA/FA and NOM is to find natural groupings of spectromicroscopic data without prior knowledge of the spectra of all components [28,29]. This was performed with principal component analysis (PCA) [30] to orthogonalize spectromicroscopic data and to eliminate analytical noise. Subsequently, cluster analysis [31] was used to classify pixels according to their spectral similarity, and then recover gradations of thickness of representative materials using the observable spectra. Furthermore, the same set of principal components in combination with carbon K-edge target spectra of isolated humic and fulvic acids were used to define possible source regions of released humic and fulvic acids in these whole rock samples.

4.2. MATERIALS AND METHODS

4.2.1. Sample Origins

Callovo-Oxfordian sedimentary rock samples representative of the future French underground laboratory located in Bure (eastern Paris Basin, departments Meuse and Haute-Marne) were taken from four burial depths (447m, 490m, 494m, 516m) of the exploration borehole EST 104 [23,24]. The formation can be classified as argillite containing 40-45% clay minerals and ~50% calcite and quartz. Detailed mineralogical studies on the composition of the clay fraction < 2μm revealed in the upper part (447 to 490m) higher amounts of 2:1 layer silicates (13-20% smectite), whereas the lower rock formations have a decreased smectite content (~2%) and show considerable amounts of kaolinite (4-9%) and chlorite (1-2%) [24,32]. In this study we present data on the smectite-containing terrestrial sedimentation environment influenced sample of the argillite (sample 447m) [33]. For comparison, one Opalinus shale sample (579.19-579.45m) of the borehole Benken provided by the Swiss National Cooperative for the Disposal of Radioactive Waste (NAGRA) was selected [34]. The total organic carbon (TOC) concentration in the clay fraction is depth independent: ~1.3 wt% in the Bure argillite and < 0.4 wt% in the Opalinus clay. The maximum burial temperature of the Bure probably never exceeded 40°C, as indicated by organic matter characterization and fluid inclusions [35,36], whereas the Opalinus shale has seen a higher maximal burial temperature of 59°C to 78°C [37]. Additionally the < 2μm size-fraction of these rock samples, separated by centrifugation and pre-treated according to ref. 38, was investigated for comparison.

4.2.2. Sample Preparation

To prevent cross-contamination with organic embedding materials (e.g., epoxy-resin) the untreated rock samples and clay fractions were embedded in sulfur and ultra-microtomed (MVA, Inc., Norcross, GA) to a thickness of 100 nm and transferred to low carbon Formvar® grids for spectroscopic investigations. Aliquots of the clay fractions used for ultra-microtoming were repeatedly humic and fulvic acid extracted using the modified [18] isolation procedure [39] of the standard method developed by the International Humic Substance Society (IHSS) [40]. 5.2 % HA and 1.3 % FA of total clay fraction associated carbon were isolated from the argillite sample 447, whereas only 1.1 % FA could be extracted from the Opalinus clay. HA/FA sample preparation for scanning transmission x-ray microscope (STXM) analysis was performed by drying a 0.5 to 1 μL droplet of the suspension on a 100 nm thick Si_3N_4 window.

4.2.3. X-Ray Absorption Near Edge Structure (XANES) Spectromicroscopy

Measurements were performed at the Scanning Transition X-Ray Microscope (STXM) of the National Synchrotron Light Source (NSLS) at Brookhaven National Laboratory (BNL) in Upton, New York: undulator beamlines X1A1 (elements C, K and Ca) and X1A2 (elements O and Fe), operated by the State University of New York at Stony Brook [41]. The principle of this method is described in detail elsewhere [41,42]. Carbon K-edge, potassium L-edge and oxygen K-edge spectra were recorded using an undulator gap of 36.8 mm, whereas for iron L-edge and calcium L-edge measurements an undulator gap of 41.7 mm was used. The Fresnel zone plate utilized for C-, K- and Ca-edge measurements had a diameter (a) of 160 µm and an outermost zone width (δ) of 45 nm, whereas for O- and Fe-edge measurements a zone plate with a = 80 µm and δ = 30 nm was used. The larger 180 µm diameter zone plate needed for wet cell measurements [43] on X1A1 gives a theoretical spatial resolution (Rayleigh criterion) of 55 nm at a focal length of 1.7 mm [44]. Energy calibration of the spherical grating monochromators was performed with the photon energy of the CO_2 gas adsorption band at 290.74 eV (X1A1) and the O_2 gas adsorption band at 532.82 eV (X1A2), respectively [45,46]. The X1A1 STXM exit and entrance slit set-up was chosen to obtain an energy resolution ΔeV of 0.1 eV at the C K-edge [47]. XANES spectra were recorded using the image stack option [48]. Only spectra with at least 1% x-ray transmittance of the incoming flux were used to avoid spectral distortion.

STXM measurements give information on the product of sample thickness d and the mass absorption coefficient µ(E), which is related to the quotient of the incident flux on the sample $I_0(E)$ and the flux detected behind the sample I(E) by $\mu(E)*d = \ln[I_0(E)/I(E)]$. Image stacks are measured as a function of photon energy and the XANES spectra were extracted after stack alignment. Image regions free of particles or separately measured XANES spectra without samples gave the $I_0(E)$ information. All carbon XANES spectra were treated according to the deconvolution procedure described in ref. 48 for semi-quantitative analysis of functional group content. In order to compare isolated HA/FA spectra with cluster regions of the whole rock microtomes in both deconvolution procedures, a Gaussian peak at 290.2 eV for carbonate was implemented as shown in Table 4.1.

4.2.4. Synchrotron FT-IR Microspectroscopy

Infrared measurements were performed at the U10B beamline (National Synchrotron Light Source) using a Spectra-Tech Continuum IR microscope coupled to a Nicolet Magna 860 FTIR. The microscope utilizes a dual remote masking aperture and matching 32x Schwartzchild objectives [50]. Area mapping was performed using the dual remote masking aperture to define the sample area and spectra were collected with 10 µm x 10 µm aperture by averaging 1024 scans in the wavenumber range 650 – 4000 cm^{-1} per point in transmission mode at 4 cm^{-1} resolution using Atlµs software (Thermo Nicolet Instruments). Background information is collected through the blank Si_3N_4 window and the final data format is absorbance.

4.2.5. Time Resolved Fluorescence Laser Spectroscopy (TRFLS)

TRFLS measurements were performed with a pulsed Nd:YAG pumped dye laser system (Continuum, Powerlite 9030, ND 6000). An optical multichannel analyser consisting of a polychromator (Chromex 250) with a 1200 lines/nm grating is used for detection of the fluorescence emission. To fade out any light scattering and background fluorescence, the emission spectra of curium(III) were recorded 1 µs after the exciting laser pulse (laser dye: Exalite 398, excitation wavelength 396.6 nm) in the wavelength range 580-620 nm in a time window of 1 ms. For lifetime

Table 4.1 Semiquantitative analysis of the average carbon K-edge by spectral deconvolution in the energy range 280 – 295 eV. Values given are in proportion (%) of the area sum of the seven π^* and the C_{aliph} mixed valence state/Rydberg transition simulated by Gaussian peaks

Sample	benzoquinone-type	$C_{arom.}$	phenol-type	$C_{aliph.}$	carboxyl-type	O-alkyl-type	carbonate	$\sum(I_{fit}-I_{meas})^2$
447 HA isolate	9	17	14	12	22	8	14	0.033840
447 FA isolate	6	17	12	16	26	12	11	0.037275
OPA FA isolate	0	8	0	24	30	17	21	0.079951
Microtome 447								
smectite region	9	17	16	16	16	12	14	0.038092
illite/MLM region	0	8	3	23	15	25	26	0.060050
carbonate region	8	15	14	16	16	13	18	0.044007
Microtome Opalinus								
illite/MLM region (cluster B)	1	12	0	25	20	16	26	0.116719
illite/MLM region (cluster C)	3	13	2	24	13	19	26	0.071879
AOM[a]-type region	0	7	4	32	26	8	23	0.104606

[a] AOM: amorphous organic matter without identification of inorganic mineral association.

measurements, the delay time between laser pulse and camera gating was scanned with time intervals of 20 μs. Samples were prepared by adding 2 x 10^{-7} mol L^{-1} Cm^{3+} to the humic acid (447 HA: 10 mg L^{-1} DOC) solution, and 1 x 10^{-7} mol L^{-1} to the two different fulvic acids (447 FA: 8.7 mg L^{-1} DOC, OPA FA: 6.6 mg L^{-1} DOC) solutions in a glove box under an argon atmosphere. Investigations were conducted at pH 5.7-6. In this range Cm^{3+} hydrolysis is negligible and only two species are formed in significant concentrations. These species are the Cm fulvic acid complex and the non-complexed Cm^{3+} ion. Therefore, spectral deconvolution is straight-forward.

4.3. RESULTS AND DISCUSSION

4.3.1. X-Ray Absorption Near Edge Structure (XANES) Spectromicroscopy

Combination of the carbon, calcium, potassium, oxygen and iron edge spectra in this study gives the possibility to differentiate nanoscopically between organic association to carbonate minerals (strong Ca K-edge and carbon K-edge π* resonance at 290.2 eV), illitic clays (potassium $L_{2,3}$-edge), smectitic clays (Ca-edge and carbon K-edge absorption without the 290.2 eV carbonate peak) and pyrite/iron oxide (Fe $L_{2,3}$-edge absorption) in the heterogeneous mineral assemblage of the whole rock microtome.

Oxfordian Argillite (447 m Depth). STXM analysis of the Oxfordian argillite microtome sample (447 m burial depth) shows a strong increase of organic content in the smectite microtome region as indicated by the optical density (OD) shown in Figure 4.1.

This smectite rich region was identified in the upper microtome sector (Figure 4.2B) because of the absence of absorption bands 1) in the energy region of the potassium L2,3-edge, and 2) at the energy of 290.2eV (π*-transitions for carbonate); and 3) a detectable absorption at the calcium K-edge representing the exchangeable interlayer cations (Figure 4.2E). In the lower part of the microtome section, two regions of generally lower overall organic carbon content can be differentiated. One region (Figure 4.2C) shows a significant absorption at the potassium L2,3-edge, a considerable calcium K-edge absorption and some iron L2,3-edge absorption (Figure 4.2E), indicative of illitic clays or smectite/illite mixed layer minerals (MLM). This region shows significantly different carbon edge spectra, with lower absorption in the aromatic carbon region and higher aliphaticity (Table 4.1). The second region (Figure 4.2D) that could be identified in the lower part of the microtome section displays very strong calcium K-edge absorption, the highest iron L_3-edge absorption and a strong absorption band at 290.2 eV in the carbon K-edge spectra indicative of carbonate. From the spectroscopic results we can deduce that this organic matter is associated with carbonate phases, most probably a mixed (Fe,Ca)-carbonate or nano-crystalline intergrowth of pure siderite and calcite end members, not resolved by the resolution of STXM (<50 nm). The carbon edge of the smectite rich regions (Figure 4.2B) and the carbonate rich region (Figure 4.2D) are surprisingly very similar in the overall features, with the sole difference of increased absorption at the C(1s) π*-transition at 290.2 eV for carbonate and the generally lower organic association with carbonate, as indicated by the optical density (Figure 4.1).

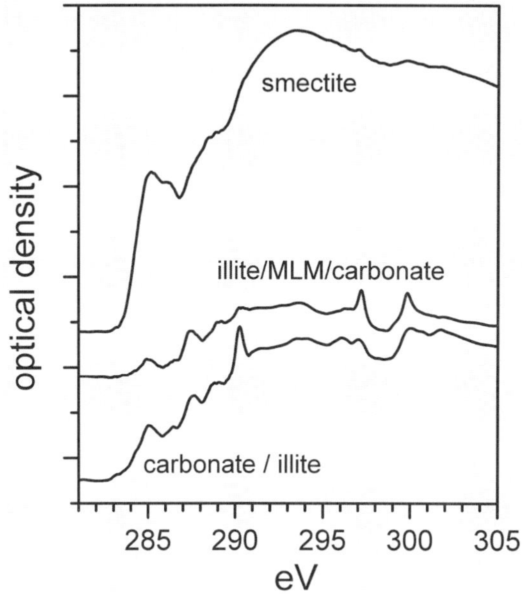

Figure 4.1 Optical density of different principal components identified in microtome section 447 of the Oxfordian argillite

Overall, these data clearly demonstrate a higher natural organic matter concentration with smectite type clay minerals. ^{14}C activity measurements in a recent Savanna system have demonstrated that clay mineralogy has a strong influence on organic matter preservation, with smectite having significantly lower organic matter turnover and higher mean residence time [51]. Therefore, the mineralogically-dependent organic carbon concentration differences observed in both the ancient Callovo-Oxfordian sediments (sample 447 m) and recent soils leads to the preliminary conclusion that only a minor organic matter overprint by diagenesis has occurred in the Bure sample. In this case, the functional group composition should change significantly with the clay mineralogy. The deconvolution of the average spectra (Table 4.1) extracted from the smectite rich region revealed a significantly higher proportion of aromatic carbon (42%) and a lower aliphatic carbon content (16%) compared to the illite/mixed layer mineral (MLM) region with 11% aromatic carbon and 23% aliphatic carbon, respectively. No particulate or amorphous organic matter (AOM) without mineral association could be observed in sample 447 under the given resolution of the STXM.

To estimate the source of soluble organic matter (that is, isolated humic and fulvic acid) in the whole rock samples, the FA/HA spectra shown in Figure 4.3C are used as target spectra to fit the first four principal components from the PCA [29]. As expected, deconvolution of the isolated HA (Table 4.1) generally showed a higher humic acid aromaticity ($\sum$(hydroquinone, C_{arom}, phenol-type) ~ 40%) compared to 35 % for the isolated FA and a lower aliphaticity of approximately 12% to 16% in FA, respectively.

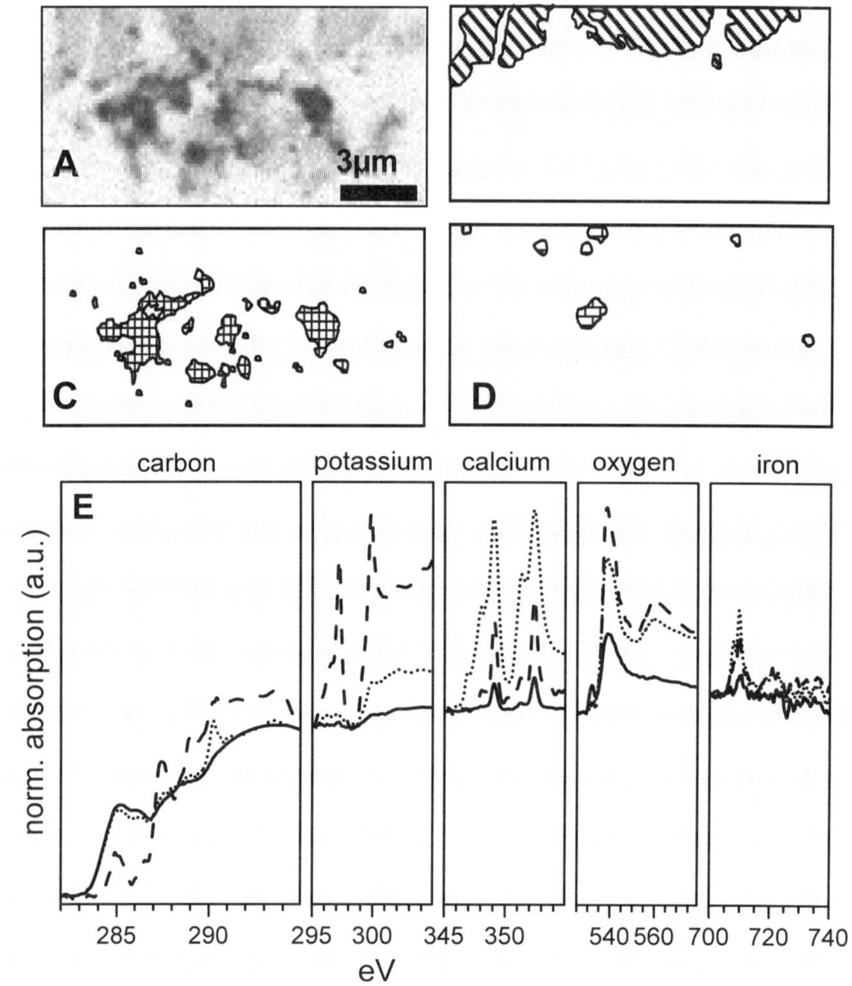

Figure 4.2 STXM analysis of whole rock microtome sample 447. In the upper part of the figure, the absorption image at 280 eV below the carbon K-edge (A) and three clusters identified using the first eight significant principle components from the PCA analysis are shown (B, C, D). Cluster analysis was performed over the energy range 280 to 740 eV with the following set-up: 4 clusters seeking procedure, eigenimage scaling factor: 0.2, first principal component removed, Euclidean distance measurement (for details see ref. 29). E: Carbon, potassium, calcium, oxygen and iron absorption spectra of the clusters found (cluster B solid line, cluster C dashed line, cluster D dotted line)

The maps of the raw target spectra (isolated FA, HA; not shown separately) and the fitted target spectra show a good correlation with light gray regions indicating high concentrations of spectral constituents and dark areas being a sign of depleted regions of target spectra. From the fitted target spectra maps (Figures 4.3A and 4.3B) and the results of the cluster analysis (Figure 4.2), the smectite rich region can be identified as a source of isolated HA in sample 447, whereas the isolated FA shows more similarities to the illite/MLM region of the microtome. However, detailed comparison of HA and FA target spectra with the fitting results shows systematic variations as well as good conformity (σ = 0.0206-0.0248). Both the isolated FA and HA have a significantly higher amount of carboxylic groups (288.6 eV) as well as $C_{arom} = C_{arom}$ or C_{arom}-H (285.2 eV), whereas a lower absorption intensity at 287.4 eV (C_{aliph}) and 286.6 eV (phenol type), respectively, is observed. Released HA or FA are most probably coming from surface layers (~ 10 nm) having slightly higher aromatic and carboxylic acid-like structure as indicated by the differences in the spectra. The decreased 287.4 eV intensity assigned to the mixed valence/Rydberg transition C-H* in the HA/FA target spectra can be explained by a decreased carbon atom number (chain length) in the alkyl pendant groups [52]. This observation, together with the significantly lower carboxyl type groups (1s-π^* transition of C=O), indicates a higher polymerization of the non extractable mineral-associated organic matter compared to the organic surface layer (HA/FA origin). Similar observations with aromatic, carbohydrate and carboxylic acid-like structures in the surface layer and increased aliphatic content in the bulk organic matter were found in studies of recent wetland soil in Saskatchewan [53]. These authors explained the observations of labile aliphatic structures in deeper layers of the organic coating by a protection mechanism involving surface layer refractory substances.

Opalinus Clay (Benken Depth 579m). Analysis of the Opalinus Clay microtome section as shown in Figure 4.4 reveals organic matter in regions where significant potassium and calcium absorption indicative of illite/mixed layer minerals (MLM) intergrown with nanocrystalline carbonate appear. Both clusters (Figures 4.4A and 4.4B) separated within this microtome region are very similar and show in the deconvolution (Table 4.1) a low aromaticity ($\sum$ (benzoquinone-type, $C_{arom} = C_{arom}$, phenol-type = 13-18%) with a low amount of C_{arom} associated oxygen containing functional groups (phenol-type 0-2%) and a high aliphaticity (24-25%). These values are in good agreement with the results found for illite/MLM region in the Oxfordian argillite 447 (see Table 4.1) and show that the illite/MLM associated organic matter of both host rock types is comparable with respect to organic functionality.

This observation raises the question of whether the illite/MLM region serves as a source of fulvic acid-like organics in the Opalinus shale. Therefore, the spectra of the isolated Opalinus FA (OPA FA, see Figure 4.5B1) were used as target spectra to fit the first four principal components of PCA [29]. From the fitted target spectral map and the results of the cluster analysis, the illite/MLM region (Figure 4.5B1) can be identified with a high certainty (σ= 0.0365) as a source of the isolated FA. Differences between the combination of the first four principal components and the target spectra (isolated FA) are a lower absorption in the phenol-type group region and a higher absorption in the C_{aliph} region of the isolated OPA FA. Additional analysis on various other regions of the microtome section (not shown in detail) revealed that besides the illite/MLM association, particulate organics or amorphous organic carbon (AOM) without significant absorption on the potassium, calcium and iron-edge also could be the source of extractable FA (Figure 4.5B2). Besides the good correlation (σ = 0.0195), significantly higher amounts of C_{aliph} and lower carboxyl and O-alkyl type groups (47% in OPA FA versus 34% in AOM) are found in the AOM, indicating a higher hydrophobicity of this material and therefore a less probable FA source.

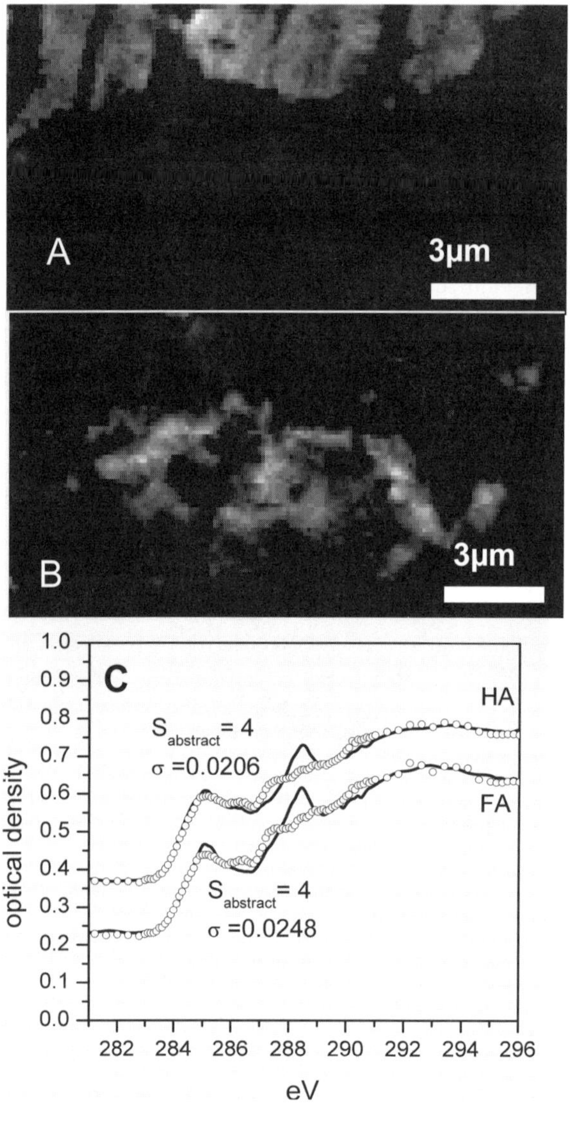

Figure 4.3 Target spectral analysis of the STXM data. In the upper part the generated maps from target spectra fits (A, B) are shown. C: shows the fit of isolated HA and FA target spectra using the first four significant principle components $S_{abstract}$ from the PCA (open circles) and the original isolated HA and FA target spectra (solid lines)

4.3.2. Synchrotron Based FT-IR Microspectroscopy

Oxfordian Argillite (Depth 447 m). The infrared spectra of the microtome section 447 is shown in Figure 4.6. A differentiation of the clay mineral assemblage as performed by X-Ray diffraction could not be accomplished from the spectra of the separated clay fraction and the rock microtome shown in Figure 4.6 [18,32]. In the region of Al-OH stretching vibrations of the aluminosilicate layers, a broad absorption at 3625 cm^{-1} and two additional significant humps at 3666 cm^{-1} and 3694 cm^{-1} could be identified. This broad hydroxyl absorption is typical of random substitution present in dioctahedral micas and smectites, where the MgAlOH absorption at 3687 cm^{-1} could not be resolved from the Al_2OH absorption at 3620-

3630 cm^{-1} [54]. The additional hump at 3666 cm^{-1} can be assigned to a weak OH-stretching feature of trioctrahedral chlorites. Kaolinite group minerals typically show three sharp high frequency features at 3697, 3669 and 3652 cm^{-1}, and the 3620 cm^{-1} inner hydroxyl group band could not be differentiated, as expected [54]. The band at 1161 cm^{-1} can be assigned to the C-C bond in aliphatics and polymeric substances or to a Si-O-C bond, which might indicate a linkage of some organic macromolecules to the clay Si-O groups [55]. The absorption bands around 1076, 1039 and 916 cm^{-1}, and the doublet around 800 cm^{-1} can be identified as Si-O or Al-OH vibrations in quartz and/or aluminosilicates, whereas the absorption band near 1415 cm^{-1} can be assigned to the asymmetrical stretch of CO_3^{-2} in carbonates [54].

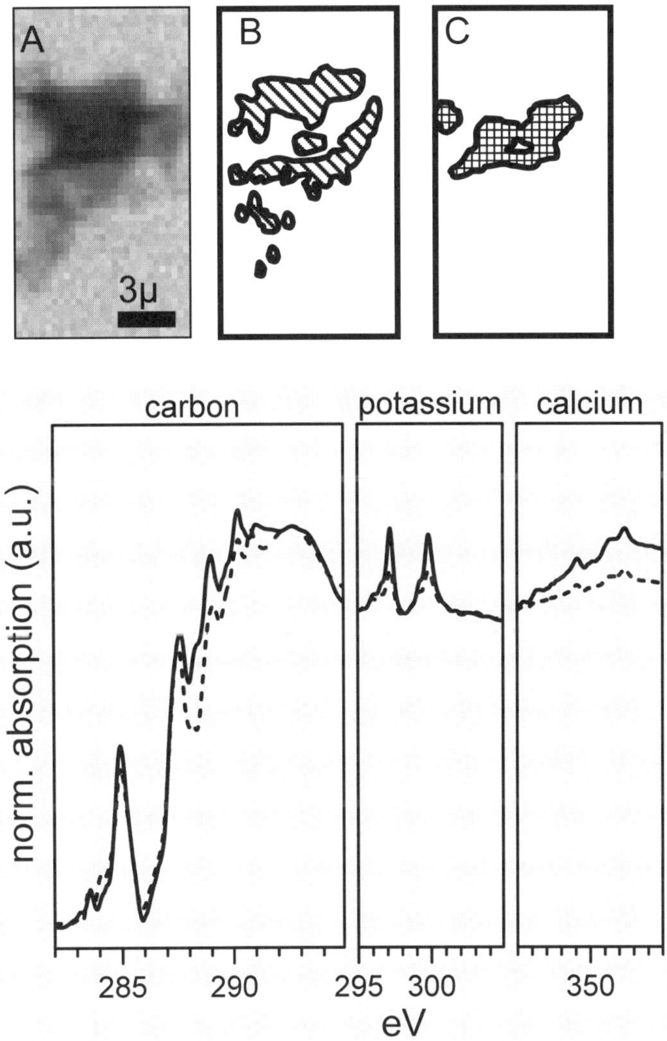

Figure 4.4 STXM analysis of whole rock microtome sample Opalinus. Upper part: the absorption image at 280 eV below the carbon K-edge (A) and two clusters identified using the first seven significant principle components from the PCA (B, C). Cluster analysis was performed over the energy range 280 to 360 eV with the following set-up: 4 clusters-seeking procedure, eigenimage scaling factor 0.2, first principal component removed, Euclidean distance measurement (for details see ref. 29). Lower part: Carbon, potassium and calcium absorption spectra of the clusters found (cluster B solid line, cluster C dashed line. No iron $L_{2,3}$- edge absorption could be detected)

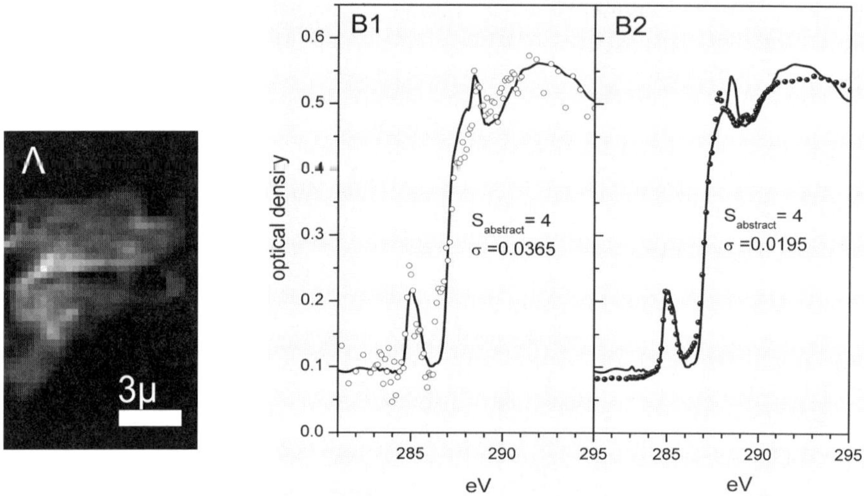

Figure 4.5 Target spectral analysis of the STXM data. The upper part (A) is the generated map from the target spectral fit. Figures B1 and B2 show the fit of isolated FA target spectra using the first four significant principal components $S_{abstract}$ from the PCA (B1: illite/MLM region; B2: amorphous organic matter region) and the original isolated FA target spectra (solid lines)

Characteristic bands for organic components include OH groups of phenols, alcohols and carboxylic OH near 3430-3383 cm^{-1}, aromatic CH groups by a shoulder at 3073 cm^{-1}, aliphatic CH_2 and CH_3 bands at 2929, 2870, 1450 and 1375 cm^{-1}, a carbonyl band of ketones, acids or esters at 1715 cm^{-1} and 1261 cm^{-1}, peptidic amide I and amide II bands at 1650 and 1540 cm^{-1} and the band at 1630 cm^{-1} for mostly aromatic C=C [5,55]. The above mentioned aliphatic CH_2 and CH_3 bands can be furthermore separated into a exclusively CH_3 absorption at 1375 cm^{-1} and a mixed linear CH_3 and cyclic CH_2 vibration at 1455 cm^{-1} [5]. To estimate the length, the degree of aliphatics and side-chain branching the technique of CH_2/CH_3 (2915-2940 cm^{-1}/2950-2975 cm^{-1}) intensity ratio measurement was applied [56]. Furthermore, data analysis was performed by an A-factor (3000-2800 cm^{-1}/3000-2800 cm^{-1}+1650-1520 cm^{-1}), and C-factor (1800-1650 cm^{-1}/1800-1650 cm^{-1}+1650-1520 cm^{-1}) calculation, which produces similar results to the classical H/C versus O/C van Krevelen plot (not shown) [57]. The smectite-rich region defined as an HA source plots in the region of type III kerogen or terrestrial derived organic material, whereas the illite/MLM region shows higher A-factor (H/C ratio) and C-factor (O/C ratio) values and falls in the region of type II kerogen typical of a marine influenced sedimentation environment. These results are compatible with the general observation of higher aliphaticity in FA compared to HA.

A clear indication of increased carboxyl group content from a significantly higher absorption at 1709 cm^{-1} in the illitic region of rock microtome sample 447 can be detected. However, compared to infrared spectra of isolated natural humic or fulvic acids [55,58] or humic-like substances generated in the presence of clays [59], this C=O band intensity of all clay associated organics is significantly reduced, indicating a loss of oxygen-containing functional groups as already shown by

XANES analysis. A second difference between regions B and C (Figure 4.6) is the band at 1161 cm^{-1}, which can be identified in the illite rich region C, but cannot be resolved in the smectite rich region B. Taking into account the C(1s) XANES target spectra analysis derived information of smectite as the potential HA source and the higher HA extraction yield compared to FA [18] of these sediments, a possible explanation of the higher HA mobilisation can be the missing Si-O-C linkage of humic macromolecules.

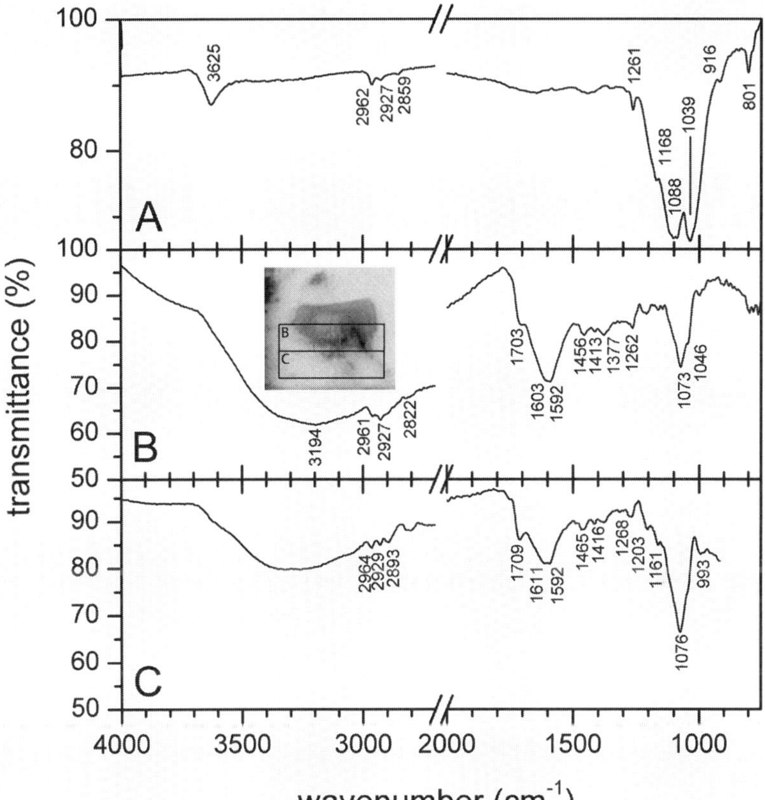

Figure 4.6 Synchrotron based infrared microspectroscopy of Oxfordian argillite sample 447m depth. A: average IR spectra of the extracted clay fraction, B: average spectra extracted from the smectite rich region B, C: average spectra of the illite/MLM region C. Size of square gives the spatial resolution used for spectra acquisition (10μm)

A direct implication of these observations is (a) a missing Si-O-C linkage of HA, (b) preferential HA release and (c) HA association with smectitic minerals could indicate that in chemically disturbed environments (high pH plume of cement corrosion) primarily the smectitic clay minerals are expected to react.

Opalinus Clay (Benken Depth 579m). The infrared spectrum of the Opalinus shale microtome section is shown in Figure 4.7. In the upper spectra (Figure 4.7), extracted from the clay-rich region A kaolinite group mineral, specific sharp high frequency features at 3696, 3669 and 3651 cm^{-1} besides the 3619 cm^{-1} inner hydroxyl group band could by identified [54]. The additional hump at 3666 cm^{-1} can be interpreted as a weak OH-stretching feature of trioctrahedral chlorites. The band observed at 1163 cm^{-1} can be assigned again to the C-C bond in aliphatics and polymeric substances or to a Si-O-C bond, which might indicate a linkage of some organic macromolecules to the clay Si-O groups [55]. The absorption bands around 1076 cm^{-1} and 1039 cm^{-1} can be identified as Si-O or Al-OH vibrations in quartz and/or aluminosilicates, whereas the absorption band near 1426 cm^{-1} can be assigned to the asymmetrical stretch of CO_3^{-2} in carbonates [54].

Characteristic bands for organic components include a broad band of phenols, alcohols and carboxylic OH near 3430-3383 cm^{-1}, aromatic CH groups by a pronounced shoulder at 3073 cm^{-1}, aliphatic CH_2 and CH_3 bands at 2960, 2922, 2857, 1460 and 1345 cm^{-1}, a pronounced carboxylic band of ketones, acids or esters at 1753 cm^{-1} and 1260 cm^{-1}, peptidic amide II bands at 1575 cm^{-1} and the band at 1631 cm^{-1} for mostly aromatic C=C [5,55]. The A-factors (H/C ratio) and C-factor (O/C ratio) data of the Opalinus shale microtome all fall near the boundary between type II and type III kerogen typical of a mixed terrestrial/marine influenced sedimentation environment. This result of the Opalinus shale deposition environment is in agreement with published data [37].

In the amorphous organic matter region B (Figure 4.7B) a very high absorption for organic OH (phenols, alcohols and carboxylic OH) around 3300 cm^{-1}, C_{arom}-H at 3069 cm^{-1}, aliphatic CH_2 and CH_3 bands (2960 to 2850 cm^{-1} and 1453 cm^{-1}), peptidic amide I and amide II bands at 1650 and 1540 cm^{-1} and the band at 1630 cm^{-1} for mostly aromatic C=C could be identified [5,55]. Striking is the rather high transmittance in the carbonyl band region of ketones, acids or esters at 1753 cm^{-1}, indicating a loss of oxygen containing functional groups and increase in the hydrophobicity/polymerization of this material.

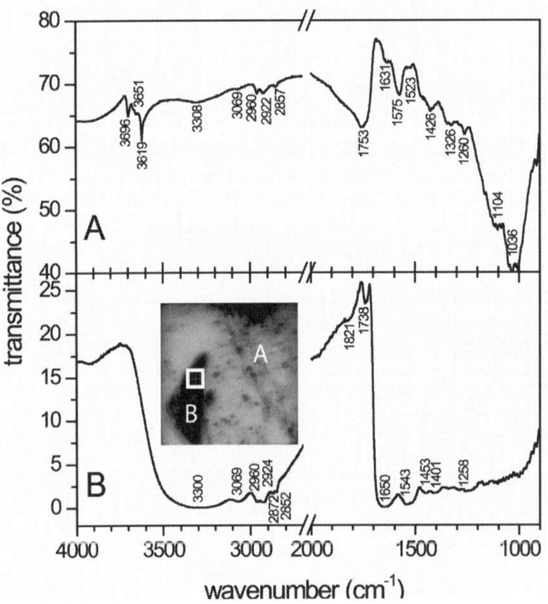

Figure 4.7 Synchrotron based infrared microspectroscopy of Opalinus clay microtome. A: average IR spectra of the clay rich region A, B: average spectra extracted from the amorphous organic matter(AOM) type material. Size of square gives the spatial resolution used for spectra acquisition (10μm)

4.3.3. Curium(III) Complexation Behavior

The trivalent actinide complexation behavior of isolated HA/FA was studied using time resolved fluorescence spectroscopy (TRFLS) on Cm(III) [60]. The emission spectra show the characteristic shape of humate and fulvate curium complexes with a peak maximum around600 nm, compared to 593.8 nm for the non-complexed Cm^{3+} aqua-ion as shown in Figure 4.8 [61]. The curium(III) fluorescence decay and derived lifetimes of the investigated humic and fulvic acid samples are basically indistinguishable and show the same time behavior as found for the Cm(III)-Aldrich humic acid complex. Whereas the 447 humic acid complexes Cm(III) almost quantitatively, both fulvic acids 447 FA and OPA FA still show free Cm(III) aqua-ions. Therefore, the relative peak areas of the two species in the different spectra were evaluated by peak deconvolution (for details see ref. 18). Quantification of the two respective Cm(III) species was accomplished by assuming the relative fluorescence intensity of the fulvate complex to be a factor of four higher than that of the non-complexed Cm^{3+} ion [61]. The Cm(III) loading capacity of sample 447 FA is calculated to be 58 %, as expected for fulvic acid at the given pH and ionic strength (between 50 and 63 % [62]). Contrary to this, the number found for the OPA FA is considerably lower (7.9 %).

The Opalinus clay derived FA shows a lower Cm(III) complex intensity than for the 447 FA. The reason for this lower Cm(III) FA fluorescence intensity is not yet clear. It may result from lower complexation strength of carboxylic groups that are adjacent to aliphatic groups. If such complexes are formed, they may show lower fluorescence intensity than those adjacent to aromatic/chromophoric structures. The lower fluorescence intensity could originate from lack of energy transfer from aliphatic structures of the FA ligand to the Cm(III) complex compared to such complexes in an aromatic/chromophoric vicinity. However, the method clearly shows that ancient (Oxfordian) sediment derived FA has comparable trivalent actinide complexation behavior to literature reported values found for groundwater isolated FAs.

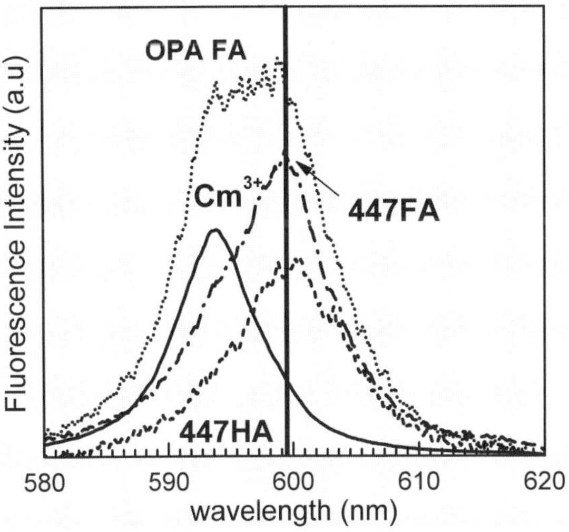

Figure 4.8 Cm(III) fluorescence spectra of isolated FA (447 FA, OPA FA) and HA (447 HA) investigated in this study. The vertical line gives the position of the humic/fulvic acid complex

4.5. CONCLUSIONS

The combination of carbon, potassium, calcium, oxygen and iron XANES spectra using PCA and cluster as well as target spectra analysis revealed a mineral dependent organic matter functionality in Jurassic low-carbon sedimentary formations, namely Oxfordian argillite and Opalinus shale. Furthermore, this method is capable of identifying the sources of isolated humic and fulvic acids in an undisturbed rock sample with nanoscopic spatial resolution. By employing synchrotron based FT-IR microspectroscopy with its high sensitivity to aliphatic carbon, it was also possible to retrieve additional information on the deposition environment. These results demonstrate that even in approximately 160 Ma old geological formations, part of the organic inventory keeps the structure/functionality of HA/FA. Investigations with time-resolved laser fluorescence spectroscopy (TRLFS) clearly showed that Oxfordian sediment derived FA-like material has not only kept its structure/functionality, but also shows comparable trivalent actinide complexation behavior, as expected for FA from literature data.

ACKNOWLEDGEMENTS

The results presented in this paper were partly collected during research supported by a Marie Curie fellowship of the European Commission programme "Nuclear Fission" under contract number FIKW-CT2002-50509. The authors are indebted to L. Griffault (ANDRA, France) and A. Gautschi (NAGRA, Switzerland) for providing core samples of the Callovo-Oxfordian argillite and Opalinus shale (Benken), respectively. We are grateful for beamtime allotment by BNL/NSLS. Spectromicroscopic data were collected using the X1A STXM developed by the group of Janos Kirz and Chris Jacobsen at SUNY Stony Brook, with support from the Office of Biological and Environmental Research, U.S. DOE under contract DE-FG02-89ER60858, and from the NSF under grant DBI-9605045. The zone plates were developed by Steve Spector and Chris Jacobsen of Stony Brook and Don Tennant of Lucent Technologies Bell Labs with support from NSF under grant ECS-9510499. The FTIR data were collected at beamline U10B of the NSLS, supported by the U.S. Department of Energy under contract DE-AC02-98CH10886.

REFERENCES

1. Hedges JI, Hu FS, Devol AH, Hartnett HE, Tsamakis E, Keil RG. Sedimentary organic matter preservation: A test for selective degradation under oxic conditions. *Am. J. Sci.*, 1999; 299: 529-555.
2. Hedges JI, Oades JM. Comparative organic geochemistries of soils and marine sediments. *Org. Geochem.*, 1997; 27: 319-361.
3. Hedges JI, Keil RG. Sedimentary organic-matter preservation - an assessment and speculative synthesis. *Mar. Chem.*, 1995; 49: 81-115.
4. Hedges JI, Keil RG. Sedimentary organic-matter preservation - an assessment and speculative synthesis - closing comment. *Mar. Chem.*, 1995; 49: 137-139.
5. Tissot BP, Welte DH, *Petroleum formation and occurence*. Hamburg: Springer, 1984.

6. Tegelaar EW, Deleeuw JW, Derenne S, Largeau, C. A Reappraisal of kerogen formation. *Geochim. Cosmochim. Acta*, 1989; 53: 3103-3106.
7. Sinninghe Damst'e JS, Rijpstra WIC, Kock-van Dalen AC, De Leeuw JW, Schenck PA. Quenching of labile functionalised lipids by inorganic sulphur species: Evidence for the formation of sedimentary organic sulphur compounds at the early stages of diagenesis. *Geochim. Cosmochim. Acta*, 1989; 53: 1343-1355.
8. Henrichs SM. Sedimentary organic-matter preservation - an assessment and speculative synthesis - a comment. *Mar. Chem.*, 1995; 49: 127-136.
9. Ding XL, Henrichs SM. Adsorption and desorption of proteins and polyamino acids by clay minerals and marine sediments. *Mar. Chem.*, 2002; 77: 225-237.
10. Bock MJ, Mayer LM. Mesodensity organo-clay associations in a near-shore sediment. *Marine Geol.*, 2000; 163: 65-75.
11. Ransom B, Bennett RH, Baerwald R, Shea K. TEM study of in situ organic matter on continental margins: Occurrence and the "monolayer" hypothesis. *Marine Geol.*, 1997; 138: 1-9.
12. Salmon V, Derenne S, Lallier-Verges E, Largeau C, Beaudoin B. Protection of organic matter by mineral matrix in a Cenomanian black shale. *Org. Geochem.*, 2000; 31: 463-474.
13. Wattel-Koekkoek EJW, van Genuchten PPL, Buurman P, van Lagen B. Amount and composition of clay-associated soil organic matter in a range of kaolinitic and smectitic soils. *Geoderma*, 2001; 99: 27-49.
14. Furukawa Y. Energy-filtering transmission electron microscopy (EFTEM) and electron energy-loss spectroscopy (EELS) investigation of clay-organic matter aggregates in aquatic sediments. *Org. Geochem.*, 2000; 31: 735-744.
15. Khalili F. Isolation and characterization of humic-acid from Jordanian oil shale. *Fuel*, 1990; 69: 151-156.
16. Krumholz LR, Harris SH, Suflita JM. Anaerobic microbial growth from components of cretaceous shales. *Geomicrobiol. J.*, 2002; 19: 593-602.
17. Claret F, Schäfer T, Bauer A, Fanghänel T. Generation of humic and fulvic acid from Callovo-Oxfordian clay under high alkaline conditions. *Sci. Tot. Environ.*, 2003; 317: 189-200.
18. Claret F, Schäfer T, Rabung T, Wolf M, Bauer A, Buckau G. Differences in properties and Cm(III) complexation behavior of isolated humic and fulvic acids derived from Opalinus clay and Callovo-Oxfordian argillite. *Appl. Geochem.*, in review.
19. Schäfer T, Artinger R, Dardenne K, Bauer A, Schuessler W, Kim JI. Colloid-borne Americium migration in Gorleben groundwater: Significance of iron secondary phase transformation. *Environ. Sci. Technol.*, 2003; 37: 1528-1534.
20. Artinger R, Schuessler W, Schäfer T, Kim JI. A kinetic study of Am(III)/humic colloid interactions. *Environ. Sci. Technol.*, 2002; 36: 4358-4363.
21. Geckeis H, Schäfer T, Hauser W, Rabung T, Geyer FW, Götz R, Missana T, Degueldre C, Möri A, Eikenberg J, Fierz T. Results of the colloid and radionuclide retention experiment (CRR) at the Grimsel test site (GTS), Switzerland-Impact of reaction kinetics and speciation on radionuclide migration. *Radiochim. Acta*, 2004; 92: in press.
22. Schäfer T, Geckeis H, Bouby M, Fanghänel T. U, Th, Eu and colloid mobility in a granite fracture under near-natural flow conditions. *Radiochim. Acta*, 2004; 92: in press.

23. Schäfer T, Claret F, Bauer A, Griffault L, Ferrage E, Lanson B. Natural organic matter (NOM)-clay association and impact on Callovo-Oxfordian clay stability in high alkaline solution: Spectromicroscopic evidence. *J. Phys. IV*, 2003; 104: 413-416.
24. Claret F, Bauer A, Schäfer T, Griffault L, Lanson B. Experimental investigation of the interaction of clays with high-pH solutions: A case study from the Callovo-Oxfordian formation, Meuse-Haute Marne underground laboratory (France). *Clay Clay Min.*, 2002; 50: 633-646.
25. Cody GD, Botto RE, Ade H, Behal S, Disko M, Wirick S. Inner-shell spectroscopy and imaging of a sub-bituminous coal: *In-situ* analysis of organic and inorganic microstructure using C(1s)-, Ca(2p)-, and Cl(2s)-NEXAFS. *Energy Fuels*, 1995; 9: 525-533.
26. Plaschke M, Rothe J, Schäfer T, Denecke MA, Dardenne K, Pompe S, Heise KH. Combined AFM and STXM in situ study of the influence of Eu(III) on the agglomeration of humic acid. *Colloids Surf. A*, 2002; 197: 245-256.
27. Schäfer T, Buckau G, Artinger R, Geyer S, Bleam WF, Wirick S, Jacobsen C. Origin and mobility of fulvic acids in the Gorleben aquifer system: Implications from isotopic data and carbon/sulfur XANES. *Org. Geochem.*, in review.
28. Jacobsen C, Feser M, Lerotic M, Vogt S, Maser J, Schäfer T. Cluster analysis of soft X-ray spectromicroscopy data. *J. Phys. IV*, 2003; 104: 623-626.
29. Lerotic M, Jacobsen C, Schäfer T, Vogt S. Cluster analysis of soft x-ray spectro-microscopic data. *Ultramicroscopy*, 2004, in press.
30. Malinowski E. *Factor analysis in chemistry*. New York: Wiley, 1991.
31. Everett B, Landau S, Leese M. *Cluster analysis.* London: Arnold Publishers, 2001.
32. Claret F. Caractérisation structurale des transitions minéralogiques dans les formations argileuses: Contrôles et implications géochimiques des processus d'illitisation. Cas particulier d'une perturbation alcaline dans le Callovo-Oxfordien Laboratoire souterrain Meuse-Haute-Marne. PhD Thesis, Université Joseph Fourrier, Grenoble, France, 2001:174 pp.
33. Landais P, Elie M. Utilisation de la géochimie organique pour la détermination du paléoenvironnement et de la paléothermicité dans le Callovo-Oxfordien du site de l´Est de la france. Etude de l´Est du bassin de Paris, Edition EDP Sciences, 1999: pp. 35-61.
34. The Swiss National Cooperative for the Disposal of Radioactive Waste (NAGRA). Project Opalinus clay - safety report - demonstration of disposal feasibility for spent fuel, vitrified high-level waste and long-lived intermediate level waste (Entsorgungsnachweis). NTB 02-05, Nagra, Wettingen (Switzerland), 2003.
35. Pellenard P, Deconinck JF, Marchand D, Thierry J, Fortwengler D, Vigneron G. Controle geodynamique de la sedimentation argileuse du Callovien-Oxfordien moyen dans l'Est du bassin de Paris: influence eustatique et volcanique. *C.R. Acad. Sci. Paris, Sciences de la terre et des planetes*, 1999; 328: 807-813.
36. Elie M, Faure P, Michels R, Landais P, Griffault L. Natural and laboratory oxidation of low-organic-carbon-content sediments: Comparison of chemical changes in hydrocarbons. *Energy Fuels*, 2000; 14: 854-861.
37. The Swiss National Cooperative for the Disposal of Radioactive Waste (NAGRA). Projekt Opalinuston -Synthese der geowissenschaftlichen

Untersuchungsergebnisse: Entsorgungsnachweis für abgebrannte Brennelemente, verglaste hochaktive sowie langlebige mittelaktive Abfälle-. NTB 02-03, NAGRA, Wettingen (Switzerland), 2003.
38. Moore DM, Reynolds RC. *X-ray diffraction and the identification and analysis of clay minerals*. New York: Oxford University Press, 1989.
39. Buckau, G. Komplexierung von Americium (III) mit Huminstoffen in natürlichen Grundwässern. PhD Thesis, Free University of Berlin, Berlin, 1991.
40. Swift RS. Organic matter characterization. In: Bartels JM ed. *Methods of soil analysis. Part 3: Chemical methods*. Madison, WI: Soil Science Society of America Book Series, 1996: pp. 1390.
41. Jacobsen C, Williams S, Anderson E, Browne MT, Buckley CJ, Kern D, Kirz J, Rivers M, Zhang X. Diffraction-limited imaging in a scanning transmission x-ray microscope. *Optics Commun.*, 1991; 86: 351-364.
42. Zhang X, Ade H, Jacobsen C, Kirz J, Lindaas S, Williams S, Wirick S. Micro-XANES: chemical contrast in the scanning transmission x-ray microscope. *Nucl. Instrum. Methods Phys. Res. A*, 1994; 347: 431-435.
43. Neuhäusler U, Abend S, Jacobsen C, Lagaly G. Soft X-Ray spectromicroscopy on solid-stabilized emulsions. *Colloid Polym. Sci.*, 1999; 277: 719-726.
44. Spector S, Jacobsen C, Tennant D. Process optimization for production of sub-20 nm soft x-ray zone plates. *J. Vacuum Sci. Technol. B*, 1997; 15: 2872-2876.
45. Hitchcock AP, Mancini DC. Bibliography of atomic and molecular inner-shell excitation studies. *J. Electron Spectrosc.*, 1994; 67: 1-132.
46. Ma Y, Chen CT, Meigs G, Randall K, Sette F. High-resolution-shell photoabsorption measurements of simple molecules. *Phys. Rev. A*, 1991; 44: 1848-1858.
47. Winn B, Ade H, Buckley C, Howells M, Hulbert S, Jacobsen C, Kirz J, McNulty I, Miao J, Oversluizen T, Pogorelsky I, Wirick S. X1A: Second generation undulator beamlines serving soft x-ray spectromicroscopy experiments at the NSLS. *Rev. Sci. Instrum.*, 1996; 67: 1-4.
48. Jacobsen CJ, Zimba C, Flynn G, Wirick S. Soft x-ray microscopy from sub-100nm regions. *J. Microsc.*, 2000; 197: 173-184.
49. Schäfer T, Hertkorn N, Artinger R, Claret F, Bauer A Functional group analysis of natural organic colloids and clay association kinetics using C(1s) spectromicroscopy. *J. Phys. IV*, 2003; 104: 409-412.
50. Dumas P, Miller L. The use of synchrotron infrared microspectroscopy in biological and biomedical investigations. *Vibrational Spectrosc.*, 2003; 32: 3-21.
51. Wattel-Koekkoek EJW, Buurman P, van der Plicht J, Wattel E, van Breemen N. Mean residence time of soil organic matter associated with kaolinite and smectite. *Europ. J. Soil Sci.*, 2003; 54: 269-278.
52. Kikuma J, Tonner BP. Photon energy dependence of valence band photoemission and resonant photoemission of polystyrene. *J. Electron Spectrosc. Relat. Phenom.*, 1996; 82: 41-52.
53. Jokic A, Cutler JN, Ponomarenko E, van der Kamp G, Anderson DW. Organic carbon and sulphur compounds in wetland soils: Insights on structure and transformation processes using K-edge XANES and NMR spectroscopy. *Geochim. Cosmochim. Acta*, 2003; 67: 2585-2597.
54. Farmer VC ed. *The infrared spectra of minerals*. Mineralogical Society Monograph 4. London: Mineralogical Society, 1974: 539 pp.

55. Naidja A, Huang PM, Anderson DW, van Kessel C. Fourier transform infrared, UV-visible, and X-ray diffraction analyses of organic matter in humin, humic acid, and fulvic acid fractions in soil exposed to elevated CO_2 and N fertilization. *Appl. Spectrosc.*, 2002; 56: 318-324.
56. Painter PC, Starsinic M, Coleman MM. Determination of functional groups in coal by fourier transform interferometry. In: Ferraro JR, Basile LJ eds. *Fourier transform infrared spectroscopy*. New York: Academic Press, 1985: 169-240.
57. Ganz, Kalkreuth. *Fuel*, 1987; 66: 708.
58. Schmeide K, Sachs S, Bubner M, Reich T, Heise KH, Bernhard G. Interaction of uranium(VI) with various modified and unmodified natural and synthetic humic substances studied by EXAFS and FTIR spectroscopy. *Inorg. Chim. Acta*, 2003; 351: 133-140.
59. Bosetto M, Arfaioli P, Pantani OL. Study of Maillard reaction products formed by glycine and D-glucose on different mineral substrates. *Clay Min.*, 2002; 37: 195-204.
60. Kim JI, Wimmer H, Klenze R. A study of curium(III) humate complexation by time resolved laser fluorescence spectroscopy (TRLFS). *Radiochim. Acta*, 1991; 54: 35-41.
61. Buckau G, Kim JI, Klenze R, Rhee DS, Wimmer H. A comparative spectroscopy study of the fulvate complexation of trivalent transuranium ions. *Radiochim. Acta*, 1992; 57: 105-111.
62. Kim JI, Czerwinski KR. Complexation of metal ions with humic acid: Metal ion charge neutralization model. *Radiochim. Acta*, 1996; 73: 5-10.

Part 2

SORPTION AND REACTIVITY

Chapter 5

WHEN SHOULD HUMIC SUBSTANCES BE TREATED AS DYNAMIC COMBINATORIAL SYSTEMS?

Cooper H. Langford and Joseph R. Melton

Department of Chemistry, University of Calgary, 2500 University Dr NW, Calgary, AB T2N 1N4, Canada

5.1. INTRODUCTION

This paper will begin with a review of the recent work in mass spectrometry and NMR that has, over the last two years, provided us with a view of humic mixtures of unprecedented intimacy. Our understanding of the significance of this shift of view from "seeing the forest from 30,000 feet" to finding ourselves viewing the diversity from "inside the canopy of this rain forest" is that the importance of weak interactions (H-bonding, hydrophobic, charge transfer) is emphasized. Humic mixtures emerge as a dynamic system of aggregational and conformational equilibria among monomers, oligomers, and polymers that produce the functional entities. When we consider binding of a metal ion to these mixtures, we must recognize that the equilibria will be disturbed.

Buffle [1] has categorized three Factors that are involved in producing the broad distributions of binding constants derived from metal ion titration onto humics. His Factor A is the presence of a variety of functional groups that can participate in metal complexation. Factor B is the electrostatic variation on humic aggregates as metal ions are progressively attached. Factor C is characterized mainly as the conformational changes that occur on progressive metal binding. Most quantitative models do not attempt to include Factor C (see, for example, Van Reimsdijk and Koopal [2]).

Recognition of the central role of aggregational equilibria leads to a major emphasis on Factor C and indicates, at least in principle, how it might be accommodated in quantitative models. The appropriate theoretical approach has been developed, especially by J.K.M. Sanders in the last few years, as the dynamic combinatorial system model [3]. The second section of this paper will present this model.

The final section will consider some current experimental information derived from study of the interaction of Cu^{2+} ion with solutions of Laurentian humic acid (LHA). Cu^{2+} complexation is studied by a kinetic method [4] exploiting the uptake of labile copper by a Chelex resin [5]. These "titration" data are compared with information about the extent of aggregation derived from dynamic light scattering, and with information about the local environment of strongly bound Cu^{2+} ions derived from paramagnetic effects on the relaxation time of water protons as measured using low field NMR relaxometry [6].

5.2. THE ENHANCED HUMIC PARADIGM

The humic literature has long recognized that humics are complex mixtures. Among the most telling evidence is the contrast observed by Piccolo et al. [7] between the size exclusion chromatograms of humic substances and conventional macromolecular substances, but it is only recently that it has been possible to specify the full dimensions of complexity for any representative sample. We now have access from Stenson et al. [8] to a very high resolution Fourier transform ion cyclotron resonance mass spectrum obtained with electrospray ionization of the well known Suwannee River fulvic acid (SRFA). 4626 empirical molecular formulas were assigned. Table 5.1 shows two series of formulas corresponding to substitution of CH_4 for O. DBE signifies double bond equivalence.

Table 5.1 Two series of molecular formulas of some neutral molecules in the 410 Dalton mass range (data from [8]).[a]

Label	Formula	DBE	Calcd. mass
1	$C_{16}H_{10}O_{13}$	12	410.0121
2	$C_{17}H_{14}O_{112}$	11	410.0485
3	$C_{18}H_{18}O_{11}$	10	410.0949
4	$C_{19}H_{22}O_{10}$	9	410.1213
5	$C_{20}H_{26}O_{9}$	8	410.1577
6	$C_{21}H_{30}O_{8+}$	7	410.2305
A	$C_{20}H_{10}O_{10}$	16	410.0274
B	$C_{21}H_{14}O_{9}$	15	410.0638
C	$C_{22}H_{18}O_{8}$	14	410.1002
D	$C_{23}H_{22}O_{7}$	13	410.1366
E	$C_{24}H_{26}O_{6}$	12	410.1729
F	$C_{25}H_{30}O_{5}$	11	410.2093

[a] substitution of CH_4 for O. DBE signifies double bond equivalence (number of rings plus double bonds)

It is apparent from Table 5.1 that data of this *seven figure resolution* allow recognition of important relationships such as replacement of CH_4 (approx mass 16) by O (16) and identification of the number of double bonds and ring closures.

Further analysis with tools such as Kendrick plots led to identification of 266 "homologous series" related by sequential addition of a standard unit such as additional oxygen. The overall range of carbon number was 14 to 58, the oxygen range 5 to 29, and the (DBE) double bond equivalence range from 6 to 33. A recent contribution from Cooper et al. [9] presents evidence from comparison with lignin for the structure shown in Figure 5.1 as a structure of one of the 4626 formulas identified. The conformation shown in Figure 5.1 is the lowest energy obtained in a Monte Carlo molecular mechanics simulation run in Spartan®. It should be emphasized that several other conformations are at comparable energy and could contribute at room temperature. The complexity of the mixture is unmistakable.

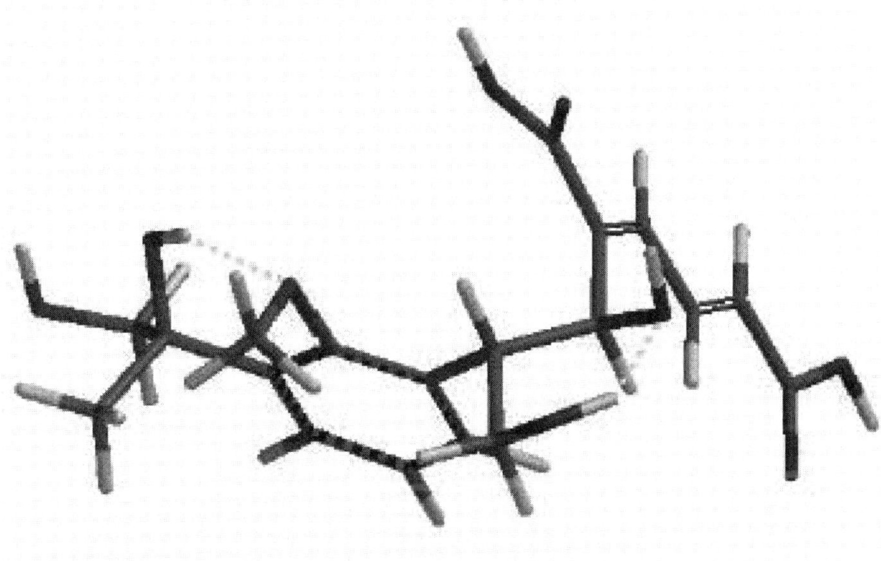

Figure 5.1 An optimized conformation (one of several at similar energies) of a structure identified from FTICR-MS as obtained using Spartan®. The structure was first treated by molecular dynamics using a Monte Carlo technique then several low energy conformations were re-optimized using atomic level parameters with PM3 semi-empirical quantum mechanics. Similar approaches were used for the structures in Figure 5.3

Evidence for complexity of molecular formula is complemented by NMR evidence for complexity of structural features. Using a range of 1- and 2-dimensional pulse sequences in a multinuclear study of Laurentian fulvic acid (LFA), Cook et al [10] were able to characterize more than 200 sharp features (long relaxation times) indicative of small molecules or very flexible and rapidly tumbling molecular fragments that are uncharacteristic of typical polymers. Also, the APT, INEPT and 2-D spectra defined structural elements up to six atoms along a chain of bonds. An indication of the coupling diagnosed is the ^{1}H TOCSY of LFA shown in Figure 5.2. The aliphatic-carbohydrate and amine-aromatic couplings are enlarged. The NMR spectra are consistent with the MS in indicating that a large fraction of the humic mixture is composed of relatively small molecules that can aggregate under the influence of weak forces.

5.3. DYNAMIC COMBINATORIAL SYSTEMS

The model for humics that emerges from our increasingly intimate view is of a complex mixture of small molecules, oligomers (*vide infra*), and polymers aggregated into functional assemblies by non-covalent forces. In these laboratories we have twice been pointed in that direction before. Aggregation of small units was the required hypothesis to explain a decrease of polydispersity ratio where Rayleigh scattering increased in data from light scattering [11].

Association of small units with larger polymers (or aggregates) seemed to be required to explain the decrease of the differential binding capacity of LFA for atrazine as LFA concentration increased. That is, the LFA mixture self competed for atrazine binding sites [12, 13]. Thus, we propose to treat humics as labile "supramolecular" assemblies susceptible to reorganization with labile equilibria among different aggregates that may be perturbed by introduction of new entities (e.g., metal ions, organic contaminants). This fits the definition of a dynamic combinatorial system.*

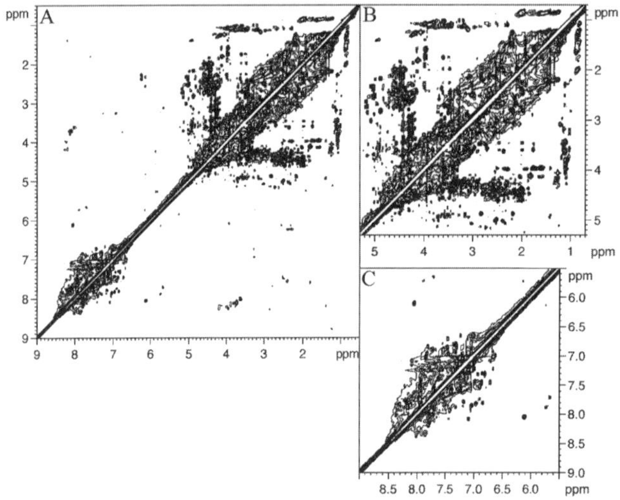

Figure 5.2 ^{1}H TOCSY spectrum of LFA. Block A is the complete spectrum, B. shows the aliphatic carbohydrate region, and C shows the aromatic amine region. Notice the extensive aliphatic-carbohydrate coupling and the relative absence of coupling of aromatic protons to either. All resolvable dots reflect long relaxation times (after Cook et al [10], reproduced by permission)

---

* According to Rowan et al. [3], a *combinatorial* library is formed by many members (M1… … … Mn), with each containing two or more elementary building blocks denoted A, B, C,…, arranged in a particular way. In a traditional combinatorial library the non-labile covalent bonds between building blocks fix the members during synthesis so there can be no post-synthetic interconversion between members such as from ABC to ACB. In contrast, in a *dynamic combinatorial library* the connections between building blocks are reversible and continuously in flux, with bonds being made and broken. Such connections may be labile covalent bonds or metal-ligand interactions or they may be non-covalent, utilizing hydrophobic, hydrogen bonding, or charge transfer complexing. The composition of a dynamic library will depend on its environment.

Scheme 5.1 gives a simple pictorial representation of the concept. The members of the library are M_1, M_2, M_3, to ....M_n. The arrows connecting them are drawn to imply that all of the members represented are present at similar concentrations at equilibrium. Now consider the consequences of adding a new entity, T_1.

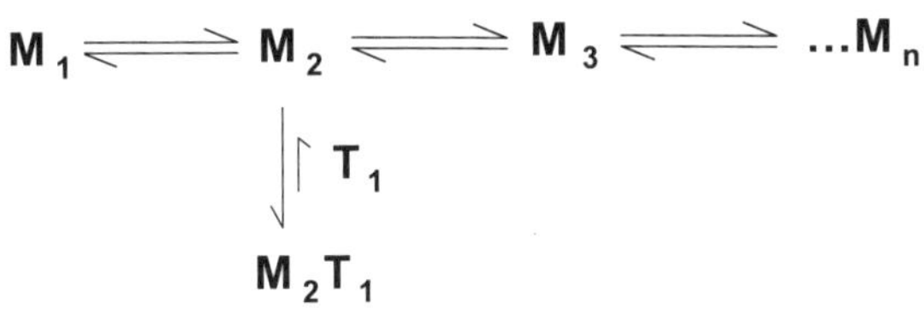

Scheme 5.1

If M_2 provides the most favorable site for binding T_1, the concentration of free M_2 will be depleted. However, if the equilibria are labile the activity (concentration) will be restored to an equilibrium value at the expense of other members of the library. If the complex M_2T_1 is significantly favored, titration of the library with T_1 will produce that complex as long as the library can sufficiently replenish M_2 to maintain the preference. However, the measured energetics of the binding process will include the energetics for formation of M_2 from other members of the library. The implications for complexation of Cu^{2+} by a humic acid system are that a favorable site could be filled over a fairly wide range of added Cu^{2+}, but the energetics of the process might be seen to vary as a consequence of the re-adjustments of equilibria in the library that forms the humic mixture.

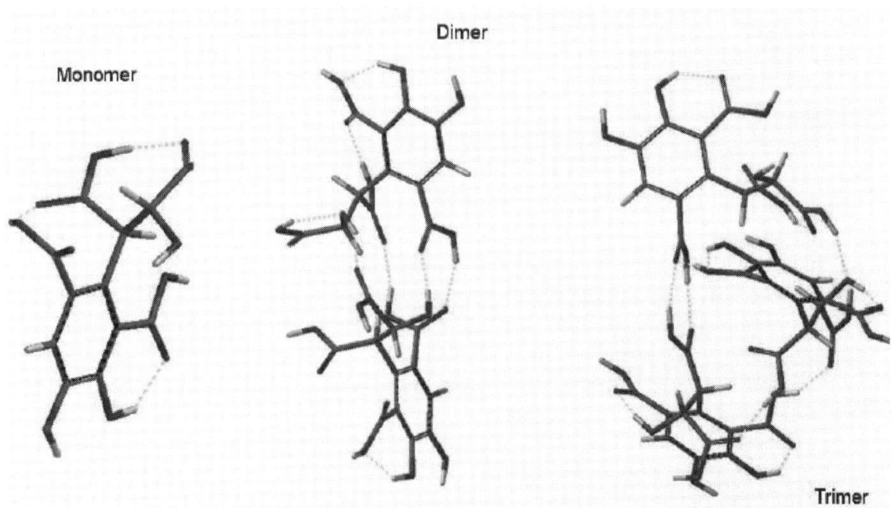

Figure 5.3 Monomer, dimer, and trimer structures of a component of a humic mixture as calculated in Spartan® following procedures of Bruccoleri et al [20]. Notice the role of H-bonding in the association. These can be thought of as modeling three members of a dynamic library

The structure shown at the left in Figure 5.3 has been assigned on the basis of high resolution MS to represent one component of a humic mixture [14]. The group of peaks around this molecular mass was repeated at twice the mass and three times the mass, suggesting a "monomer-dimer-trimer" series. Interestingly, following size separation on a size exclusion column, all three of these mass groupings were found in the mass spectra of each of the fractions, although in different ratio.

We exploit this intriguing observation by offering a model of a small part of the dynamic library of a humic in Figure 5.3. The structures shown were calculated using Spartan®. They are the energetically most favorable conformations found from a Monte Carlo molecular mechanics calculation on the monomer followed by a PM3 semi-empirical quantum mechanics optimization. Again, it is important to note that the monomer conformation used is only one of several very close in energy. The three calculated structures offer us a concrete model to aid visualization of a few members of the huge dynamic combinatorial library making up a humic.

5.4. COPPER(II) BINDING BY LHA

The experiments described below examined solutions with copper(II) concentrations ranging from 1 x 10^{-6} M to 5 x 10^{-4} M and LHA concentrations from 25 to 250 mg L^{-1}. The pH was buffered at 4.0 or 7.0. A quantitative estimate of copper bound to LHA was required for interpretation of the NMR relaxometry experiments. Because bound copper that is labile on a time scale approaching the NMR timescale will be able to exchange waters with the solvent on essentially the same basis as free copper, we need a measure of kinetically "non-labile" (NMR timescale) copper(II). We adopted the stripping ligand approach of Gutzman and Langford [4], which adds an excess of a strongly complexing ligand and measures the time course of capture of copper from the solution and the humic complexes. The method chosen follows Charkrabarti et al. [5] in using a 400 mesh Chelex 100 resin as a separate phase to avoid ambiguities arising from the adjunctive pathway of the displacement reaction [5]. Table 5.2 contains a summary of partition of copper concentrations into labile and non-labile.

5.5. LASER LIGHT SCATTERING

The first test that dynamic combinatorial modeling must face is whether or not size distributions of aggregates reflect changes induced by the redistribution in the library upon addition of a complexing metal ion. A powerful probe is dynamic light scattering. We emphasize that the limitation it suffers is that it can lead to overestimation of the largest aggregates. Representative data are presented here. The qualitative trends are consistent with the model.

The shift in molecular diameter is shown in Figure 5.4 at pH 7.0 in terms of the summed contributions for successive ranges (0-50, 51-120, 121-250, 250-400 and 401-9000 nm) as Cu(II) concentrations rise from 0 to 10^{-5} to 10^{-4} M. As expected, increasing concentrations of LHA shift aggregates to higher mass. The size range found is similar to that reported by Manning et al. [15]. At pH = 7.0, addition of even small amounts of Cu(II) shifts the mass distribution to a peak in the 175 nm neighborhood similar to the maximum seen for higher LHA concentrations. Smaller shifts to higher diameter are seen throughout at the higher LHA concentrations. Similar results are obtained at pH 4.0 with the important exception that a more protonated dilute LHA exhibits a peak near 120 nm even in the absence of Cu^{2+}. This shifts to approximately 175 nm on addition of Cu^{2+}, again with smaller shifts to higher diameter induced by Cu^{2+} in solutions of higher LHA concentration. A notable feature is the "onset" of coagulation at the ratio 10^{-4} M Cu^{2+} to 25 mg L^{-1} LHA at pH 7.0.

Table 5.2 Kinetic partition of Cu(II) concentrations. The first component is kinetically fast and the second is slow. The fast component is taken as the "free" (NMR timescale) Cu(II) and the slow component is taken to be the bound copper

pH	LHA conc. (mg/L)	Cu^{2+} conc. (M)	1st component (%)	SD (%)	2nd component (%)	SD (%)
4	50	5.00E-06	38.3	1.71	60.1	0.99
		1.00E-05	60.3	4.75	36.8	3.95
		5.00E-05	83.1	1.14	31.9	0.87
		1.00E-04	97.4	1.55	3.8	1.55
	150	5.00E-06	34.5	2.02	65.2	1.14
		1.00E-05	47.0	2.06	42.0	1.49
		5.00E-05	69.7	0.45	30.3	0.41
		1.00E-04	84.9	2.32	15.8	2.38
	250	5.00E-06	28.1	1.44	70.8	7.96
		1.00E-05	44.1	12.02	56.1	11.6
		5.00E-05	62.0	3.28	24.9	2.69
		1.00E-04	76.3	2.78	23.4	2.35
7	50	5.00E-06	0.0	0.00	99.6	2.65
		1.00E-05	27.9	1.68	72.1	0.78
		5.00E-05	43.7	1.99	55.8	1.50
		1.00E-04	86.7	3.04	15.4	3.13
	150	5.00E-06	0.00	0.00	98.7	1.56
		1.00E-05	23.3	1.38	76.3	10.3
		5.00E-05	36.3	1.51	63.5	1.27
		1.00E-04	45.1	1.93	54.1	1.01
	250	5.00E-06	0.0	0.00	99.7	2.02
		1.00E-05	0.0	0.00	99.9	0.43
		5.00E-05	22.4	0.76	77.6	0.03
		1.00E-04	39.8	1.91	59.6	0.99

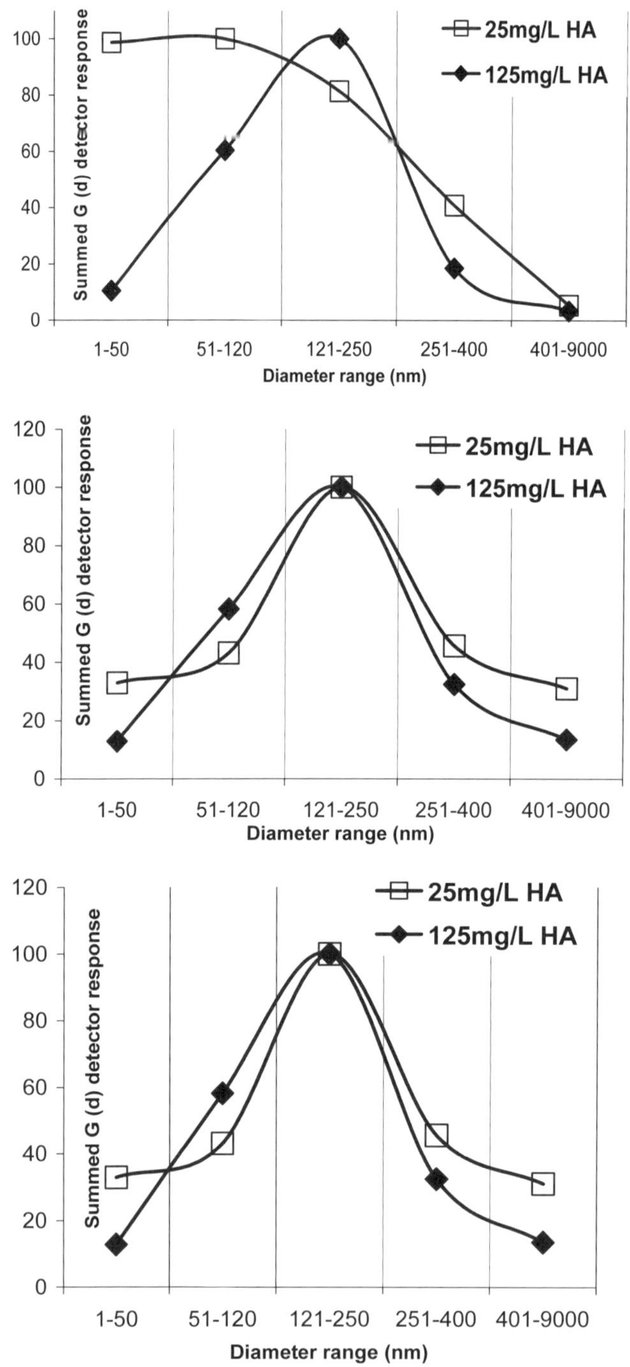

A. [Cu(II)] = 0 M

B. [Cu(II)] = 10^{-5} M

C. [Cu(II)] = 10^{-4} M

Figure 5.4 Molecular mass distribution at pH 7. Distributions are grouped as points for 1-50, 51-120, 121-250, 251-400, and 400-9000 Da. The vertical axis is the summed B detector response

5.6. WATER PROTON NMR RELAXATION TIMES

Mildvan and Cohn [16] recognized that metal ions coordinated to proteins might still have coordinated water with protons in fast exchange with bulk water so that a single average solvent proton signal would be observed. Paramagnetic metal ions influence the observed average relaxation time $T_2(obs)$. The fast exchange case follows the Bloch equation, which for only two environments (diamagnetic and paramagnetic) is as in Eq. 5.1, where P_D and P_P are the probabilities that the protons

$$1/T_2(obs) = P_D/T_2(D) + P_P/T_2(P) \qquad (5.1)$$

are in the diamagnetic and paramagnetic environments, respectively, and $T_2(D)$ and $T_2(P)$ are the transverse relaxation times in the diamagnetic and paramagnetic environments, respectively. Complexation of a paramagnetic ion to a protein was recognized to affect average relaxation times as a result of one of, or a combination of, two main factors. Attachment to the protein can reduce the number of bound waters on the ion and reduce P_p. Alternatively, attachment of the ion to the large protein may reduce the tumbling time with respect to the magnetic field experienced by the ion. This will tend to shorten $T_2(P)$. A paramagnetic ion coordinated to a multidentate site on the protein can experience a significantly reduced effect on the water proton relaxation as a consequence of reduced P_P. In contrast, a paramagnetic ion coordinated to a site where it retains several coordinated water molecules with exchangeable protons can have an enhanced effect on water proton relaxation as a consequence of reduced ("faster") $T_2(P)$.

There has been little exploitation of these effects to probe the nature of sites for paramagnetic ions bound to humics. Deczky and Langford [17] reported that coordination of Fe(III) to a fulvic acid matched the first case of reduction of number of exchangeable coordinated water molecules, thereby increasing average relaxation times compared to free Fe(III). In the case of Cu(II) and fulvic acid the results were opposite, suggesting reduction in the tumbling time of Cu(II) attached to the fulvic mixture, resulting in reduced relaxation times.

This report describes initiation of systematic study of Cu-LHA solutions using the efficient low field relaxometry technique measuring T_2 with the widely exploited Carr-Purcell-Meiboom-Gill (CPMG) pulse sequence [18]. While the fulvic acid mentioned above [17] had little effect itself on relaxation times, LHA is observed to cause a reduction of $T_2(obs)$. This no doubt results from the fact that water interior to aggregates and bound on the surfaces experiences longer tumbling times and more efficient relaxation as it does in soils [6]. It is confirmed by EPR measurements that it is not a function of LHA paramagnetism. A dry LHA solid gave an EPR signal that was reduced on equilibration with water saturated air and rendered nearly unobservable by creating a mud with water weight equal to 40% of the LHA weight. The EPR signal was lost in noise in solution under the same conditions where a fulvic acid signal was recorded. The decrease in $T_2(obs)$ is linear in LHA concentration. In these circumstances, the Bloch equation can be written as shown in Eq. 5.2.

$$1/T_{2obs} = (1/T_{2W}+P_{HA}/T_{2HA}) + (P_{CuF}/T_{2CuF}+P_{bound}/T_{2bound}) \qquad (5.2)$$

In Eq. 5.2, T_{2W} is the relaxation time in bulk water. In dilute solution the probability of a proton being in the bulk environment is little reduced and can be set to 1. P_{HA} is the probability that water is found in the LHA environment and T_{2HA} is the corresponding relaxation time. P_{CuF} is the probability that a water molecule is found in the coordination sphere of a free Cu^{2+}(or an NMR labile one) and T_{2CuF} is the corresponding relaxation time. P_{bound} is the probability that a copper ion is complexed to LHA (a site non-labile on the NMR time scale) and T_{2bound} is the corresponding relaxation time.

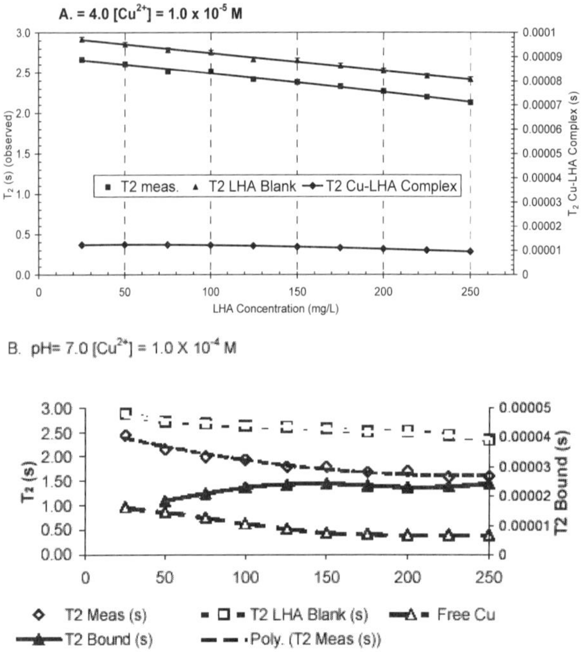

Figure 5.5 Measured T_2 values for: B. 1.0 x 10^{-5} M Cu^{2+} at pH = 4.0, and A. 1.0 x 10^{-4} Cu^{2+} at pH = 7.0 as a function of LHA concentration with accompanying measurements of the "blanks" for LHA solutions. The values of T_2 for bound Cu^{2+} are calculated from Eq. 5.2

The objective is to find T_{2bound} as a probe of the nature of the bound Cu^{2+} environment. The first two terms in Eq. 5.2 can be obtained from the curve of T_{2obs} as a function of LHA concentration‡. T_{2CuF} is estimated from known T_{2W} and the relaxation times of Cu^{2+} in water in the absence of LHA. P_{CuF} and P_{bound} are estimated from the kinetic partition of Cu^{2+} into labile and bound (non-labile) using the kinetic stripping experiments introduced above (Table 5.2). The interesting outcome is that for large LHA to metal ratios we find that estimated T_{2bound} is approximately constant, probably within the substantial experimental uncertainty. This is illustrated by the curves in Figure 5.5 with data for 10^{-5} M Cu^{2+} at pH 4.0 and 10^{-4} M Cu^{2+} at pH 7.0. As expected, the values are a function of pH with the relaxation time of water bound to Cu being near 1 x 10^{-5} s at pH 4.0 and above 2 x 10^{-5} s at pH 7.0. For all Cu^{2+} concentrations, the values at high LHA to metal ratio are near 1 x 10^{-5} s at pH 4 and near 2 x 10^{-5} s at pH 7.0. At lower LHA to metal ratio there is some upward deviation. This may reflect complexation at weaker sites or

‡ Note that a small humic paramagnetism contribution would be accommodated in this signal.

systematic error in our kinetic partition of "NMR" free and bound metal.

The consistency within the experimental error (that arises both from taking the small difference of the two experimental curves and from the kinetic partitioning of free and bound copper) is consistent with the expectations of a dynamic combinatorial model. At reasonably high LHA to Cu^{2+} ratio we can expect a preferred member of the set, or in a system of this complexity a set of closely related members, to form the binding site(s). Under these circumstances, changes of the aggregational distribution are not linked to significant change of the bound Cu^{2+} environment since there is no water bound to Cu^{2+} in the complex.

On balance, these results suggest that the dynamic combinatorial model may provide fruitful insights into contaminant binding by humic mixtures. Since the labile aggregational equilibria picture has so much fundamental support in humic chemistry, it appears to be fruitful to try to continue to extend the application of the dynamic combinatorial model to the functioning of humic mixtures.

5.7. MATERIALS AND METHODS

5.7.1. Materials

LHA [12,13] was extracted from the Laurentian organic rich podzol obtained from the forest preserve of Laval University, Quebec. Following base extraction under N_2, the extract was acidified to pH = 2. The recovered solid was subjected to repeated washing with 1.0 M HCl under N_2 to remove metals. After 10 washings, residual Fe was 1.7 ppm (in solid LHA) and Na was undetectable in a flame. The sample is > 99% in the proton form. Stock solutions were prepared in containers pre-cleaned with 0.1% EDTA solution and ultrapure water. Appropriate weights of LHA were dissolved in NaOH solutions at pH 10 and pH was then adjusted with either sodium acetate (BDH) (pK_a = 4.76) buffers or sodium dihydrogen phosphate (BDH) (pK_a = 7.20) buffers. Stock solutions were stored in pre-cleaned Nalgene® HDPE screw top bottles that were wrapped in Al foil. Stock solutions were used within one month. Copper stock solutions were prepared from copper sulfate (MCB Reagents).

5.7.2. Stripping Ligand Kinetics

An experimental cell was made from glass (10 cm tall and 6 cm in diameter). An outlet on the bottom edge allowed solution to flow into a fitted Acrodisc® Supor® membrane syringe filter (Pall Gelman). The filter, either 0.45 µm or 1.2 µm pore size, was attached by Tygon® tubing to the ICP pump and from there into the ICP nebulizer. The cell was placed on a heat-insulated stirrer with a Teflon™ coated stir bar placed inside the cell. The cell was continuously stirred over the course of the experiment at a moderate rate. The ion exchange resin was a Chelex 100 "molecular biology grade" 200-400 mesh size in the sodium form, with a resin size range of 75 – 150 µm (Bio-Rad). The resin was pre-treated to adjust to proper pH by soaking in buffer solutions of the appropriate pH till the supernatant over the resin was within ± 0.2 pH units of the sample. The resin was weighed prior to runs to ensure a 1% (w/w) loading and used moist. The resin was weighed and used moist, since drying has been found to affect the metal binding properties [5]. Loading of 1% (equivalent to a Chelex to metal ratio of 100 or higher, depending on the metal concentration within the sample) was used.

Calibration of the ICP was accomplished with 1.005 mg/L and 10.05 mg/L standards, prepared fresh daily from a certified 1005 mg/L standard (SCP Science). Prior to the introduction of the Chelex resin, several readings were taken of the sample solution to get an accurate measure of the total solution copper concentration. Once a stable copper concentration was recorded, the resin was introduced and measurements were begun immediately. Measurement was then repeated every 30 seconds until a total of approximately forty were completed.

The data were smoothed by fitting to a sixth degree polynomial and an approximate Laplace transform calculated according to the procedure given by Olson and Shuman [19]. The peaks of the Laplace transform provide an objective measure of the number of kinetic components. The peak frequencies provide approximate rate constants and the areas give approximate concentrations. The values from the Laplace treatment were then used as inputs to a non-linear regression of the sum of first order rate processes using the Marquardt algorithm carried out on the unsmoothed data. In all present cases, a two-component fit was satisfactory with a negligible time independent term.

5.7.3. Light Scattering

The light scattering cells were rigorously cleaned prior to use to ensure no dust or metal contamination. All cleaned cells were protected from dust prior to analysis. LHA solutions were filtered with 1.2 μm Acrodisc® 32 mm Supor® membrane syringe filters prior to analysis. 0.45 μm filters were not used for the LHA solutions due to literature reports of HA of diameters up to 436 $\pm$ 36 nm [15], and a desire not to bias the samples towards the smaller diameter HA species. LHA filtered and non-filtered solutions were checked by absorbance to ensure no LHA was lost to the filter. No difference between the solutions was detected. All copper solutions were filtered with 0.45 μm nylon syringe filters (MSI). The LHA and copper were mixed 35 minutes prior to analysis and capped to avoid contamination. Prior to analysis, the cells were wiped with KimWipes® and introduced into the light scattering apparatus. A diffraction matching liquid was pumped around the cell for 1 minute to ensure no air bubbles, and then left to become quiet for a further minute prior to analysis. The samples were analyzed at room temperature.

A Hamatsu HC 120 photomultiplier with a BI-9025 AT system made by Brookhaven Instruments fitted with a BI-200 SM collector was used for all runs. Software used to collect the light scattering data was BI-2P v. 4.3 manufactured by Brookhaven. The detector was set to 90° to minimize effects of flare, dust and misalignment. The viscosity was 1.005, refractive index 1.333, and λ = 632.8 nm. The detector aperture was set to 400 nm.

All results were analysed using a CONTIN model. The measured baseline was used and the software-suggested particle size range was accepted. The results were transferred to a spreadsheet. Particle sizes were summed from 1-50, 51-120, 121-250, 251-400, and 401-9000 nm. The distributions were then plotted.

5.7.4. NMR Relaxometry

Sample cells were cylindrical glass vessels 3.5 cm in diameter and 5 cm in height. The cells had round glass lids that were secured in place with Teflon™ tape. Auto pipettes were used to deliver 9.90 mL of buffered LHA solution and 0.10 mL of the copper solution. The cell was sealed and placed in a 30 $\pm 1^{o}$C water bath for 35 minutes prior to analysis. A standard of 10^{-4} M copper solution was run as a measure of day to day instrument fluctuation.

The low-field NMR experiments were performed on a CoreSpec-1000™ (NUMAR Corp.) relaxometer at a frequency of 1 MHz (field strength of 0.024 T) with a Teflon™ tube sample holder thermostated at 30°C. The T_2 relaxation time distributions were recorded using the CPMG pulse sequence. Spectra were optimized at an inter-echo spacing of 0.6 ms, with a signal to noise (S/N) ratio of at least 225. A recovery time of 5500 ms was allowed for a series of 5000 echoes. System gain was set to 400 with the number of trains (discrete pulses) at 16. Each run took three minutes and fifty-six seconds. Prior to analysis the instrument was tuned to ensure accurate performance.

The data were processed using EchoFit v.3.02 (NUMAR Corporation), a software package created for relaxometry that utilizes non-negative least squares methods to create the T_2 distributions. The T_2 geometric mean was used for further data analysis. Independent analysis demonstrated that a single exponential provided a good fit to the data.

ACKNOWLEDGEMENTS

We thank Dr. Apostolos Kantzas for guidance with the relaxometry experiments and Dr. Guojun Liu for assistance with light scattering. This work was funded by a grant from the Natural Sciences and Engineering Research Council of Canada.

REFERENCES

1. Buffle J. *Complexation reactions in aquatic systems*. Chichester, UK: Ellis-Horwood, 1988.
2. Van Reimsdijk WH, Koopal LK. In: Van Riemsdijk WH ed. *Environmental particles*. Chelsea MI: Lewis, 1992: Chap. 12.
3. Rowan SJ, Luckman PS, Reynolds DJ, Sanders JKM. Engineering diversity into dynamic combinatorial libraries by use of a small flexible building block. *New J. Chem.*, 1998; 1015-1018.
4. Gutzman DW, Langford CH. Kinetic studies of metal ion speciation. *Anal. Chim. Acta,* 1993; 283: 773-783.
5. Chakrabarti CL, Lu Y, Gregoire DC, Back MH, Schroeder WH. Kinetic studies of metal speciation using Chelex cation exchange resin: Application to cadmium, copper, and lead speciation in river water and snow. *Environ. Sci. Technol.*, 1994; 28: 1957-1967.
6. Manalo FP, Kantzas A, Langford CH. Soil wettability as determined from using low-field nuclear magnetic resonance. *Environ. Sci. Technol.*, 2003; 37: 2701 – 2706.
7. Piccolo A, Conte P, Cozzolino A. Differences in high performance size exclusion chromatography between humic substances and macromolecular substances. In: Ghabbour EA, Davies G eds. *Humic substances: Versatile components of plants, soil and water.* Cambridge: Royal Society of Chemistry, 2000: 111-125.

8. Stenson A, Marshall AG, Cooper, WT. Exact masses and chemical formulas of individual Suwannee River fulvic acids from ultrahigh resolution electrospray ionization fourier transform ion cyclotron resonance mass spectra. *Anal Chem.,* 2003; 75: 1275.
9. Cooper WT, Milligan L, Chanton J, Dittmar T, Filley T. Ultrahigh resolution mass spectrometry of aquatic humic substances: they really are geopolymers. *Humic Substances Seminar VII*, Northeastern University, Boston, March 17-19, 2004.
10. Cook RL, McIntyre, DD, Langford CH, Vogel HJ. A comprehensive liquid-state heteronuclear and multidimensional NMR study of Laurentian fulvic acid. *Environ. Sci. Technol.,* 2003; 37: 3935–3944.
11. Underdown AW, Langford CH, Gamble DG. Copper(II) titration of fulvic acid ligand sites with theoretical, potentiometric, and spectrophotometric analysis. *Anal. Chem.,* 1981; 53: 2139-2140.
12. Wang Z-D. Stoichiometric interactions of atrazine and hydroxyatrazine with chemically- characterized Laurentian soil and its key components. Ph.D. thesis Montreal, QC: Concordia University, 1989.
13. Wang Z-D, Pant BC, Langford CH. Spectroscopic and structural characterization of a Laurentian fulvic acid: notes on the origin of colour. *Anal. Chim. Acta*, 1990; 232: 181-188.
14. These A, Reemtsma T. Molecular formulas of fulvic acid as derived by size exclusion chromatography – electrospray ionization mass spectrometry. *Humic Substances Seminar VII*, Northeastern University, Boston, March 17-19, 2004.
15. Manning TJ, Bennett T, Milton D. Aggregation studies of humic acid using multiangle laser light scattering. *Sci. Total Environ.,* 2000; 257: 171-176.
16. Milldvan AS, Cohn M. Aspects of enzyme mechanisms studied by nuclear spin relaxation induced by paramagnetic probes. *Adv Enzymol.,* 1970; 2: 910.
17. Deczky K, Langford CH. Application of water nuclear magnetic resonance relaxation times to study of metal complexes of the soluble soil organic fraction fulvic acid. *Can J. Chem.*, 1978; 56: 1947-1951.
18. Sanders J, Hunter B. *Modern NMR spectroscopy*. NY: Oxford University Press, 1993: 248-249.
19. Olson DL, Shuman MS. Kinetic spectrum method for analysis of simultaneous first-order reactions and application to copper(II) dissociation from aquatic macromolecules. *Anal. Chem.*, 1983; 55: 1103-1107.
20. Bruccoleri A, Sorenson T, Langford C. Molecular modeling of humic structures. In: Ghabbour EA, Davies G eds. *Humic substances: Structure, models, and functions*. Cambridge, UK: The Royal Society of Chemistry, 2001: 193-208.

Chapter 6

MASS SPECTROSCOPIC AND CAPILLARY ELECTROPHORESIS STUDIES OF HUMIC ACIDS INTERACTIONS WITH VARIOUS COMPOUNDS AND XENOBIOTICS

Ma. de Lourdes Pacheco H.[1] and Josef Havel[2]

[1] Universidad Autónoma del Estado de Hidalgo, Centro de Investigaciones Químicas, Carretera Pachuca-Tulancingo Km. 4.5, C.P. 42076, Pachuca, Hidalgo, México
[2] Department of Analytical Chemistry, Faculty of Science, Masaryk University, Kotlářská 2, 611 37 Brno, Czech Republic

6.1. INTRODUCTION

Humic acids (HAs) are the main components of soil organic matter. They are widespread in nature but still of unknown structure [1]. Knowledge of HAs structure might help to explain their remarkable properties. One of the main problems to be addressed is the notion that because HAs are complex supramolecules, many analytical problems must be solved before HAs can be fully exploited. For example, it is difficult to understand how their aggregation proceeds when the characteristics of humic substances themselves and their primary compounds are still unclear. Nevertheless, HAs have been documented to interact with nutrients, toxic metals, radionuclides and the halogens, among others. The effect on toxic metals and micronutrients is unpredictable and often paradoxical, sometimes making them more available to organisms while also acting as sequestering agents so as to reduce their toxicity or beneficial value. But due to their complexity, the characterization of HAs and their many environmental processes (for example, metal complexation) requires a versatile assortment of analytical tools to better understand them.

Mass spectrometry has been used with the aim of elucidating HAs structures [2-7], but most of the analyses have been made after applying thermal [8] or chemical [9] degradation techniques. Examples of these combined-technique studies are those of Schulten et. al. [10], who analyzed HAs degradation products by applying pyrolysis coupled to gas chromatography-mass spectrometry (GC-MS), and of Fievre et. al. [11], who reported the analysis of humic substances by liquid chromatography coupled to MS. Virtually all early mass spectrometric studies of humic substances using electron ionization techniques were characterized by the extensive fragmentation produced, yielding mass spectra with fragment ions at almost every nominal mass or m/z value [5]. Recently, Havel et al. [6] reported mass spectra where it was demonstrated that HAs can be ionized without fragmentation. Nevertheless, much more work with standards and fractionated samples is needed to better understand what affects the apparent molar mass distribution of HAs in matrix-assisted laser desorption ionization (MALDI) mass spectra. Recently, a considerable breakthrough has been made by applying soft ionization techniques with high resolution, in this case electrospray ionization (ESI) with Fourier transform ion cyclotron resonance mass spectrometry (FT-ICR) [4,12-15]. As for fulvic acids, it was found that HAs are mostly derived from lignin [13,15] and several thousand compounds were identified and chemical formulas of individual fulvic acids were determined [15].

Capillary electrophoresis (CE) is an electromigration separation technique based on electrophoresis performed in very narrow (25-100 μm) capillaries. Electrophoresis is defined as the migration of analytes within an electrolyte solution under the influence of an electric field. CE has proven to be a useful technique for the separation and characterization of proteins, peptides, nucleic acids, as well as polysaccharides and HAs [2,3,6,17-19,20-22,23-25]. It was a CE separation technique with which HAs were shown to be mixtures of many compounds. Because the pK_a values of HAs are in the range 3-5, HAs are negatively charged at pH>3 and should migrate under such conditions towards the anode. However, having sufficiently high electroosmotic flow (EOF), they migrate towards the cathode. Selection of the buffer used is crucial, and good reviews can be found on this subject [18,21]. We have recently found that the best buffer is a mixture of borate, EDTA and TRIS [21] and this buffer was therefore used in the present work.

Despite all the MS and CE studies and attempts to fractionate HAs, the search for new methodologies to further characterize HAs from various sources and origins continues. Therefore, the purpose of this work was to study HAs from different sources (peat, coal, soil, leonardite) and of different origin by MALDI-TOF mass spectrometry and CE with the aim of finding out if they have something in common or if HAs of different origin are substantially different. Additionally, we set out to study the interactions of various cations (including alkali metals, transition metal ions and organic cations) with HAs to gain further information about ion-HA interactions. Finally, we sought to develop a methodology for the study of interactions between HAs and compounds of environmental interest in order to explain or at least better elucidate the possible mechanisms of HAs interactions with charged molecules.

6.2. MATERIALS AND METHODS

6.2.1. Materials

All reagents used were of analytical grade. Sodium hydroxide, sodium chloride, 1,1,1-Tris-hydroxymethylaminomethane (Tris), ethylendiamintetraacetic acid (EDTA), hydrochloric acid, potassium ferrocyanide and potassium ferricyanide were obtained from Lachema (Brno, Czech Republic). Boric acid, sodium tetraborate, glacial acetic acid and potassium acetate were from Merck (Darmstadt, Germany). Copper(II) nitrate and mercury(II) nitrate were purchased from Sigma-

Aldrich (Steinheim, Germany). Fluorescein and 5-sulphosalicylic acid were from Loba-Chemie (Wien, Austria) and mesityl oxide used as an EOF marker was from Fluka (Buchs, Switzerland). Parathion insecticide (50%) was obtained as a commercial sample "Foley-50-E" produced by Agricultura Nacional (Mexico) and malathion insecticide (47%) "Helios" produced by Laboratorios Helios (Mexico).

Deionized water double-distilled from a Heraeus Quartzschmelze (Hanau, Germany) quartz apparatus was used to prepare all solutions.

Humic Acids. In order to study characteristics of HAs from different countries and sources (soil, peat, leonardite and coal), a collection of samples was examined. For this purpose, several companies and laboratories provided us with HAs, and for comparison studies standard humic acids from the International Humic Substances Society (IHSS) were also used in this work. The sources and places of origin of the HAs studied are shown in Table 6.1.

Table 6.1 Humic acids used in the present study

Name	Source	Obtained from
Chemapex standard	Coal	CHEMAPEX Co. [26], Czech Republic
Cananeia	Soil	Sao Paulo soil [27], Brazil
Fluka I		[28]
Humitron 60S	Leonardite	GBM Co. [29], Mexico
Humic acid	Leonardite	IMERCA Co. [30], Guatemala
Humic acid	Coal	ARTECH Co. [31], USA
Grama	Semiarid soil	Cordoba [32], Argentina
Peat Standard	Peat	International Humic Substances Society [33]
Soil standard	Soil	International Humic Substances Society [33]
Leonardite standard	Leonardite	International Humic Substances Society [33]
China HA	Coal	CHEMAPEX Co. [26], Czech Republic

Preparation of Aqueous HAs Solutions. Stock solutions of 1000 mg L^{-1} of HAs were prepared by dissolving the corresponding mass in 360 μL of 0.1M NaOH and diluted to a total volume of 10 mL. For CZE analysis, the stock solutions were diluted 10 times with water. All HAs working solutions were analyzed after approximately 10 days of preparation because it was found that after this time the electropherograms were reproducible as reported previously in Pokorná et al. [3]. HA-xenobiotic solutions were prepared by mixing the HA solution (100 mg L^{-1}) with the required amount of xenobiotic. The mixture was shaken for 1 min, injected and analyzed. All measurements of the mixed solutions were done on the same day.

6.2.2. Methods

Mass Spectrometry. Mass spectra of HAs were measured in an Axima Kratos-CFR (Shimadzu) mass spectrometer controlled by Kompact 1.2.0 software and equipped with a nitrogen laser (wavelength 337 nm, pulse duration 4 ns, energy pulse 300 μJ) from Laser Science Inc. (Franklin, MA, USA). The energy of the laser was changed in the range of 0-180 (arbitrary units). Up to 1000 laser shots were used for each sample, the signals of which were averaged and smoothed. O-phenanthroline and insulin were used for mass calibration.

Mass spectra were measured in various modes depending on the sample analyzed (linear positive, linear negative or reflectron negative modes), but generally spectra with the most peaks were obtained in the linear positive mode. For all measurements, 1 μL of 1000 mg L^{-1} HA solution was placed on a target spot and dried in a stream of air at ambient temperature.

Capillary Electrophoresis. Experiments were carried out with a Beckman (Model P/ACE) System 5500 (Palo Alto, CA, USA) equipped with a diode array detection (DAD) system, automatic injector, fluid-cooled column cartridge and System Gold Data station. Fused silica capillary tube of 37 cm (30.5 cm length to the detector) x 75 μm I.D. was used. The normal polarity mode of the CZE system (cathode at the side of detection) was applied unless indicated otherwise. All sample solutions were filtered using 0.2 μm pore size filters (Sigma) before being injected and analyzed. The pH of the buffer solution was measured with a combined glass OP-08080 electrode and a Precision Digital pH-meter OP-208/1 (Radelkis, Budapest, Hungary). Standard Radiometer and/or Radelkis buffer solutions were used for calibration. Activation and conditioning of the capillary tube was as previously reported [20].

6.3. RESULTS

6.3.1. Mass Spectrometry

MALDI Mode. The use of several matrices was examined, but with no success because ionization was low and mass spectra were very poor. Useful HAs mass spectra were observed only with ferulic acid as the matrix. In this case, several groups of signals were observed with a difference between the peaks of approximately m/z 190 Da. Several experiments showed that most of these signals were due to the ferulic acid matrix. However, it is interesting to note that mass spectra of ferulic acid after the laser beam action are very similar to the mass spectra of HAs. It is not excluded that with a ferulic acid matrix, some "humic-like" polymers and/or aggregates similar to HAs are formed. These clusters of ferulic acid interfere. Therefore, further analyses of HAs were carried out in the laser desorption ionization (LDI) mode. In this sense, the LDI mode is a "self matrix" approach and the results obtained are in good agreement with those using ESI [4,12-15].

LDI Mode. Examples of LDI-TOF mass spectra of different HAs are shown in Figures 6.1a and 6.1b. The LDI mass spectrum of Chemapex HA in an m/z interval of 500-1000 Da is shown in Figure 6.2. The main and characteristic group of peaks for all HAs occurs in the m/z range 750-950 Da.

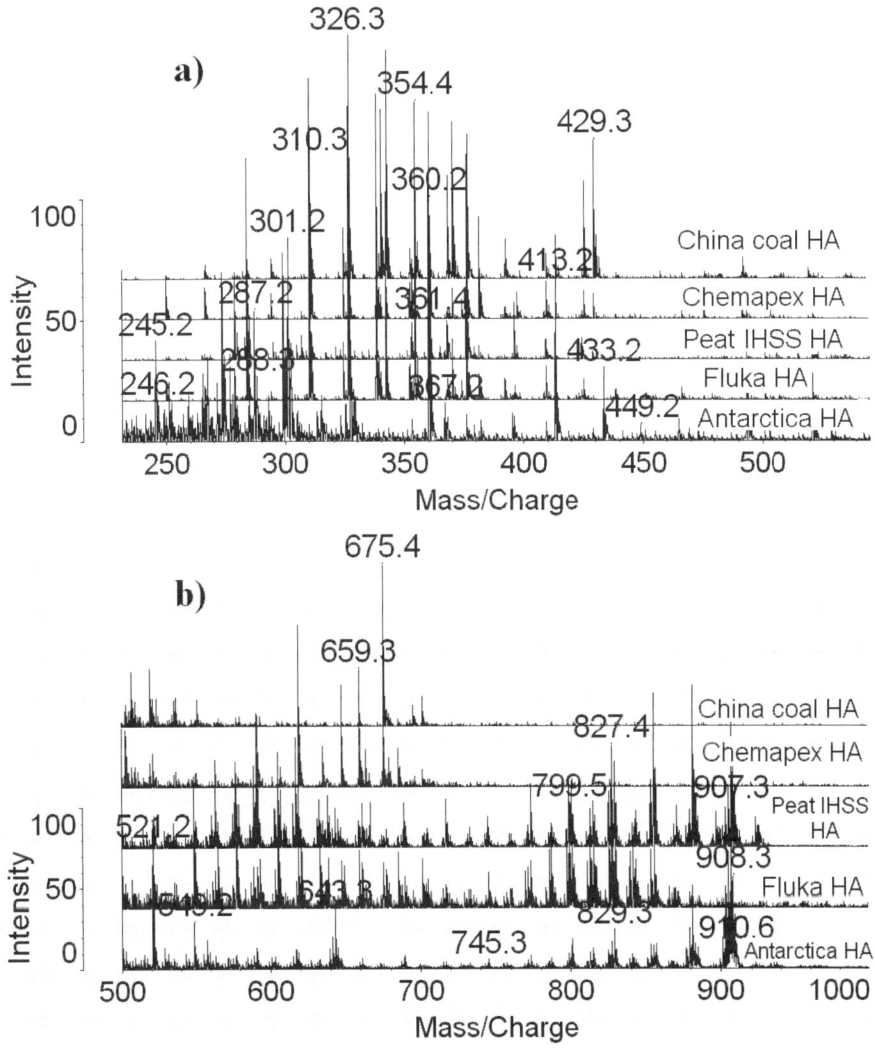

Figure 6.1 LDI-TOF MS of several HAs (1000 mg L^{-1}) measured after10 days of preparation. Linear positive mode. Scale from a) m/z 200-500 Da and b) m/z 500-1000 Da

Two amplifications of the mass spectra obtained with 150 units of laser energy in the intervals m/z 720-920 Da and 820-840 Da are shown in Figure 6.3, It is evident from Figure 6.3b that the differences between the isotopic patterns of the peaks are equal to one, which is evidence that the group of peaks around m/z $\approx$ 830 Da are singly charged ions and so the m/z values directly indicate molar mass. In addition, the difference between the groups of peaks (Figure 6.3a) is almost exactly 14.00 Da. We can reasonably suggest that this difference corresponds to the -CH_2- group. Thus, the group of peaks around the most intense peak at m/z 829.77 Da most probably represents at least 12-14 compounds, likely with the same structure containing different numbers of -CH_2- groups. For example, such compounds might be waxes. Other possibilities are discussed in the literature [15]. In addition, the rather complicated isotopic patterns of the individual peaks indicate that overlap of several compounds that differ by only 1 or two protons occurs (see the details in Figure 6.3b).

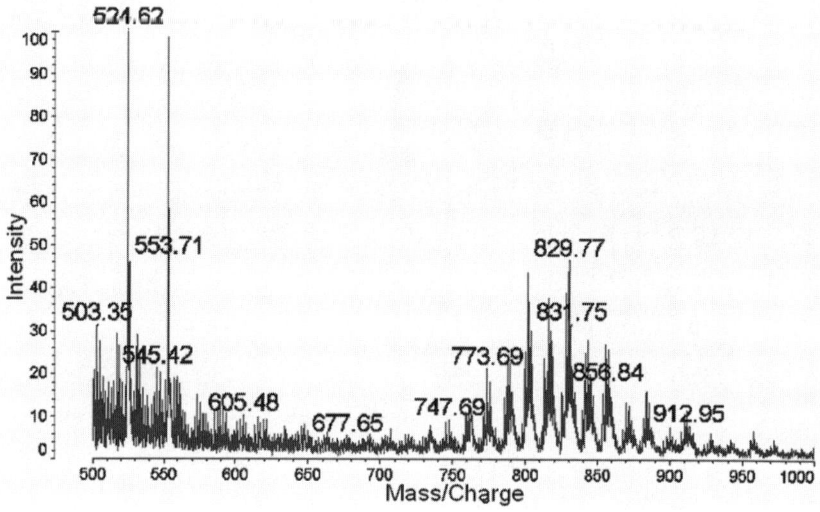

Figure 6.2 LDI- TOF mass spectra of Chemapex HA at laser energy 150

6.3.2. Capillary Electrophoresis

Based on previous studies [21] with a buffer composed of borate, Tris and EDTA, several experiments were conducted to compare the fingerprints of various HAs. Figure 6.4 shows the electropherograms of HAs from different sources and origins with borate-Tris-EDTA buffer at pH 8.5. Figure 6.4 shows that the HAs from soil, peat and leonardite contain the same compounds but in different distributions and concentrations.

On the basis of the similar CE patterns of the HAs under study, Chemapex HA [26] was selected for further examination because it has been well characterized [22]. Our intention was to continue our previous work [20] and follow HA interactions with both organic and inorganic compounds in more detail. A variety of different model compounds were studied with the aim of finding some markers to better understand HAs-xenobiotics interactions.

Fluorescein and 5-Sulphosalicylic acid. Negatively charged fluorescein and 5-sulphosalicylic molecules (containing carboxylic groups) are easily detected by CE. Figure 6.5 shows the effect on HA fingerprints of adding different amounts of organic solutes to an HA solution. The HAs separation pattern is not substantially changed with fluorescein additions (Figure 6.5a), suggesting that fluorescein does not interact with HA under the experimental conditions. On the other hand, 5-sulphosalicylic acid, which has a phenol group *ortho* to the carboxylic group (Figure 6.5b), most probably interacts with all HA fractions and the migration time of the whole HA peak envelope is decreased. Free 5-sulphosalicylic acid is not observed as it moves more slowly towards the detector (migration time more than 30 min) than HA.

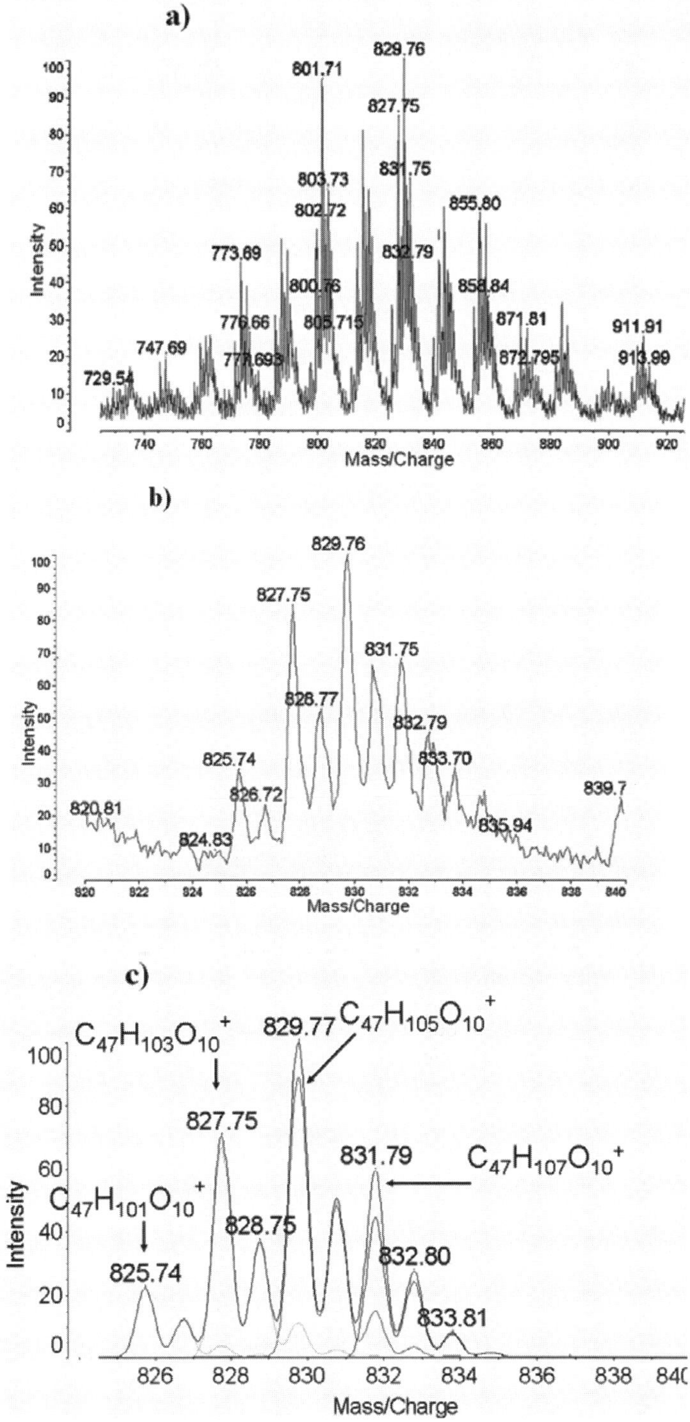

Figure 6.3 Amplifications of LDI-TOF MS of Chemapex HA (1000 mg L^{-1}) measured after 10 days of preparation. Scale from a) m/z 700-930 Da, b) m/z 820-840 Da. 1 μL 1000 mg L^{-1} HA. Linear positive mode, c) Model of mass spectra concerning peaks in the range m/z 820-840 Da. Overlapping of $C_{47}H_{101}O_{10}^{+}$, $C_{47}H_{103}O_{10}^{+}$, $C_{47}H_{105}O_{10}^{+}$, and $C_{47}H_{107}O_{10}^{+}$ ions is assumed

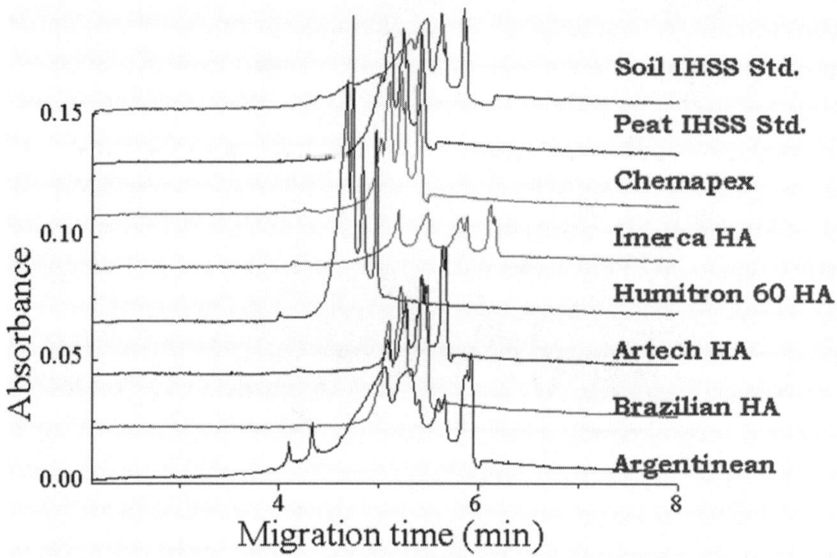

Figure 6.4 Electropherograms of various HAs (100 ppm) in borate (90mM)–Tris (90mM)–EDTA (1mM) buffer, pH 8.5. HAs 100 ppm, $\lambda = 214$ nm, normal polarity, 20 kV, 20°C, 20 seconds hydrodynamic injection

Organophosphorous Insecticides. Malathion and parathion complexation with HAs was followed by capillary zone electrophoresis (CZE). Figures 6.6a and 6.6b show the Chemapex HA electropherograms when adding increasing amounts of malathion and parathion insecticides, respectively. Even though these insecticides are negatively charged at the pH of the buffer (8.5), the interactions HA-parathion and HA-malathion must be strong because the CE fingerprints and migration times of the HA in the presence of these xenobiotics are different.

Quite interesting is that a small fraction of HA at 7-8 min migrates faster when the insecticide is added. On increasing the malathion concentration, another signal appears with almost the same migration time (Figure 6.6a). A new, intense peak also appears at about 2.3 min. It is probable that a) malathion displaces some small fraction trapped in the HAs structure (that is, encapsulated in a supramolecule) and is replaced by malathion, or b) an HA-malathion aggregate (peak at $\approx$ 2.3 min) forms and the peak intensifies after the malathion additions. A similar effect of supramolecular interaction in the system HAs-congo red has been observed previously [20].

On the other hand, with parathion additions this small HA fraction (around 7-8 min) just moves and diminishes in intensity. As for the HA main group of peaks (around 2-3 min), malathion and parathion change the Chemapex HA fingerprint substantially and a considerable number of peaks appear. This may be due to structural rearrangement of the HA because of the insecticide interactions. However, it was not possible to continue analyzing further additions because of precipitation.

Nitrate. Figure 6.7 shows amplification of the HA electropherograms resulting from increasing additions of potassium nitrate. It is clear from Figure 6.7 that the negative charge of the main group of HA peaks (about 5 min) increases because the electropherograms are slightly shifted to longer migration times. Therefore, interaction between negatively charged HAs and nitrate anions can be presumed. In previous studies it was demonstrated that chloride also is strongly complexed by HAs [20].

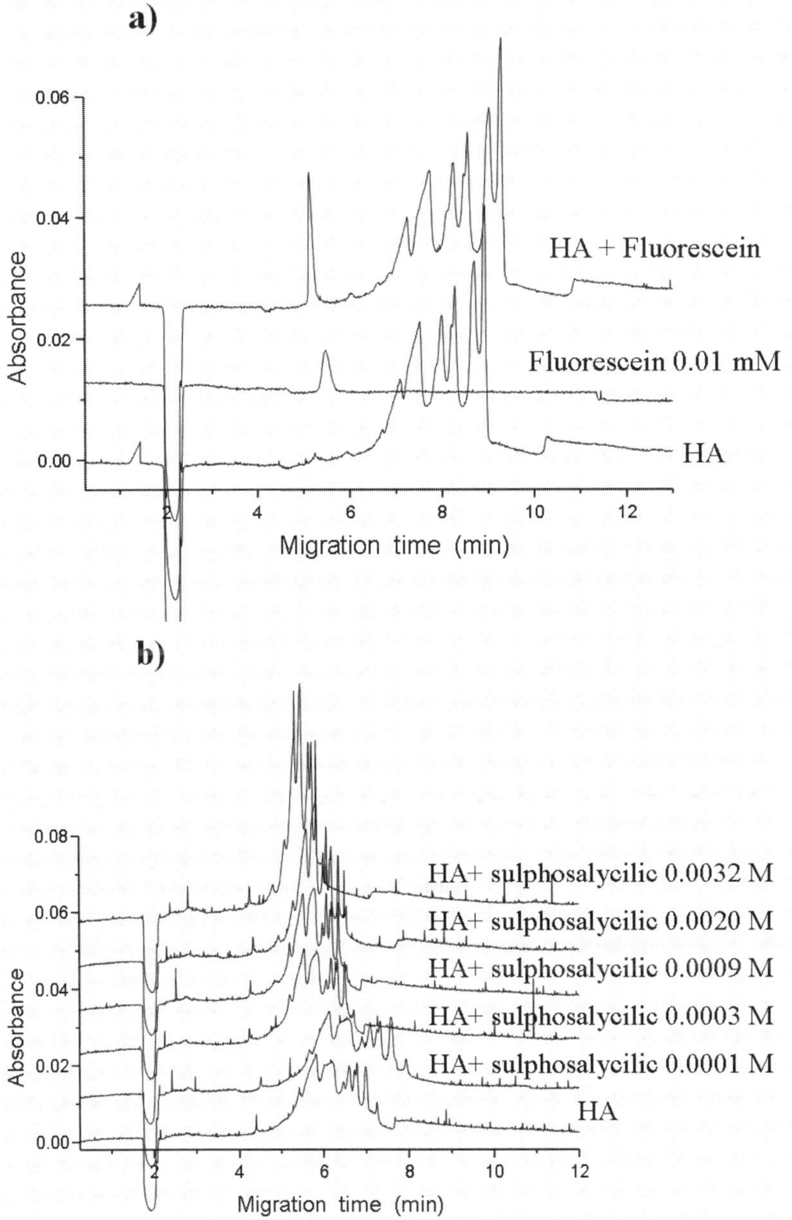

Figure 6.5 Electropherograms of Chemapex HA at increasing additions of different organic molecules a) fluorescein and b) 5-sulphosalicylic acid. Same separation conditions as in Figure 6.4

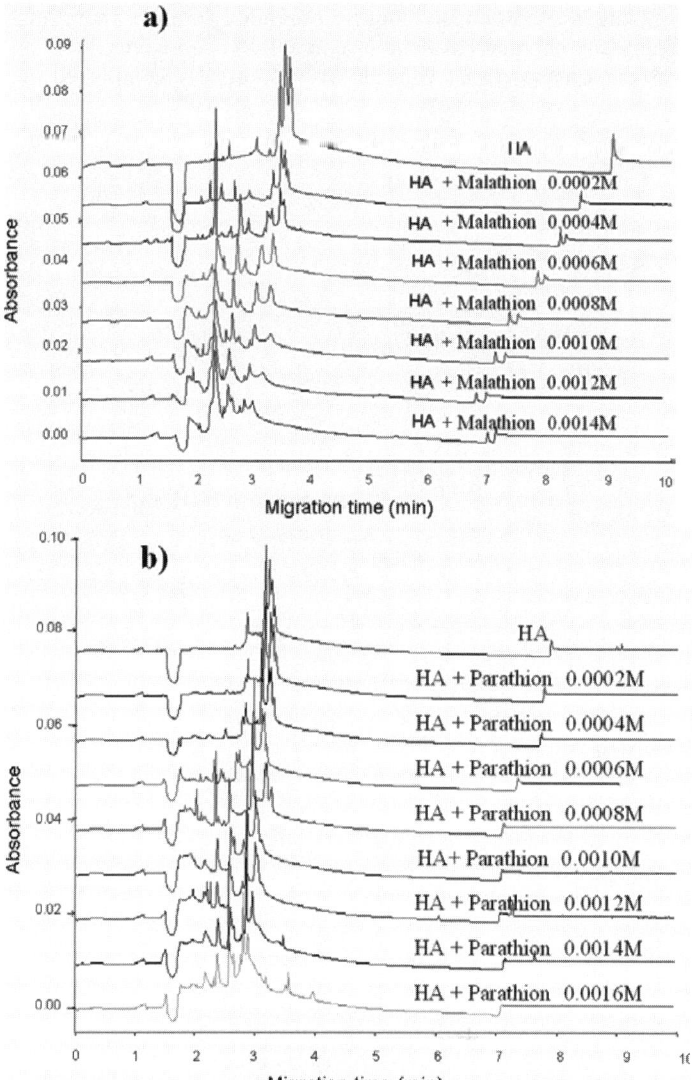

Figure 6.6 Electropherograms of Chemapex HA in borate-Tris-EDTA buffer, pH 8.5 and increasing additions of insecticides a) malathion and b) parathion. Separation conditions the same as in Figure 6.4

Cyanides. Because of the surprisingly marked interaction of the nitrate anion with Chemapex HA, some other experiments adding ferro- and ferricyanide to HAs solutions were performed. The results indicate strong binding of these kinds of anions with negatively charged HAs molecules. In addition, Figure 6.8 demonstrates the redox properties of HAs starting with a ferricyanide, $Fe(CN)_6^{3-}$, solution. After various additions of HA, the final product was the ferrocyanide ion, $Fe(CN)_6^{3-}$. The signal of HA is not observed for the same reason as with 5-sulphosalicylic acid.

Sodium cation. The interaction of HAs with alkali metal cations is reported in the literature. Humic acids are usually studied as sodium salts and the charge of humates is partially compensated by sodium cations. Therefore, we tried to complex sodium by adding a specific complexing agent, that is, dibenzo-18-crown-6 ether (18C6). Figure 6.9 shows the effect of adding 18C6 crown ether to the HA solution, with the aim of complexing the Na^+ present. On the contrary, the interactions with HA become stronger. That is, the sodium-18C6 crown complex interacts more strongly than the sodium cation itself.

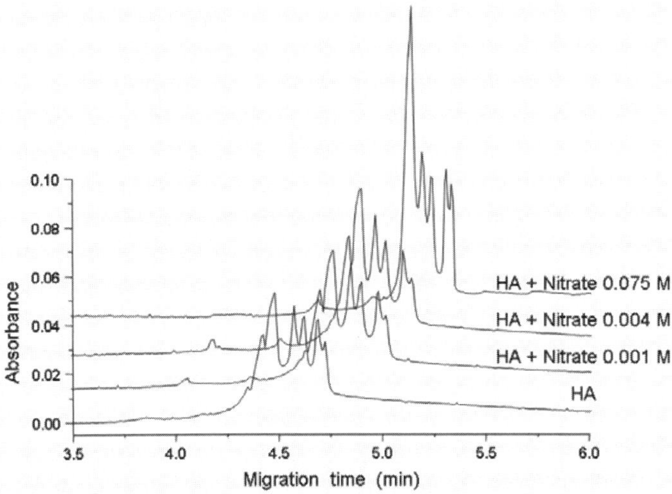

Figure 6.7 Electropherograms of Chemapex HA with increasing additions of potassium nitrate. Separation conditions the same as in Figure 6.4

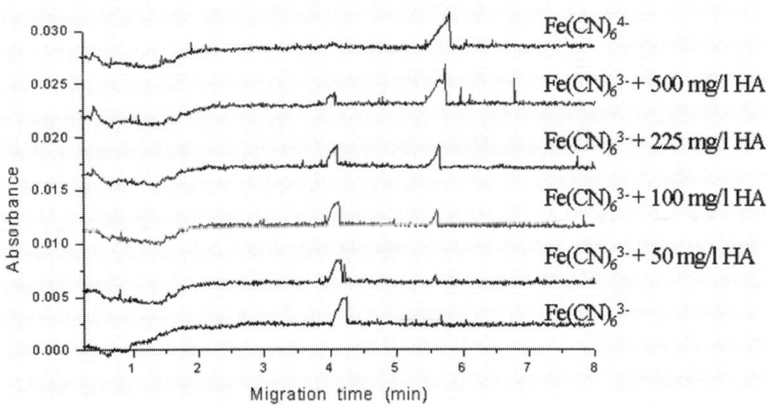

Figure 6.8 Electropherograms for additions of HA to a solution of 0.01M $Fe(CN)_6^{3-}$. Reversed polarity and other conditions as in Figure 6.4

Metal ions. Copper(II). It is well known that practically all metal ions interact with HAs. In previous work dealing with uranium(VI)–HAs interactions [23] it was shown that CE is a suitable technique for study of metal-ligand equilibria [34]. Electropherograms showing the interaction HA-Cu(II) are shown in Figure 6.10. When HA is added to a copper(II) solution (Figure 6.10a), the peak of free Cu^{2+} diminishes and the signal of humics decreases substantially. On the other hand, when Cu^{2+} is increasingly added to an HA solution (Figure 6.10b), the electropherograms show that the interaction is so strong that the HA fingerprint is totally modified at low, even sub-equimolar concentrations of copper(II).

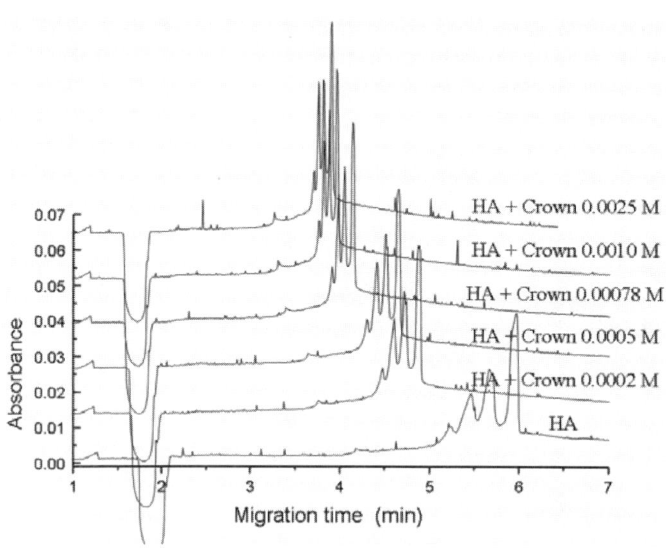

Figure 6.9 Electropherograms for various additions of 18C6 crown ether to bind Na^+. Separation conditions the same as those in Figure 6.4

a)

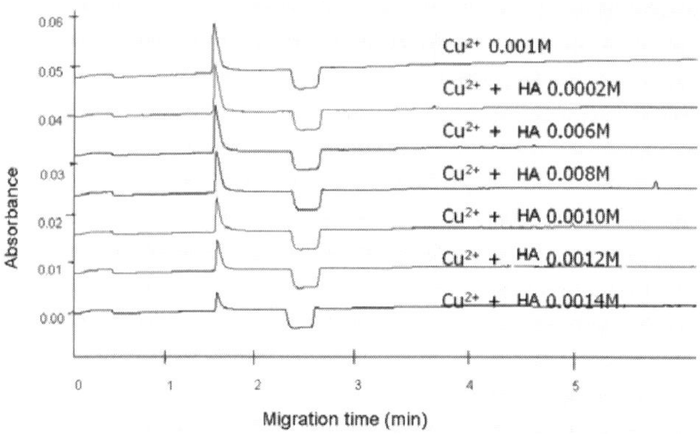

b)

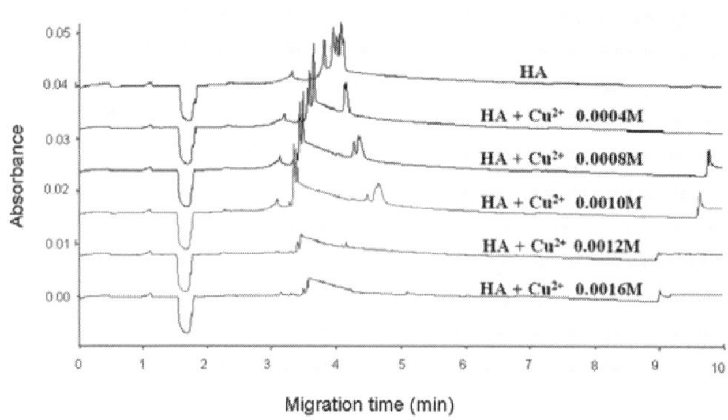

Figure 6.10 Electropherograms for a) various additions of Chemapex HA to a 1 mM Cu^{2+} solution and b) additions of Cu^{2+} to a Chemapex HA solution. Separation conditions the same as in Figure 6.4

Mercury(II). Results concerning HA-Hg(II) complexation are shown in Figure 6.11. The complex formed is so strong that the Chemapex HA electropherograms are substantially modified.

6.4. DISCUSSION

6.4.1. Mass Spectrometry

We abandoned the experiments conducted with common MALDI matrices because even the most useful matrix (ferulic acid) interacts with HA (Figure 6.1). On the other hand, reproducible and descriptive mass spectra for several HAs were obtained in the LDI mode. The mass spectra of HAs of different origin show quite similar patterns (Figure 6.2), indicating the presence of low molar mass substances in all the HAs samples. In most cases, two groups of signals around m/z 300-400 and 700-900 Da were observed, with the latter group the more intense (Figure 6.3a). The detailed mass spectra (shown, for example, in Figures 6.2 and 6.3) contain signals with a difference of m/z 14 or 28 Da. This is attributed to different numbers of $-CH_2-$ groups in HAs molecular structures. However, several other explanations are possible [15].

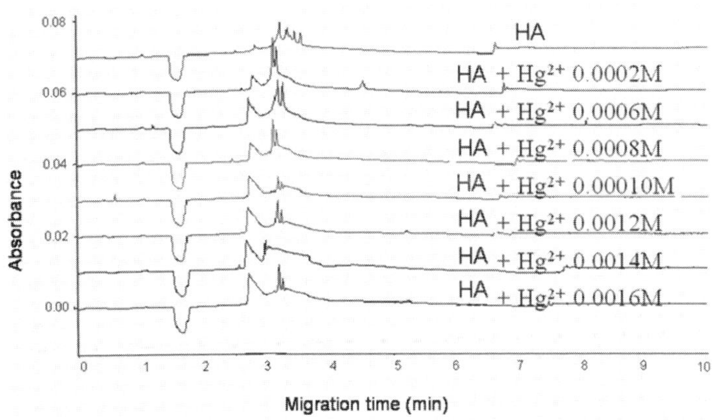

Figure 6.11 Electropherograms for various additions of a Hg^{2+} solution to Chemapex HA solution. Separation conditions the same as those in Figure 6.4

Based on the results obtained with LDI-TOF MS and also on the basis of the isotopic patterns observed for compounds present in HAs (Figure 6.3), a molecular formula was suggested. The simulation was performed by computer modeling of HAs mass spectra with AXIMA Kratos software. Several fragments containing carbon, oxygen and hydrogen were tested to simulate the experimental mass spectra (Figures 6.1-6.3). We concluded that the stoichiometric formula of HAs can resemble $C_{40}H_{38}O_{18}Na$, or $C_{40-n}H_{38-2n}O_{18}Na(CH_2)_n$, n = m + 2, m + 3, m + 4, m + 6, and so on. However, there are other possibilities (see above).

As shown above, the most frequent difference between the peaks is m/z 14 Da. Therefore, the proposed HA components should differ in the number of $-CH_2-$ groups, and because they are singly charged cations (detected in linear positive mode), some sodium adducts are proposed. Figure 6.12 shows theoretical mass spectra for a mixture of 5 compounds where the singly charged molecular ion corresponds to m/z 829.1773 Da and the other peaks represent molecules with the same structure but with the gain and loss of one or two $-CH_2-$ fragments. The agreement with experimental spectra is quite good considering the relatively low experimental mass resolution.

Mass spectra like those in Figure 6.3c can be obtained by assuming alternative formulas for the most intense $C_{47}H_{101}O_{10}^+$ ion, for example, $C_{40}H_{100}O_{14}Na^+$, $C_{40}H_{61}O_{16}^+$, $C_{41}H_{65}O_{17}^+$, $C_{42}H_{69}O_{16}^+$, $C_{43}H_{73}O_{15}^+$, $C_{44}H_{77}O_{14}^+$, $C_{47}H_{105}O_{10}^+$ ions with monoisotopic m/z peak values equal to 829.71, 829.39, 829.42, 829.46, 829.49, 829.53, and 829.77 Da. Observe that m/z for $C_{40}H_{100}O_{14}Na^+$ and $C_{47}H_{105}O_{10}^+$ are almost identical (within the m/z precision of $\pm$ 0.06 Da).

The same approach as for the most intense peak at m/z 829.76 Da (Figure 6.3a) is also valid for the other groups of peaks in the range m/z 720 – 920 Da. These groups of peaks differ by m/z 14.00 Da and so at least 40 different compounds are ionized. This observation agrees with our previous results [3] and also with the results of Kujawinski et al. [4]. Several groups are working to resolve the structures of HAs components with mass spectrometry and other methods.

6.4.2. Capillary Electrophoresis

CE is a suitable way of detecting chemical interactions between HAs components and different compounds due to its high sensitivity and relatively fast analysis. It follows from Figure 6.4 that electrophoretic study of HAs can be accomplished with the borate-Tris-EDTA buffer we developed recently [21]. Most CE literature on HAs describes the typical unresolved "humic hump" [19, 24]. As demonstrated here, "the hump" can be at least partially separated into smaller fractions. Thus, the hump most probably represents supramolecular aggregates. Also from Figure 6.4, the CE patterns are almost the same for various HAs, independent of the source or country of origin. This confirms our MS results that similar (or the same) low molar mass substances are present in all the HAs samples used in this study. Similar results were also obtained for Antarctic organic matter [25].

Fluorescein and 5-Sulphosalicylic Acid. Fluorescein and 5-sulphosalicylic acid are simple, representative organic molecules. Both contain a carboxylic group. Figure 6.5 shows the fingerprints for addition of each substance to a Chemapex HA solution. Addition of fluorescein does not change the fingerprint much, but it does slightly transform the electropherogram by resolving the signals (Figure 6.5a). Perhaps the HA-fluorescein interaction is weak because of steric effects.

On the other hand, 5-sulphosalicylic acid clearly modifies the HA fingerprints (Figure 6.5b). This molecule may interact with particular aggregating HA moieties. Figure 6.5b shows that the resolution of the signals is lost when the 5-sulphosalicylic acid concentration increases. Also, the migration time of the peaks slightly decreases as when the molar mass increases due to aggregation. The decrease in migration time may be due to the increased molar mass of the {HA-5-sulphosalicylic acid} aggregate. This conclusion agrees with the well-known facts that humic acids aggregate with hydroxycarboxylic acids and that they also self-aggregate in aqueous solution [35].

Organophosphorus Insecticides. Figure 6.6a shows the effect on the HA migration time when increasing amounts of the insecticide malathion were added to HA. The main group of HA peaks (around 4 min) is shifted towards the electro-osmotic flow (EOF). This means that the negative charge of HA is decreased. Also, the fraction observed initially at around 12 min is shifted slightly to the left, perhaps due to the higher molar mass with the encapsulated insecticide. Similar results were observed with the insecticide parathion, which also modified the electropherograms, suggesting strong interactions. However, the interactions of these insecticides are selective with some HA fractions and are different for malathion and parathion.

Nitrate. The electropherograms in Figure 6.7 show that adding potassium nitrate solution to an HA solution causes the migration time of the main signals to increase (more negatively charged molecules). This means that even though K^+ ions from KNO_3 are present in the solution, the stronger interaction is between HA and nitrate. Chloride and nitrate both interact with HA fractions.

Therefore, we suggest that the interactions of HAs with anions in some cases can be stronger than those with cations. This kind of interaction of negatively charged HA with another anion can be explained by considering the possibility that HA supramolecules contain macropolycyclic polyamines [36]. Such macrocycles, known as "anion binders," are able to sequester anions such as Cl^-, NO_3^- and CN^-. A certain HAs fraction might contain such polyamines able to complex anions. However, other kinds of sequestering cannot be excluded, including complexing with amino acids.

Cyanides. We reported previously that ferrocyanide and ferricyanide anions interact with HAs [20]. Figure 6.8 demonstrates the redox properties of HAs. The experiments carried out to follow change of the $Fe(CN)_6^{3-}$ signal when a solution of HA is added show that HA reduces ferricyanide and another signal appears that corresponds to $Fe(CN)_6^{4-}$. The HA signal is not observed for the same reason as for 5-sulphosalicylic acid.

Sodium Cation. It is well known that sodium salts influence the solution behavior of HAs. The HAs studied in this work were dissolved in sodium hydroxide, so they are sodium salts. We tried to decrease the effect of sodium cations by adding a strong complexing agent. The interaction of HAs with the sodium cation was followed by adding dibenzo-18C6 crown ether to an HA solution. Our idea was to complex the free sodium cation present in the HA solution and avoid its interaction with the HA. Surprisingly, Figure 6.9 shows that the effect was the opposite. Increasing the crown ether concentration caused the migration time of HA fractions to decrease, which means that the negative charge of HA decreases. The possible explanation is that the Na^+-crown complex interacts more strongly with the HA than the sodium cation itself. Therefore, the migration time is decreased due to the higher molar mass of the HA-$\{Na\text{-crown}\}^+$ macromolecule and/or the HA negative charge is more compensated than with sodium cations.

Heavy Metal Ions. We used the Medusa program [37] to obtain the distribution diagram of the aqueous system Cu(II)-acetate. The result was that Cu(II) in acetate buffer (pH 4.2) is mostly in the form of cationic species, that is, as Cu^{2+} and $Cu(acetate)^+$ complexes. Therefore, when a mixture of HA and Cu(II) was injected and analyzed by CE in acetate buffer (Figure 6.10a), on increasing the concentration of HA the peak of Cu^{2+} (1.6 min) was diminished. Therefore, HAs interact with both $Cu(acetate)^+$ and free Cu^{2+} ion.

In summary, complexation of Cu(II) with humic acids takes place even at pH 4.2 and so strongly that at slight excess of HA the peak of positively charged Cu(II) almost disappears. Figure 6.10b gives insight on Cu(II) complexation by HA in alkaline solution. In spite of the fact that Cu(II) is strongly complexed with EDTA at pH 8.5, the HA separation pattern (Figure 6.10b, curve 1) changes substantially after adding Cu(II) at sub-equimolar amounts (first curves, Figure 6.10b). While the first HA peaks (with no copper) are shifted to lower migration times, the last group of HA peaks are shifted towards longer migration times when the Cu(II) concentration is increased. Finally, at equimolar conditions (concentration of HA ~ Cu (that is, Figure 6.10b, curve 5), the peaks almost disappear, most probably due to Cu-humate precipitation that was observed after this point.

It is apparent especially from Figure 6.10b (curves 1-4) that Cu(II) reacts differently with different HA fractions and complexes with different stability are formed. This can explain the great discrepancy of Cu-HA stability constants in the literature. Our conclusion agrees with the indication from Lubal et al. [38] that the best model for potentiometric data is one with different copper(II) complexes with different HA components. In addition, it is also shown here (as with uranium(VI) [23]) that this is a general phenomenon: metal ions do not form single complexes with humic acids. Rather, they form different complexes with different HA fractions that can be separated by capillary electrophoresis.

Most of the published stability constants are average values of the constants for different complexes with individual HA fractions. As the HA fraction percentages in HA of different origin are different, HA-metal ion stability constants differ almost for each HA.

Concerning the HA-Hg(II) interaction (Figure 6.11), similar results to those with Cu(II) were observed but with a slight difference: Hg(II) is selectively bound only by some specific component of the HA sample, because the main group of peaks around 3 min shows that only 2 peaks stay without any alteration while the rest of the signals disappear with increasing Hg(II) concentration. In other words, complexation of HAs with Hg(II) is specific. Only one main predominant signal was observed before total metal complex precipitation.

All the experiments described confirm that HAs in aqueous solutions react with almost every species, neutral, positive or negatively charged. Figure 6.12 summarizes possible interactions between HAs and xenobiotics, which can be of various kinds: charge neutralization, charge transfer, π-π interactions, Van der Waals forces, hydrogen bonds, extraction of ion associates, supramolecular entities formation, aggregation, and so on. All these interactions are due to the polyelectrolyte character of HAs and the large number of functional groups present in all HAs molecules.

6.5. CONCLUSIONS

The fingerprints of HAs from various origins (peat, soil, coal and Leonardite) were obtained by LDI-TOF mass spectrometry. It was found that several m/z values are the same for all the HAs (even from different continents), suggesting that some compounds are the same and are present in HAs with very different origins. The isotopic patterns of ionized HA molecules were obtained. By mass spectral modeling it was possible to propose empirical formulas agreeing well with the main group of peaks. Formulas proposed for the main peak at m/z 879.26 are $C_{47}H_{101}O_{10}^{+}$, $C_{40}H_{100}O_{14}Na^{+}$, $C_{40}H_{61}O_{16}^{+}$, $C_{41}H_{65}O_{17}^{+}$, $C_{42}H_{69}O_{16}^{+}$, $C_{43}H_{73}O_{15}^{+}$, $C_{44}H_{77}O_{14}^{+}$ and $C_{47}H_{105}O_{10}^{+}$. Much higher mass resolution and careful further investigations are needed to distinguish between them. It was also found that several compounds with different numbers of $-CH_2-$ groups are present in HAs, while several ions with a m/z difference of 2 Da overlap in mass spectra. The results are in good agreement with those achieved by high resolution ESI FT-ICR [4,13-15].

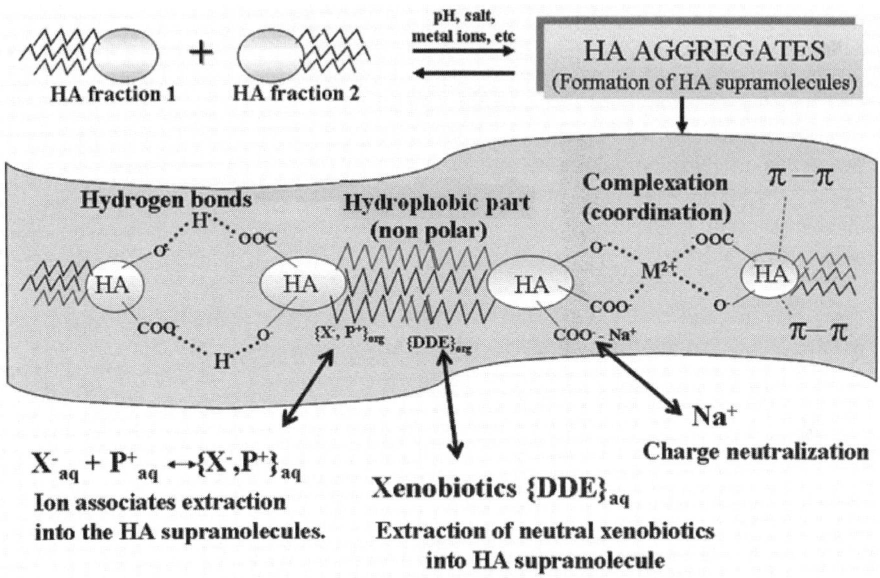

Figure 6.12 Review of various possibilities concerning interactions of humic acids themselves (aggregation) in the form of supramolecules and interactions of HAs with compounds observed in this work

With the optimal CE conditions reported before [21], the interactions of HAs with several compounds and ions were demonstrated successfully. Capillary electrophoretic analysis of HAs of different origin helps to demonstrate the presence of several similar fractions that are perhaps of the same or very similar nature in most of the HAs under study. We conclude that HAs might contain the same compounds in soil, peat and leonardite originated HAs. This conclusion is supported by the mass spectrometric results described above.

Complexation of insecticides by HAs is strong even if both species (HA and insecticide) are negatively charged under the working conditions. The interactions of the insecticides were selective with only some HA fractions. Strong complexation of nitrates and of ferro- and ferricyanides with HAs was also demonstrated. Concerning the interactions between negatively charged humic acids and inorganic anions like Cl^-, NO_3^-, $Fe(CN)_6^{3-}$ and $Fe(CN)_6]^{4-}$, it is suggested that macropolycyclic polyamine structures are present in humic acid supramolecules and that they may be responsible for strong anion binding. However, further study and confirmation by other methods is needed.

HAs electropherograms are strongly influenced by sodium cations. This effect was found to be more strongly expressed if Na^+ is complexed with 18C6 crown ether, perhaps because the more hydrophobic $\{Na(crown\ 18C6)\}^+$ cation interacts more strongly with negatively charged groups of humics. As for the other metals, we observed that HAs-metal complexes can be positive, neutral or negatively charged, depending on the concentration of both reactants and on the HA:metal ratio. Our work shows that Cu(II), Hg(II) and uranium(VI) react differently with different HA fractions and that several complexes with different stability are formed.

In general, the interactions were found to be of different kinds depending on the structure of the reactant molecule and its charge (neutral, positive or negative). Special care must be taken when studying these kinds of interactions because counter-ions added to the solution can be complexed more strongly than the metal itself. The buffer components should not react with the metal in order to leave it free to react with the HA. Thus, the counter-ion, the buffer and pH all play important roles in HA-metal interactions, irrespective of the analytical technique used.

ACKNOWLEDGEMENTS

This work is a part of a research program within a grant from the Ministry of Education and Youth of the Czech Republic, No. J07/98: 143100011.

REFERENCES

1. Davies G, Ghabbour EA. Understanding life after death. *Chemistry Industry,* 1999; 7: 426.
2. Pokorná L, Gajdošová D, Havel J. Characterization of humic acids by capillary zone electrophoresis and matrix assisted laser desorption/ionization time of flight mass spectrometry. In: Ghabbour EA, Davies G eds. *Understanding humic substances: Advanced methods, properties and applications.* Cambridge: Royal Society of Chemistry, 1999: 107-112.
3. Pokorná L, Gajdošová D, Mikeska S, Homoláč P, Havel J. The stability of humic acids in alkaline media. In: Ghabbour EA, Davies G eds. *Humic substances: Structures, models and functions.* Cambridge: Royal Society of Chemistry, 2001: 133-139.
4. Kujawinski EB, Freitas MA, Zang X, Hatcher PG, Green-Church KB, Jones RB. The application of electrospray ionization mass spectrometry (ESI-MS) to the structural characterization of natural organic matter. *Org. Geochem.,* 2002; 33: 171-180.
5. Haberhauer G, Bednar W, Gerzabek M.H, Rosenberg, E. Application of MALDI–TOF-MS to the characterization of fulvic acids. In: Ghabbour EA, Davies G eds. *Humic substances: Versatile components of plants, soil and water.* Cambridge: Royal Society of Chemistry, 2000: 143-152.
6. Havel J, Fetsch D, Peña-Méndez EM, Lubal P, Havliš J. Recent developments in humic acid characterization by CE and MALDI-TOF mass spectrometry. In: Swift RS, Spark KM eds. *Understanding and managing organic matter in soils, sediments and waters.* Adelaide, Australia: IHSS, 2001: 77-84.
7. Pfeifer T, Klaus U, Hoffman R, Spiteller M. Characterization of humic substances using atmospheric pressure chemical ionisation and electrospray ionisation mass spectrometry combined with size-exclusion chromatography. *J. Chromatogr. A.,* 2001; 926: 151-159.
8. Schulten HR. In: Boutton TW, Yamasaki, S eds. *Mass spectrometry of soils.* New York: Dekker, 1995.
9. Allard B, Templier J, Largenau C. Formation of melanoidin-like artifact

macromolecular material during the usual isolation process. *Org. Geochem.,* 1997; 26: 691-696.

10. Schulten HR, Plage B, Schnitzer M. A chemical structure for humic substances. *Naturwiessenschaften*, 1991; 78: 311-321.
11. Fievre ST, Marshall AG, Cooper WT. High resolution Fourier transform ion cyclotron resonance mass spectrometry of humic and fulvic acids using laser desorption and electrospray ionization. *Energy Fuels*, 1997; 11: 554-560.
12. Brown TL, Rice JA. Effect of experimental parameters on the ESI FT-ICR mass spectrum of fulvic acid. *Anal. Chem*., 2000; 72: 384-390.
13. Kujawinski EB, Hatcher PG, Freitas MA. High-resolution Fourier transform ion cyclotron resonance mass spectrometry of humic and fulvic acids: Improvements and comparisons. *Anal. Chem.*, 2002; 74: 413-419.
14. Stenson AC, Landing WM, Marshall AG, Copper WT. Ionization and fragmentation of humic substances in electrospray ionization Fourier transform-ion cyclotron resonance mass spectrometry. *Anal. Chem.*, 2002; 74: 4397-4409.
15. Stenson AC, Marshall AG, Copper WT. Exact masses and chemical formulas of individual Suwannee River fulvic acids from ultrahigh resolution ESI FT-ICR mass spectra. *Anal. Chem.*, 2003; 75: 1275-1284.
16. Bragato G, Mori A, De Nobili M. Capillary electrophoretic behaviour of humic substances from Sphagnum peats of various geographical origin: Relation with the degree of decomposition. *Europ. J. Soil Sci.,* 1998; 49: 589-601.
17. Pompe S, Heise KH, Nitsche H. Capillary electrophoresis for a "finger-print" characterization of fulvic and humic acids. *J. Chromatogr. A*, 1996; 723: 215-220.
18. Fetsch D, Havel J. Capillary zone electrophoresis for the separation and characterization of humic acids. *J. Chromatogr. A,* 1998; 802: 189-197.
19. Schmitt-Kopplin P, Hertkorn N, Garrison AW, Freitag D. Kettrup A. Influence of borate buffers on the electrophoretic behavior of humic substances in capillary zone electrophoresis. *Anal. Chem,.* 1998; 70: 3798- 3810.
20. Pacheco ML, Peña-Méndez EM, Havel J. Supramolecular interactions of humic acids with organic and inorganic xenobiotics studied by capillary electrophoresis. *Chemosphere*, 2003; 51: 95-108.
21. Pacheco ML, Havel J. Capillary zone electrophoresis of humic acids from the American continent. *Electrophoresis*, 2002; 23: 268-277.
22. Pokorná L, Gajdošová D, Mikeska S, Havel J. Analysis and characterization of a standard coal derived humic acid. In: Ghabbour EA, Davies G eds. *Humic substances: Structures, models and functions.* Cambridge: Royal Society of Chemistry, 2000: 299-307.
23. Pacheco ML, Havel J. Capillary zone electrophoretic study of uranium(VI) complexation with humic acids. *J. Radioanal. Nucl. Chem*., 2001; 248: 565-570.
24. Pokorná L, Pacheco ML, Havel J. Highly reproducible capillary zone electrophoresis of humic acids in cyclodextrin or oligosaccharide-modified background electrolytes. *J. Chromatogr. A*, 2000; 723: 345-350.
25. Gajdošová D, Novotná K, Prošek P, Havel J. Separation and characterization of humic acids from Antarctica by capillary electrophoresis and matrix-assisted laser desorption ionization time-of-flight mass spectrometry: Inclusion complexes of humic acids with cyclodextrins. *J. Chromatogr. A,* 2003; 1014: 117-127.

26. CHEMAPEX s.r.o., http://www.chemapex.cz
27. Personal communication from Dra. Maria O. Rezende, Chemistry Institute of Sao Carlos, Sao Paulo University, CP 780-13560-970, Sao Carlos, S.P., Brazil.
28. FLUKA I HA, Fluka Chemie AG, Buchs, Switzerland.
29. GBM, Grupo Bioquímico Mexicano, S.A. de C.V., Blvd. Dr. Jesus Valdes Sanchez No. 2369, Saltillo, Coahuila C.P. 25290, Mexico.
30. IMERCA, Importacion y Mercadeo, S.A,Av. 36-40, Col. Las Charcas, Guatemala, C.A.
31. ARTECH, Inc. 14100 Park Meadow Dr., Chantilly, VA 22051, USA.
32. Personal communication with Ceppi, SB et al., Department of Natural Resources, Faculty of Agricultural Sciences, University National of Cordoba, CC 509-5000-Cordoba, Argentina.
33. See http://www.ihss.gatech.edu
34. Havel J, Janoš P. Evaluation of capillary zone electrophoresis equilibrium data using the CELET program. *J. Chromatogr. A,* 1997; 786: 321-33.
35. Fetsch D, Hradilová M., Peña Méndez EM, Havel J. Capillary zone electrophoresis study of aggregation of humic substances. *J. Chromatogr. A,* 1998; 817: 313-323.
36. Lehn JM. *Supramolecular chemistry*. Weinheim, Germany: VCH, 1995.
37. MEDUSA and HYDRA; http://www.inorg.kth.se
38. Lubal P, Široký D, Fetsch D, Havel J. The acidobasic and complexation properties of humic acids. Study of complexation of Czech humic acids with metal ions. *Talanta,* 1998; 47: 401–412.

Chapter 7

PHOTOSENSITIZED GENERATION OF SINGLET OXYGEN AND ITS QUENCHING BY HUMIC SUBSTANCES

Andrea Paul,[1] Steffen Hackbarth,[2] Elke Zwirnmann,[1] Beate Röder[2] and Christian E.W. Steinberg[1]

[1] Leibniz Institute of Freshwater Ecology and Inland Fisheries, Müggelseedamm 301, 12561 Berlin, Germany
[2] Humboldt University Berlin, Institute of Physics, Newtonstr. 5, 12489 Berlin, Germany

7.1. INTRODUCTION

Reactions of $^1O_2$ are of high current interest because of their importance for processes in many photo-oxidations of chemical and biological systems, such as reactions used in photo-dynamic therapy [1]. Singlet oxygen is a powerful oxidant, which among others plays an important role in the photo-degradation of dyes and polymers [2,3].

Humic substances represent the main fraction of dissolved organic carbon in natural waters that absorbs solar radiation, and hence are mainly responsible for aquatic photochemistry. Irradiated HSs generate not only $^1O_2$ [4-7], but also free radicals and hydrated electrons [8], hydroxyl radicals [6,9], superoxide [6,10], hydrogen peroxide [6,11] and reactive HS triplet states [12,13]. Scheme 7.1 [14] serves to illustrate various reactions starting after photo-excitation of HSs that lead to the formation of reactive species. Generation of reactive oxygen species can increase the normally low rate of oxidation of several organic compounds in natural waters [15,16], change the redox state [17,18] and cause oxidative stress to organisms [19-21]. Photolytic processes of dissolved organic matter are involved both in the formation and in the degradation of HSs by photo-mineralization [22,23]. Aquatic organisms benefit from the release of fatty acids, carbonyl compounds and amino acids, which serve as organic nutrients for heterotrophic bacteria [24,25]. Although it is not yet clear which of these reactive species are responsible for the processes listed above, $^1O_2$ is known to be a selective oxidant that reacts with electron-rich olefins, conjugated dienes, sulfides and phenols [13,26]. In living cells, both impairment and activation of signaling processes by

$^1O_2$ result from the interaction of $^1O_2$ with biomolecules and their oxidative modification [27-29]. In aquatic systems, $^1O_2$ is suggested to be responsible for the indirect photochemical degradation of organic pollutants [30-32] and allelochemicals, such as cyanotoxins [33].

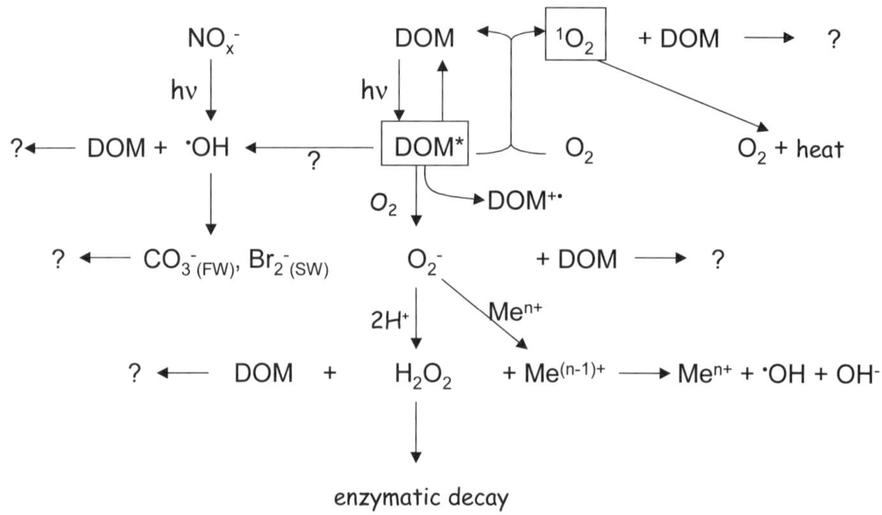

Scheme 7.1 Reactive oxygen species produced after light absorption (hν) by dissolved organic matter (DOM) modified according to Kieber et al. [14]

Previous attempts to determine $^1O_2$-quantum yields (Φ_Δ) of HSs indirectly by chemical trapping using 2,5-dimethylfuran were recently followed by direct steady state and time-resolved detection of singlet oxygen phosphorescence. The observed values for Φ_Δ ranged between 0.5 and 3% [14], in one case even 6% (see Tables 7.2 and 7.3 below). Action spectra that indicate an increase of Φ_Δ with decreasing excitation wavelengths have been published by Haag et al. [5] and Scully et al. [34]. Based on fractionation of HSs by gel permeation chromatography, Haag et al. [35] suggested that the lower molar mass fractions seem to be more efficient in generating $^1O_2$. This result was supported by Sandvik et al. [36], who investigated the photoreactivity of dissolved organic matter along salinity transects in the Mississippi and Atchafalaya River plumes. Although the photo-inductive capacities of various HSs correlated well with their ability to produce singlet oxygen, no apparent correlation was found between structural characteristics evaluated by IR, elemental analysis (including metals) and solid-state ^{13}C NMR [32]. The latter results are contradictory to flash photolysis studies [37], which suggested that quenching of excited triplet states of HSs should be increased by complexed paramagnetic metals. Recent investigations even point to a HS-mediated photo-oxidation of Mn(II) by $^1O_2$ and OH-radicals as a significant geochemical pathway in natural water [38]. These results may serve to illustrate the complexity of and contradictions within the photochemistry of HSs.

In aqueous solutions, $^1O_2$ rapidly returns to its ground state due to physical quenching by water. This process and the quenching of $^1O_2$ by dissolved organic matter are considered to be the main pathways of deactivation in the absence of any further targets (Scheme 7.1). Consequently, HSs are not only sources of reactive oxygen species, but also act as effective quenchers, as was investigated by Hessler et al. [39]. Further spectroscopic evidence for the ability of HSs to quench $^1O_2$ arose from studies by Sandvik et al. [36] and Bilski et al. [40], but these studies did not provide quenching constants. The apparent quantum yield, which can be determined by chemical or spectroscopic methods, only yields the fraction of singlet oxygen that escapes from the complex HSs macromolecules into the D_2O solution, whereas part of the photosensitized $^1O_2$ is quenched both physically and chemically by the HS itself. Consequently, not only the presence of chromophores and functional groups that allow for the formation of $^1O_2$, but also the presence of groups or structures acting as quenchers, determine the apparent value of Φ_Δ. In order to obtain further knowledge of these photochemical processes, comparison of data on sensitization and quenching of $^1O_2$ [41,42] will be useful.

7.2. MATERIALS AND METHODS

7.2.1. Materials

Chemicals. D_2O, Rose Bengal (Fluorescein 3,4,5,7-tetrachloro-2′,4′,5′,7′-tetraiodo dianion) and salts for buffers (sodium hydrogen phosphate, anhydrous and sodium dihydrogen phosphate) were purchased from Aldrich and Merck, respectively.

Humic Substances. Humic acids (HAs) and fulvic acids (FAs) were purchased from the IHSS [43] and Fredriks Research Products [44], respectively. Most of the natural organic matter (NOM) samples were collected by reverse osmosis and were characterized in the NOMiNiC project (Natural Organic Matter in the Nordic Countries) [45,46]. A detailed description of source, origin and elemental analysis of HSs in the NOM "Fuchskuhle" is given in [41,47]. HS1500 [48], a synthetic HS, was produced by Weyl GmbH [49]. Abbreviations, sources and origins of HSs are listed in Table 7.1. For all $^1O_2$-measurements, stock solutions of 0.4 - 4 mg mL^{-1} HSs in phosphate buffers (PBS), pH approximately 7.5, were prepared 24 h in advance, stirred for 2 h at room temperature and kept at 4°C in darkness.

Humic Substances Characterization. The molar mass distribution of HSs was determined by gel permeation chromatography using a Toyopearl HW 50 S resin (250 x 20 mm), eluted with 29 mmol L^{-1} phosphate buffer (pH 6.5) at a flow rate of 1 mL min^{-1}. Molar masses at peak maximum of humic substances (M_p) were calculated after calibration with polydextrans (M_p: 830, 4400, 9900, 21400, 43500) from Polymer Standards Corp. (Mentor, Ohio), and saccharides (raffinose, maltose, glucose), glycerin and methanol from Merck [50]. Aromaticity (SAC/DOC in L mg^{-1}m^{-1}) was taken as the ratio between the spectral absorption coefficient (SAC in m^{-1}, at 254 nm) and the organic carbon (DOC in mg L^{-1}) and was used as a measure of aromatic and unsaturated structures in HSs [51].

7.2.2. Time Resolved Detection of Singlet Oxygen

Time-Resolved Detection of Singlet Oxygen Phosphorescence. Excitation of samples was performed with a Nd-YAG laser pumped OPO (BMI), and a 5 x 5 mm² Ge-Pin diode with cooled preamplifier (Northcoast) was used for detection. The measured decay curves were fitted mono-exponentially by a least squares method (Figure 7.1), yielding the value for the initial amplitude and the decay time, Eq. 7.1.

Table 7.1 Abbreviations, origin and sources of natural and synthetic humic substances

abbreviation	HS	origin/catchment/ country	source
SuRi-NOM	Suwannee River NOM	aquatic, blackwater river, USA	IHSS [43]
Skj-(f)NOM/ Skj-(s)NOM	Skjervatjern NOM fall 1999/ spring 2000	aquatic, small dystrophic lake, Norway	NOMiNiC [46]
Svart-(f)NOM/ Svart-(s)NOM	Svartberget NOM fall 1999/ spring 2000	aquatic, small stream, boreal pristine, Sweden	NOMiNiC [46]
Val-(f)NOM/ Val-(s)NOM	Valkea-Kotinen NOM fall 1999/ spring 2000	aquatic, small acidic headwater lake, Finland	NOMiNiC [46]
Bir-(f)NOM/ Bir-(s)NOM	Birkenes NOM fall 1999/ spring 2000	aquatic, acidified small stream, Norway	NOMiNiC [46]
Hiet-(f)NOM/ Hiet-(s)NOM	Hietajärvi NOM fall 1999/ spring 2000	aquatic, large oligotrophic lake, Finland	NOMiNiC [46]
Heo-(s)NOM/ Heo-(f)NOM	Hellerudmyra NOM spring 1996/ fall 1996	aquatic, small tarn, Norway	E. Gjessing [45]
Fuku-(f)-NOM/ Fuku-(s)-NOM	Fuchskuhle NOM fall 2001/ spring 2002	aquatic, small dystrophic lake, Germany	IGB [47]
SuRi-HA	Suwannee River HA Standard	aquatic, blackwater river, USA	IHSS [43]
Peat(S)-HA	Peat HA Standard	terrestrial, peat, USA	IHSS [43]
Peat(R)-HA	Peat HA Reference	terrestrial, peat, USA	IHSS [43]
Leon-HA	Leonardite HA Standard	terrestrial, lignite, USA	IHSS [43]
Sum-HA	Summet Hill Soil HA Reference	terrestrial, soil, tussock grass area, New Zealand	IHSS [43]
Soil-HAII	Soil HA Standard	terrestrial, prairie soil, USA	IHSS [43]
Lau-HA	Laurentian HA	terrestrial, soil, Canada	Fredriks Res. Prod. [44]
SuRi-FA	Suwannee River FA Standard	aquatic, blackwater river, USA	IHSS [43]
Lau-FA	Laurentian FA	terrestrial, soil, Canada	Fredriks Res. Prod [44]
Soil-FAII	Soil FA Standard II	terrestrial, prairie soil, USA	IHSS [43]
HS1500	HS1500	synthetic, derived from autoxidation of quinone	Weyl GmbH [48, 49]

$$f(t) = Ie^{\frac{-t}{\tau}} \tag{7.1}$$

The amplitude (I) is directly proportional to the amount of generated singlet oxygen, whereas the decay time (τ) gives information about quenching processes of singlet oxygen. Under the precondition that both sensitizer and reference are adjusted to the same optical absorbance at the excitation wavelength, singlet oxygen quantum yields Φ_Δ can be calculated by comparing the amplitudes of the samples with that of the reference (I^{ref}), Eq. 7.2.

$$\Phi_\Delta = \Phi_\Delta^{ref} \frac{I}{I^{ref}} \tag{7.2}$$

Here, we used the excitation range of 355–550 nm with Rose Bengal (Φ_Δ^{ref} =75% in D_2O [52]) as reference sensitizer. All samples were adjusted to an absorbance of 0.2 at the excitation wavelength.

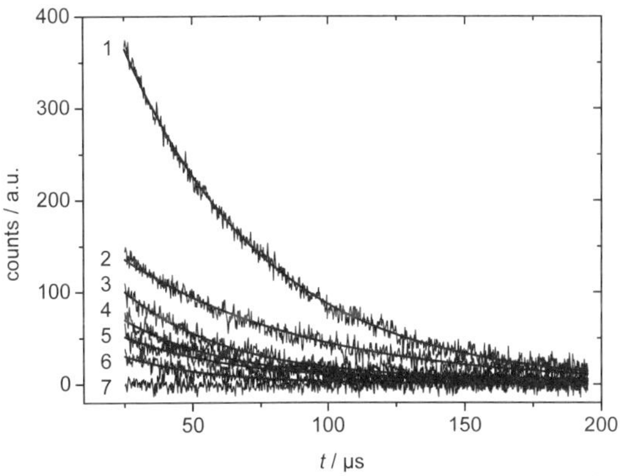

Figure 7.1 Time resolved phosphorescence of $^1O_2$ from HSs in D_2O. Excitation wavelength 480nm. Measured curves and mono-exponential fits in the temporal range 25-195 µs. 1 – Soil-FAII, 2 – Peat(S)-HA, 3 – Bir-s, 4 – Bir-f, 5 – SuRiHA, 6 – Val-s, 7 – phosphorescence quenched by NaN_3 [41].– Reproduced by permission of The Royal Society of Chemistry (RSC) on behalf of the European Society for Photobiology and the European Photochemistry Association

Excitation Spectra were determined for the samples SuRi-HA, SuRi-NOM, Peat(S)-HA, Bir-f, Bir-s, Soil-FAII, Hiet-f and Svart-s. For each excitation wavelength a new sample was prepared and adjusted to an absorbance of 0.20. The present operation mode of the Nd-YAG laser system working with a SHG-crystal plus OPO enabled a tuning range between 450–550 nm, and without OPO excitation at 355nm.

Quenching Experiments. In order to investigate the quenching behavior of HSs, phosphorescence decays of singlet oxygen, photosensitizer generated by Rose Bengal in D_2O with absorbance 0.2 at the excitation wavelength 514 nm were measured. Rose Bengal dissolved in D_2O (pH approximately 7.5) yielded a singlet oxygen decay time (τ_0) of 64 ±3 µs, which agrees well with $^1O_2$ lifetimes reviewed by Wilkinson et al. [1]. In order to calculate k_Q from Stern-Volmer plots, aliquots of HS-stock solutions with final HSs concentrations up to 8 x 10^{-4} g mL^{-1} were injected into a cuvette containing Rose Bengal solution. With increasing HSs concentration, both decay times (τ) and amplitudes of the signals decreased (Figure 7.1).

7.3. RESULTS

7.3.1. Quantum Yields of $^1O_2$

Quantum yields (Φ_Δ) of $^1O_2$ generation, calculated on the basis of time-resolved phosphorescence measurements and literature data obtained from both steady state phosphorescence and chemical trapping experiments, are summarized in Tables 7.2 and 7.3. Because Φ_Δ shows a dependence on the excitation wavelength (λ_{exc}) (cf. section 7.3.2), only Φ_Δ measured at similar λ_{exc} could be compared.

Table 7.2 Quantum yields of singlet oxygen generation by humic substances obtained at $\lambda_{exc} \geq 400$ nm

$\lambda_{exc} = 480$ nm, Φ_Δ determined by time-resolved phosphorescence measurement [41]		$\lambda_{exc} \geq 400$ nm, Φ_Δ determined by chemical trapping experiments	
sample:	Φ_Δ %	sample:	Φ_Δ %
SuRi-NOM	0.23	$\lambda_{exc} = 434$ nm	
Skj-(f)NOM	0.22	Aldrich HA [31]	0.1
Skj-(s)NOM	0.29		
Svart-(f)NOM	0.26	$\lambda_{exc} = 400$ nm	
Svart-(s)NOM	0.30	Lake Erie [34]	0.12
Val-(f)NOM	0.16	Sarpes Bay [34]	0.07
Val-(s)NOM	0.30	Smithwick Bog [34]	0.03
Bir-(f)NOM	0.16		
Bir-(s)NOM	0.91	$\lambda_{exc} = 405$ nm	
Hiet-(f)NOM	0.59	Fluka HA [5]	0.5
Hiet-(s)NOM	0.86	Black Lake HA [5]	1.3
Heo-(s)NOM*	0.16	Lake Baldegg DOM [5]	2.3
Heo-(f)NOM*	0.13		
Fuku-(f)-NOM*	0.51	$\lambda_{exc} = 436$ nm	
Fuku-(s)-NOM*	0.48	Fluka HA [5]	0.48
SuRi-HA	0.31	Black Lake HA [5]	1.2
Peat(S)-HA	0.83	Lake Baldegg DOM [5]	2.0
Peat(R)-HA	0.81		
Leon-HA	0.55	$\lambda_{exc} = 546$ nm	
Sum-HA	0.67	Fluka HA [5]	0.5
Soil-HAII	0.75	Black Lake HA [5]	0.51
Lau-HA	1.08	Lake Baldegg DOM [5]	1.1
SuRi-FA	0.57		
Lau-FA	0.26		
Soil-FAII	2.69		
HS1500	≤0.06		

Data were taken from Paul et al. [41], Aguer et al. [31], Scully et al. [34], and Haag et al. [5].* this work.

Table 7.3 Quantum yields of singlet oxygen generation by humic substances obtained at $\lambda_{exc} < 400$ nm

$\lambda_{exc} = 366$ nm, Φ_Δ determined by chemical trapping experiments		$\lambda_{exc} = 280$-380 nm, Φ_Δ determined by steady state phosphorescence measurement [36]	
sample:	Φ_Δ %	sample:	Φ_Δ %
SuRiFA pH 7.0 [37]	1.0	Mississippi, 1995, T2C >3000 Da	1.7
Bansee FA, pH 7.0 [37]	2.1	Mississippi, 1995, T2D >3000 Da	1.7
Brunnensee FA pH 8.0 [37]	1.6	Mississippi, 1995, T2E >3000 Da	1.7
BM7 FA pH 8.0 [37]	1.0	Atchafalaya, 1995, T1A >3000 Da	2.7
Zillhamer See FA pH 7.0 [37]	1.7	Atchafalaya, 1995, T1B >3000 Da	1.9
Fluka HA [5]	0.5	Atchafalaya, 1995, T1D >3000 Da	2.0
Black Lake HA [5]	1.0	Atchafalaya, 1995, T1E >3000 Da	1.6
Lake Baldegg DOM [5]	2.6	Gulf of Mexico, BGOM >1000 Da	3.9
$\lambda_{exc} = 365$ nm		Mississippi River, BMISS >1000 Da	6.1
Ranker–FA [31]	0.55	Apalachicola River, APP >3000 Da	1.4
Ranker-HA [31]	0.36	Aldrich HA	1.0
Aldrich-HA [31]	0.20	Mississippi, 1996, MD >3000 Da	2.0
Rendzina HA [31]	0.11	Mississippi, 1996, MC >3000 Da	1.8
Podzol A1 IIA [31]	0.039	Mississippi, 1996, MA >3000 Da	1.7
Podzol Bh HA [31]	0.039	Blackwater River	3.8
$\lambda_{exc} = 360$ nm		Poultney River	1.1
Lake Erie [34]	2.2	Otter Creek	2.4
Sharpes Bay [34]	0.7	Mud Creek	1.3
Smithwith Bog [34]	0.4	Alburg Dune Park	0.5

Data were taken from Frimmel et al. [37], Haag et al. [5], Aguer et al. [31], Scully et al. [34], and Sandvik et al. [36].

For $\lambda_{exc} = 280$-380 nm (Table 7.3), the values of Φ_Δ varied between 0.04 and 6.1% and for $\lambda_{exc} > 400$ nm between 0.06 and 2.7%, respectively (Table 7.2). In all studies, low Φ_Δ were found for synthetic HSs. Although a general trend of higher Φ_Δ with lower molar mass of HSs was indicated, linear regression analysis yielded no correlation of Φ_Δ ($\lambda_{exc} = 480$nm, time-resolved phosphorescence detection) with either molar mass ($R = 0.177$, $R^2 = 0.031$) or aromaticity ($R = 0.116$, $R^2 = 0.013$) for the data in Tables 7.2 and 7.4. The comparison between NOM, HA and FA for the SuRi-samples yielded Φ_Δ of 0.23%, 0.31% and 0.57%, respectively. The same pattern was distinguished for soil-HA and soil-FA with Φ_Δ of 0.75% and 2.7%, respectively, whereas the Lau-HA and Lau-FA behaved in an opposite way with 1.1% and 0.26%, respectively.

In order to check whether the presence of (paramagnetic) metals has an influence on the generation of $^1O_2$, quantum yields were plotted against the amounts of metals in HSs. For this purpose, we collected data on Fe (ferromagnetic, paramagnetic, antiferromagnetic properties possible in HSs), Mn (paramagnetic properties in HSs), Cu (diamagnetic as element, but complexed or in aggregates, paramagnetic), and Al (diamagnetic) in HSs [31,41,51,53-54]. Correlations between Φ_Δ and the Fe, Mn, Al, and Cu as well as the sum of metal contents were weak and not significant, as it is shown in Figures 7.2A and 7.2B for two independent datasets. The weak correlations may be explained by factors such as localization, redox state and spatial geometry of the metals inside HSs influencing the quenching of excited triplet states, rather than metal concentrations alone.

Table 7.4 Quenching constants (k_Q) in D_2O obtained from time-resolved phosphorescence measurements, molar masses at peak maximum (M_p) and aromaticity (A) of HS in PBS from gel permeation chromatography

Sample	k_Q, L mg^{-1} s^{-1}	M_p, Da	A, L mg^{-1}m^{-1}
SuRi-NOM	14.1	4871	5.0
Skj-NOM	15.6	5455	5.5
Svart-NOM	14.8	4871	5.6
Fuku-NOM	10.6	4073	4.0
SuRi-HA	35.6	7766	9.5
Peat(S)-HA	34.8	13098	11.85
Sum-HA	46.7	9454	11.2
Soil-HAII	38.0	9770	13.1
SuRi-FA	12.4	5458	4.7
Lau-FA	11.6	8721	5.6
Soil-FAII	15.1	7639	6.0
HS1500	32.1	11586	8.8

Data taken from Paul et al. [41].

7.3.2. Wavelength Dependence of $^1O_2$

Due to the fact that HSs represent a complex mixture of various molecules, the contribution of the absorption of the $^1O_2$ generating species to the overall absorption of HSs varies with wavelength. Consequently, a dependence of Φ_Δ on the excitation wavelength can be expected. Wavelength dependent quantum yields of singlet oxygen formation are shown in Figures 7.3A and 7.3B. All aquatic NOM samples yielded a similar spectral shape, which is characterized by an increase from 550 to 450 nm and a weak slope from 450 to 355 nm. Depending on the excitation wavelength intervals, the maximum was found between 450 nm (Figure 7.3A) and 406 nm (Figures 7.3B, data from Haag et al. [5]).

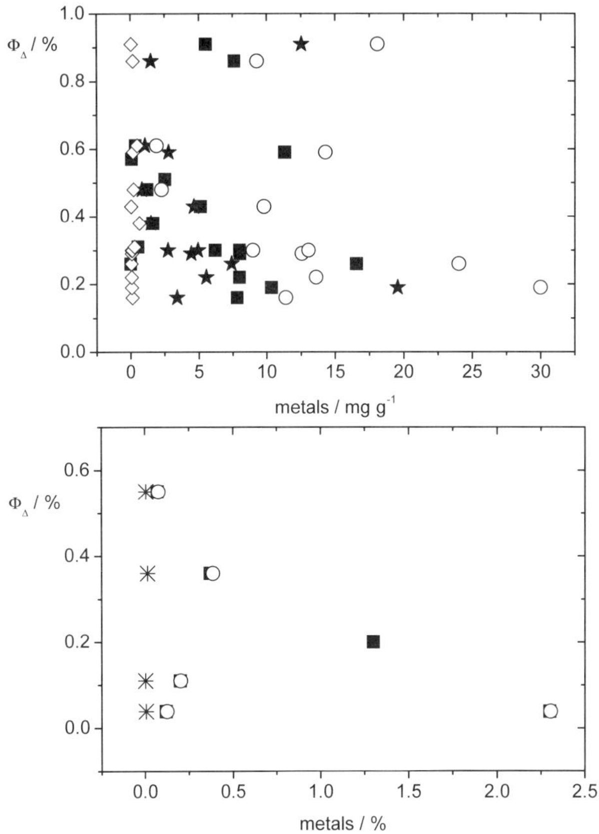

Figure 7.2 Plot of singlet oxygen quantum yields (Φ_Δ, means ± accuracy 12.5%) vs. metals in HS. **A:** Plot of (Φ_Δ) vs. Fe (■), Cu (◇), Al (★) and sum of metals in HS (○). Values for Φ_Δ were adapted from Paul et al. [41], and for metals from Riise and Salbu, [53], Cook and Langford [54] and the NOMiNiC project [51]. **B:** Plot of Φ_Δ vs. percentage of Fe (■), Mn (✳) and sum of Fe and Mn (○) in HSs. Data were adapted from Aguer et al. [31]

Interestingly, HSs from peat (Peat-S) and soil (soil-FAII) exhibited a similar spectral shape in the range 450-355 nm, but contrary to the aquatic samples, no further increase of Φ_Δ at 355 nm was found and the presence of a maximum at approximately 450 nm was more evident. Soil-FAII exhibited extraordinary high Φ_Δ over the whole spectral range. Lowest Φ_Δ were found for SuRi-HA, revealing an intermediate spectral shape.

These results indicate that there are mainly two groups of chromophores in HSs that contribute to $^1O_2$ formation: one group which is present in both fractionated HS and in NOM absorbs in the spectral range 355-550 nm with a maximum around 450 nm. A second group of chromophores is responsible for the increase of Φ_Δ in the UV-range and which dominates the excitation spectra of aquatic NOM. The latter groups of chromophores are either not contained in soil/peat HSs or they were removed or destroyed during the process of HA/FA fractionation. The data of Scully et al. [34] for three aquatic sites hint at a further increase of Φ_Δ in the spectral range between 400 to 300 nm.

Singlet Oxygen Production Rates. $d[^1O_2]/dt$ were calculated on the basis of the absorption coefficients of HSs [41] and values for underwater irradiation intensities of Zepp and Cline [55]. The amount of generated singlet oxygen is given by the product of the number of absorbed photons and the wavelength-dependent singlet oxygen quantum yield ($\Phi_{\Delta\lambda}$), Eq. 7.3, where I_{HS} is the number of photons absorbed by the humic substance.

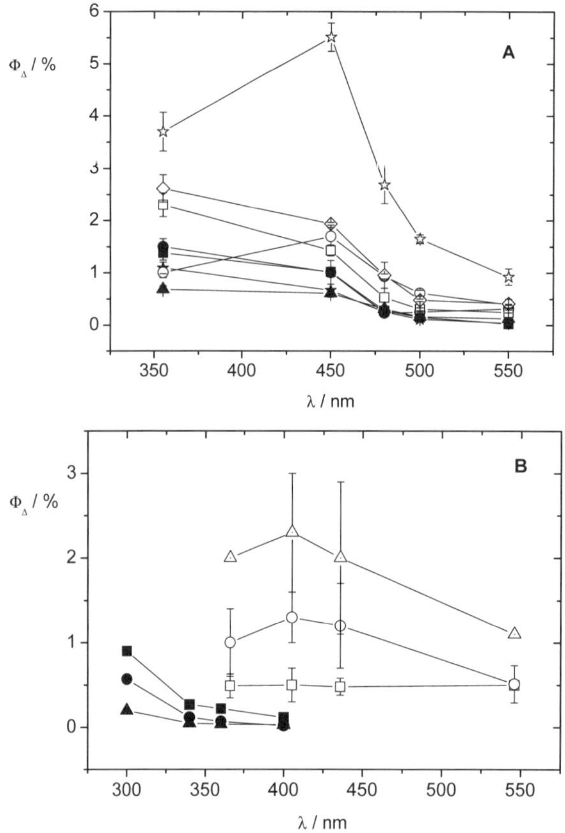

Figure 7.3 A Wavelength dependence of singlet oxygen quantum yields (Φ_Δ, means ± accuracy 12.5%) (excitation spectra), HS: Hiet-f (■), SuRi-NOM (●), SuRi-HA (▲), Sva-s (★), Peat(S)-HA (○), Bir-s (□), Bir-f (◇) and Soil-FAII (✭) [41],– Reproduced by permission of The Royal Society of Chemistry (RSC) on behalf of the European Society for Photo biology and the European Photochemistry Association. **B** Data from Haag et al. [5] (open symbols) and Scully et al. [34] (black symbols) - Lake Erie (■), Sharpes lake (●), Lake Smithwick (▲), Fluka-HA (□), Black lake-HA (○), Lake Baldegg (△)

$$\frac{d[^1O_2]}{dt} = I_{HS} \cdot \Phi_{\Delta\lambda} \tag{7.3}$$

A detailed description of the calculation of I_{HS} is provided by Paul et al. [41]. Briefly, I_{HS} is proportional to the product of concentration and absorption coefficient of HSs, photon flux density and optical pathlength. For a short optical pathlength, i.e. for a depth of only 1 cm and an HS concentration of 1 mg L^{-1}, wavelength-dependent production rates of $^1O_2$ are presented in Figure 7.4. Due to the spectral shape of sunlight (irradiation for midday, 40°N sunlight) [55], the maximum of $^1O_2$ generation was in the visible range between 406 – 450 nm. In contrast to the excitation spectra in Figure 7.3A, there were pronounced differences between single HSs from surface waters. Here, almost identical low production rates for all aquatic samples and high production rates for terrestrial HSs were obtained.

7.3.3. Quenching of $^1O_2$

The observed decay time of singlet oxygen phosphorescence is strongly solvent dependent due to quenching by solvent molecules (k_S) [1]. Compared to k_S, the rate constant of phosphorescent deactivation (k_L) can be neglected. In the presence of an additional quencher (Q), the decay time of singlet oxygen emission can be further reduced by chemical reactions (k_C), physical quenching (k_P), or a combination of both. The rate constants for both physical and chemical quenching depend on the concentration of the quencher [Q]. The observed decay time τ then is given by Eq. 7.4.

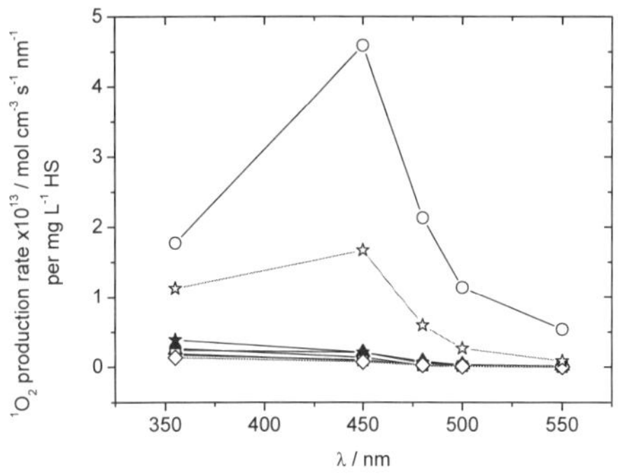

Figure 7.4 Wavelength dependence of $^1O_2$ production rates for HSs in water calculated for the surface according to Paul et al. [41], using the Z-values from Zepp and Cline for midday, summer sunlight, latitude 40°N [55]. HS: Hiet-f (■), SuRi-NOM (●), SuRi-HA (▲), Sva-s (★), Peat(S)-HA (O), Bir-s (□), Bir-f (◇) and Soil-FAII (✯).– Reproduced by permission of The Royal Society of Chemistry (RSC) on behalf of the European Society for Photobiology and the European Photochemistry Association

$$\tau = \frac{1}{k_S + k_L + (k_C + k_P)[Q]} \quad (7.4)$$

$$k_Q = k_C + k_P \quad (7.5)$$

The sum of both quenching mechanisms is k_Q (Eq. 7.5), which can be determined by comparing the decay times of singlet oxygen in the same solvent with (τ) and without quencher (τ_0), Eq. 7.6.

$$\frac{\tau_0}{\tau} = \frac{k_S + k_L + k_Q[Q]}{k_S + k_L} = 1 + \tau_0 \cdot k_Q[Q] \quad (7.6)$$

In contrast to steady state methods, any interaction of the quencher with excited states of the sensitizer, as expected with HSs [39], does not influence the result because the decay time of singlet oxygen remains unchanged. Therefore, the time-resolved detection of $^1O_2$ phosphorescence yields results that are not influenced by inner filter effects that are expected to occur at higher HSs concentrations due to the competing absorption of Rose Bengal and HSs at the excitation wavelength.

The decay time of the phosphorescence τ decreased with increasing HSs concentration. In Figure 7.5, τ_0/τ (that is, the ratio of decay times without and with HSs, respectively) is plotted versus the HSs concentration in the cuvette. The slope is the quenching constant k_Q. For all HSs investigated, linear Stern-Volmer plots were obtained (namely HA, FA, NOM, and synthetic HSs). As indicated by data in Table 7.4, the calculated quenching constants k_Q ranged from about 11 to 47 L mg^{-1} s^{-1}. Whereas FA and NOM were characterized by low k_Q, HA and HS1500 were more efficient $^1O_2$ quenchers. In agreement with the results of Hessler et al. [39], our data suggest that FA and HA from soil are in most cases stronger quenchers than those from aquatic origins (Table 7.5).

Table 7.5 $^1O_2$ production rates of HS in D_2O (pH approximately 7.5) calculated according to Zepp and Cline [55] for midday, summer sunlight, latitude 40°N

sample	$d[^1O_2]/dt \times 10^{11}$ /mol mL s^{-1}
Bir-f	0.16
Bir-s	0.12
Sva-s	0.34
Hiet-f	0.16
SuRi-NOM	0.23
SuRi-HA	0.28
Peat(S)-HA	4.77
Soil-FAII	1.84

Data adapted from Paul et al. [41].

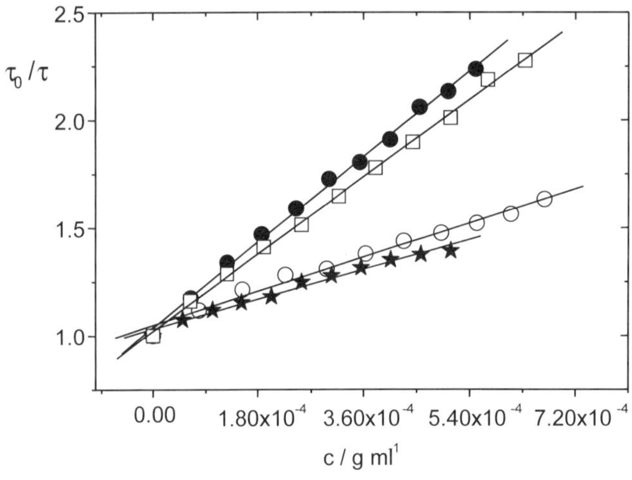

Figure 7.5 Stern-Volmer plots for the quenching of $^1O_2$ photosensitized by Rose Bengal in buffered D_2O by HS, (□) HS1500, (●) SuRi-HA, (○) SuRi-NOM and (★) SuRi-FA. Data were taken from Paul et al. [41]

7.4. DISCUSSION

7.4.1. Comparison of Aquatic and Terrestrial HSs

In contrast to former experiments, both the large dataset and the high accuracy of time-resolved phosphorescence measurements [41] allow comparison of the generation of $^1O_2$ as a measure of photoreactivity of HSs of various ecological origins. For example, comparison of HSs from surface waters, peat, soil and coal in Table 7.2 in general suggests low Φ_Δ for HSs that were derived from surface waters and high Φ_Δ for those which were derived from terrestrial ecosystems or peat. Nevertheless, three exceptions were observed: the soil HS Lau-FA with an atypically low Φ_Δ, and the aquatic HS Bir–(s)-NOM and Hiet–(s)-NOM with atypical high Φ_Δ. The latter effect can be explained by the seasonal differences between NOM, as will be discussed in section 7.4.2.

Although Lau-FA has been extensively characterized [54,56,57], there is no explanation as to why, despite triplet quantum yields of 28-60% [57], a singlet oxygen quantum yield of only 0.57% was measured. Obviously, the triplet of Lau-FA is extensively deactivated by processes such as electron transfer, triplet-triplet annihilation or self-quenching [1]. Another possibility to explain the low Φ_Δ of Lau-FA is the fact that the energy transfer to molecular oxygen is based on exchange interactions and needs a close contact between sensitizer and O_2. Therefore, any steric hindrance will reduce this contact and decrease Φ_Δ. It seems unlikely that quenching of $^1O_2$ by Lau-FA accounts for the low Φ_Δ, since Lau-FA exhibited one of the smallest quenching constants among the FAs investigated here (Table 7.4).

Whereas Φ_Δ (λ_{exc} = 480 nm) obtained for the aquatic Bir–(s)-NOM and Hiet–(s)-NOM (Φ_Δ= 0.86 and 0.91%) overlap with Φ_Δ from terrestrial samples (0.55-2.69%), $^1O_2$-production rates for "non-aquatic" HSs exceeded those of aquatic HSs by one order of magnitude (cf. Figure 7.4 and Table 7.5). As an ecological consequence of these results, an increase of photoreactivity is expected to follow the transfer of terrestrial HSs into lakes and rivers. However, it is not yet clear how long the increased photoreactivity can influence aquatic ecosystems. The question arises as to which processes may account for the transformation of more reactive terrestrial matter into relatively less reactive matter?

7.4.2. Seasonal Variation of Φ_Δ from Aquatic Sampling Sites

In all cases, the direct comparison of samples from fall and spring (Figure 7.6) displayed higher Φ_Δ for the spring samples. The largest differences between spring and fall were found for the sampling sites Birkenes and Hietajärvi. Here, the differences of Φ_Δ between spring and fall remained over the whole spectral range used for excitation (Figure 7.3A). In the NOMiNiC project [46], correlations between site characteristics and several structural and chemical parameters for 5 Scandinavien sampling sites (Birkenes, Hietajärvi, Valkea-Kotinen, Skjervatjern and Svartberget) were performed. The differences in the catchment characteristics did not explain why the observed effect was highly pronounced in these samples [41]. Nevertheless, some of the correlations pointed to the presence of coloured, oxygen rich structures with low molar mass as a precondition for high Φ_Δ.

The seasonal variations of Φ_Δ indicate that during summer and winter, modifications of HSs occurred, modulating photoreactivity. It can be speculated that during fall, winter and spring, processes like degradation of litter by microorganisms and an increased input of terrestrial material by excessive rainfalls may provide HSs structures to surface waters that in turn may increase the formation of "fresh" and highly photoreactive HSs. Instead, during summertime, photochemical modification and photomineralization of NOM may be the dominating processes, resulting in the destruction of those structures that are responsible for $^1O_2$ sensitization.

7.4.3. Quenching of Singlet Oxygen by HSs

Within the studied range of quencher concentration, linear plots of τ_0/τ (i.e. the ratio of decay times without addition of HSs and with HSs, respectively) *vs.* HSs concentration were obtained. The calculated values of k_Q principally can be attributed to the HSs classes HA (k_Q approximately 35-40 L mg^{-1} s^{-1}), NOM (k_Q approximately 11-16 L mg^{-1} s^{-1}) and FA (k_Q approximately 11-15 L mg^{-1} s^{-1}). It is interesting to note that the synthetic HS1500 had a quenching constant of 32 L mg^{-1}

s^{-1}, which was similar to those of HA. As a result of linear regression analysis, a weak correlation of k_Q with the molar mass at the maximum of the HS-fraction, determined by gel permeation chromatography ($R = 0.72$, $R^2 = 0.51$, $p = 0.004$), and a strong correlation with the aromaticity ($R = 0.94$, $R^2 = 0.88$, $p<0.0001$) was found [42]. In general, no correlation was found between Φ_Δ and k_Q, and there was no correlation between metal contents of HSs and k_Q.

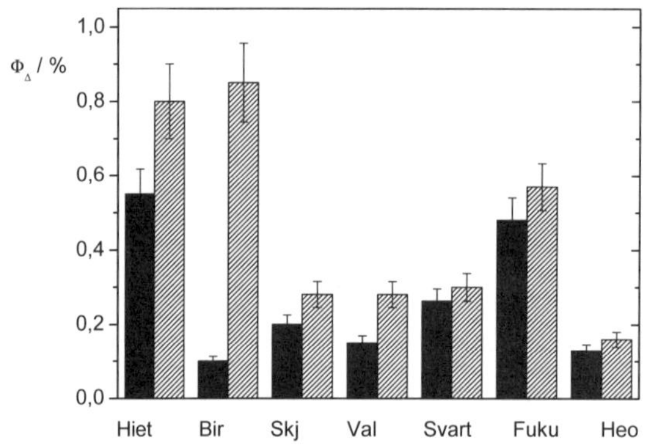

Figure 7.6 Seasonal variations (grey columns: spring 2000 samples, black columns: fall 1999 samples) of Φ_Δ (means ± accuracy 12.5%) for six Scandinavian and one German sampling site

Linear regressions of the k_Q published by Hessler et al. [39] for two HSs from ground water, one brown water fulvic acid (FA) and one podzol FA with the molar masses estimated from the gel-chromatograms (cf. Table 7.6), yielded a correlation with $R = 0.99$, $R^2 = 0.98$, $p = 0.01$. The corresponding results for these independent datasets clearly indicate that in contrast to the formation of $^1O_2$, where Φ_Δ displayed no dependence on parameters such as molar mass or aromaticity, but the quenching of $^1O_2$ by HSs did. Aromaticity is an operationally parameter derived from gel permeation chromatography. It relates to the absorbance at 254 nm at the maximum of the HS-fraction, divided by the DOC-concentration of HSs. It is well known that the only shoulder at 254 nm in the otherwise unstructured absorption spectrum of HSs is caused by $\pi\pi^*$ transitions. Hence, the quenching of $^1O_2$ by HSs is directly related to the sum of aromatic and unsaturated compounds associated with the high-molar mass fraction in HSs. Therefore, it can be concluded that not only the sizes, but also the structural compositions of HSs accounts for the quenching of $^1O_2$. At present, no differentiation is possible between the contributions of physical and chemical quenching to the observed overall quenching, but it can be expected that oxidative modification of HSs by $^1O_2$ is an important process in sunlit natural water.

Table 7.6 Quenching constants (k_Q) in D_2O and molar masses (MW) from gel permeation chromatography

abbreviation	Origin	Sample type	k_Q, $L mg^{-1} s^{-1}$	MW, Da
HO3-FA	Hohlosee	Brown-lake water	22	1417
E3-FA	Soil sample	Podzol	20	1711
BOC 3/3.5	Bachholder Feld	Ground water	13	2268
FB1-1FA	Fuhrberger Feld	Ground water	15	2033

Data taken from Hessler et al. [39].

7.5. CONCLUSIONS

Our work adds to current knowledge on singlet oxygen formation and quenching by dissolved HSs in several important respects: 1) both singlet oxygen quantum yields and $^1O_2$-production rates indicate that more reactive compounds are found in HSs of non-aquatic origin. The reason for this observation may either arise from the different ways of HSs formation, different compounds that contribute to HSs, or other modification pathways of HSs in terrestrial and aqueous surroundings. 2) the spectral properties of aquatic and non-aquatic HSs were not identical. Our results point to two different classes of chromophores that are responsible for $^1O_2$ formation. The first one has an absorption maximum between 400-450 nm, which is abundant in aquatic and terrestrial HSs. A second one has an absorption maximum in the UV-range of solar radiation, which is typical for aquatic HSs. One can speculate that small structural compounds, which are formed as a result of photolysis of dissolved HSs in aquatic systems, contribute to the observed increase of Φ_Δ at short excitation wavelengths. 3) In the RO-isolates of seven surface waters, a general trend of higher Φ_Δ in spring samples compared with fall samples was observed. This effect can be understood as the result of seasonally dependent processes that lead to the accumulation of more reactive HSs during winter time and of less reactive material in summer time, respectively. 4) The postulated tendency that low molar mass HSs have high $^1O_2$ quantum yields has been supported, but there was no significant correlation between molar mass and Φ_Δ. Nevertheless, the quenching properties of HSs seem to be closely connected with molecular parameters of HSs, such as molar mass and aromaticity. 5) There was no significant correlation between quenching constants, Φ_Δ, metal contents and Φ_Δ or quenching constants and metal contents.

ACKNOWLEDGEMENT

The authors thank R. Vogt and E. Gjessing for kindly providing humic substances.

REFERENCES

1. Wilkinson F, Helman WP, Ross AB. Rate constants for the decay and reactions of the lowest electronically excited singlet state of molecular oxygen in solution. An expanded and revised compilation. *J. Phys. Chem. Ref. Data*, 1995; 24: 663-1021.
2. Lissi EA, Encinas MV, Lemp E, Rubio MA. Singlet oxygen $O_2(^1\Delta_g)$ bimolecular processes. Solvent and compartmentalization effects. *Chem. Rev.*, 1993; 93: 699-723.
3. Gorman AA, Rodgers MAJ. CRC *Handbook of Organic Photochemistry*, Vol. II, JC Scaiano ed. Boca Raton, FL: CRC Press, 1989: 229-247.
4. Zepp RG, Wolfe NL, Baughmann GL, Hollis RC. Singlet oxygen in natural waters, *Nature*. 1977; 267: 421–423.
5. Haag WR, Hoigné J, Gassman E, Braun AM. Singlet oxygen in surface waters – Part II: Quantum yields of its production by some natural humic materials as a function of wavelength. *Chemosphere*, 1984; 13: 641–650.
6. Hoigné J, Faust BC, Haag WR, Scully Jr. FE, Zepp RG. Aquatic humic substances as sources and sinks of photochemically produced transient

reactants, In: Suffet IH, MacCarthy P eds. *Aquatic humic substances: Influence on fate and treatments of pollutants*. Adv. Chem. Series, 1989; 219: 363–381.

7. Frimmel FH, Bauer H, Putzlen J, Murasecco P, Braun AM. Laser flash photolysis of dissolved aquatic humic material and the sensitized production of singlet oxygen. *Environ. Sci. Technol.*, 1987; 21: 541–545.
8. Zepp RG, Braun AM, Hoigné J, Leenheer JA. Photoproduction of hydrated electrons from natural organic solutes in aquatic environments. *Environ. Sci. Technol.*, 1987; 21: 485–490.
9. Vaughan P, Blough NV. Photochemical formation of hydroxyl radical by constituents of natural waters. *Environ. Sci. Technol.*, 1998; 32: 2947–2953.
10. Goldstone JV, Voelker BM. Chemistry of superoxide radical in seawater: CDOM associated sink of superoxide in coastal waters. *Environ. Sci. Technol.*, 2000; 34: 1043–1048.
11. Scully NM, McQueen DJ, Lean DRS. Hydrogen peroxide formation: The interaction of ultraviolet radiation and dissolved organic carbon in lake waters along a 43-75°N gradient. *Limnol. Oceanogr.*, 1996; 41: 540–548.
12. Zepp RG, Schlotzhauer PF, Sink RM. Photosensitized transformations involving electronic energy transfer in natural waters: Role of humic substances. *Environ. Sci. Technol.*, 1985; 19: 74–81.
13. Richard C, Vialaton D, Aguer J-P, Andreux F. Transformation of monuron photosensitised by soil extracted humic substances: energy or hydrogen transfer mechanism? *J. Photochem. Photobiol. A: Chemistry*, 1997; 111: 265–271.
14. Kieber DJ, Peake BM, Scully NM. Reactive oxygen species in aquatic ecosystems. In: Helbling EW, Zagarese H eds. *UV effects in aquatic organisms and ecosystems.* Cambridge: Royal Society of Chemistry, 2003: 251-288.
15. Hoigné J. Formulation and calibration of environmental reaction kinetics: Oxidation by aqueous photooxidants as an example. In: Stumm W ed. *Aquatic chemical kinetics*. New York: Wiley, 1990: 43-70.
16. Mill T. Predicting photoreaction in surface waters. *Chemosphere*, 1999; 38: 1379-1390.
17. Voelker BM, Sedlak DL, Zafiriou OC. Chemistry of superoxide radical in seawater: Reactions with organic Cu complexes. *Environ. Sci. Technol.*, 2000; 24: 1036-1042.
18. Emmenegger L, Schonenberger RR, Sigg L, Sulzberger B. Light-induced redox cycling of iron in circumneutral lakes. *Limnol. Oceanogr.*, 2001; 46: 49-61.
19. Gjessing ET, Källquist T. Algicidal and chemical effect of UV radiation of water containing humic substances. *Wat. Res.*, 1991; 25: 491–494.
20. Lund V, Hongve D. Ultraviolet irradiated water containing humic substances inhibits bacterial metabolism. *Wat. Res.*, 1994; 28: 1111–1116.
21. Liltved H, Landfald B. Effects of high intensity light on ultraviolet-irradiated and non-irradiated fish pathogenic bacteria. *Wat. Res.*, 2000; 34: 481–486.
22. Strome DJ, Miller MC. Photolytic changes in dissolved humic substances. *Verh. Internat. Verein. Limnol.*, 1981; 20: 1248–1254.
23. Salonen K, Vähatalo A. Photochemical mineralization of dissolved organic matter in lake Skjervatjern. *Environm. Intern.*, 1994; 20: 307–312.
24. Bertilson S, Tranvik L. Photochemical transformation of dissolved organic matter in lakes. *Limnol. Oceanogr.*, 2000; 45: 753–762.
25. Bertilsson S, Allard B. Sequential photochemical and microbial degradation of

refractory dissolved organic matter in a humic freshwater system. *Arch. Hydrobiol. Spec. Issues Advanc. Limnol.*, 1996; 48: 133–141.

26. Larson RA, Weber EJ. Reaction mechanisms of environmental organic chemistry. *Carbohydrate Polymer*, 1996; 29: 293-294.
27. Klotz L-O, Kröncke K-D, Sies H. Singlet oxygen-induced signalling effects in mammalian cells. *Photochem. Photobiol. Sci.*, 2003; 2: 88-94.
28. Baker A, Kanofsky JR. Quenching of singlet oxygen by biomolecules from L1210 leukamia cells. *Photochem. Photobiol.*, 1992; 55: 523-528.
29. Piette J. Biological consequences associated with DNA oxidation mediated by singlet oxygen. *J. Photochem. Photobiol. B: Biology*, 1991; 11: 241-260.
30. Boreen AL, Arnold WA, McNeill K. Photodegradation of pharmaceuticals in the aquatic environment: A review. *Aquat. Sci.*, 2003; 65: 320-341.
31. Aguer J-P, Richard C, Andreux F. Effect of light on humic substances. *Analysis*, 1999; 27: 387–390.
32. Aguer J-P, Richard C, Andreux F. Comparison of photoinduced properties of commercial, synthetic and soil-exctracted humic substances. *J. Photochem. Photobiol. A: Chemistry*, 1997; 103: 163–168.
33. Welker M, Steinberg CEW. Rates of humic substances photosensitized degradation of microcystin-LR in natural waters. *Environ. Sci. Technol.*, 2000; 34: 3415–3419.
34. Scully NM, Vincent WF, Lean DS, Cooper WJ. Implications of ozone depletion for surface water photochemistry: Sensitivity of clear lakes. *Aquat. Sci.*, 1997; 59: 20–274.
35. Haag WR, Hoigné J. Singlet oxygen in surface waters. 3. Photochemical formation and steady state concentrations in various types of water. *Environ. Sci. Technol.*, 1986; 20: 341-348.
36. Sandvik SL, Bilski P, Pakulski JD, Chignell CF, Coffin RB. Photogeneration of singlet oxygen and free radicals in dissolved organic matter isolated from the Mississipi and Atchafalaya river plumes. *Marine Chem.*, 2000; 69: 139–152.
37. Frimmel FH, Bauer H, Putzlen J. Laser flash photolysis of dissolved aquatic humic material and the sensitized production of singlet oxygen. *Environ. Sci. Technol.*, 1987; 21: 541-545.
38. Nico PS, Anastasio C, Zasoski RJ. Rapid photo-oxidation of Mn(II) mediated by humic substances. *Geochim. Cosmochim. Acta*, 2002; 66: 4047-4056.
39. Hessler DP, Frimmel FH, Oliveros E, Braun AM. Quenching of singlet oxygen $O_2(^1\Delta_g)$ by humic substances. *J. Photochem. Photobiol. B: Biology*, 1996; 36: 55-60.
40. Bilski P, Burghart JG, Chignell CF. Photochemical characterization of water samples from Minnesota and Vermont sites with malformed frogs: potential influence of photosensitization by singlet molecular oxygen ($^1O_2$) and free radicals on aquatic toxicity. *Aquatic Toxicology*, 2003; 65: 229-241.
41. Paul A, Hackbarth S, Vogt RD, Röder B, Burnison BK, Steinberg CEW. Photogeneration of singlet oxygen by humic substances: Comparison of humic substances of aquatic and terrestrial origin. *Photochem. Photobiol. Sci.*, 2004; 3: 273-280.
42. Hackbarth S, Paul A, Zwirnmann E, Steinberg CEW, Röder B. Humic substances as sources and sinks of singlet oxygen. In preparation.
43. IHSS, International Humic Substance Society. See web page < http://www.ihss. gatech.edu/ >. Accessed 25.05.04.
44. Fredriks Research Products, Laurentian soil, the base for great humic and fulvic acid. See web page <http://home.concepts-ict.nl/~fredriks/> . Accessed

11.07.03.

45. Gjessing ET, Egeberg PK and Håkedal J. Natural organic matter in drinking water – The "NOM-Typing Project", Background and basic characteristics of original water samples and NOM isolates. *Environment International*, 1999; 25: 145-159.
46. Vogt RD, Akkanen J, Andersen DO, Brüggemann R, Chatterjee B, Gjessing ET, Kukkonen JVK, Larsen HE, Luster J, Paul A, Pflugmacher S, Starr M, Steinberg CEW, Schmitt-Kopplin P, Zsolnay Á. Key variables governing the functional characteristics of dissolved natural organic matter (DNOM) in Nordic forested catchments. *Aquat. Sci.*, 2004; 66: 1-16.
47. Sachse A, Babenzien D, Ginzel G, Gelbrecht J, Steinberg CEW. Characterization of dissolved organic carbon (DOC) in a dystrophic lake and adjacent fen. *Biogeochem.,* 2000; 54: 279–296.
48. Lange N, Kühnert M, Haase A, Höke H, Seubert B. Untersuchungen zum Resorptionsverhalten eines niedermolekularen synthetischen Huminstoffes nach einmaliger oraler Applikation an der Ratte. *Dtsch. Tierärztl. Wschr.*, 1996; 103: 113–115.
49. Weyl GmbH: Vorläufiges Datenblatt RHS1500. 1990, Ausgabe 10., EP-B 0281678 v. 10.02.1987, EP 0537427 v. 31.07.1992.
50. Perminova IV, Frimmel FH, Kovalevskii DV, Abbt-Braun G, Kudryavtsev AV, Hesse S. Development of a predictive model for calculations of molar mass of humic substances. *Wat. Res.,* 1998; 31: 872-881.
51. Vogt RD, Andersen DO, Bishop K, Clarke N, Gadmar T, Gjessing E, Lundstrøm U, Starr M. Natural organic matter in the Nordic Countries (NOMiNiC): 1. TOC intercalibration 2. Physico-chemical characteristics of DOM. NT Techn Report 479. NordTest, Espo, Finland. Available at <*http://www.nordtest.org/register/techn/tlibrary/tec479.pdf*>, 2001: 150-153.
52. Bilski P, Dabestani R, Chignell CF. Influence of cationic surfactant on the photoprocesses of eosine and Rose Bengal in aqueous solution. *J. Phys. Chem.*, 1991; 95: 5784–5791.
53. Riise G, Salbu B. Major and trace elements in standard and reference samples of aquatic humic substances determined by instrumental neutron activation analysis (INAA). *Sci. Total Environ.*, 1989; 81/82: 137-142.
54. Cook RL, Langford CH. Structural characterization of a fulvic acid and a humic acid using solid-state Ramp-CP-MAS ^{13}C nuclear magnetic resonance. *Environ. Sci. Technol.*, 1998; 32: 719-725.
55. Zepp RG, Cline DM. Rates of direct photolysis in the aquatic environment. *Environ. Sci. Technol.*, 1977; 11: 359-366.
56. Cook RL, McIntyre DD, Langford CH, Vogel HJ. A comprehensive liquid-state heteronuclear and multidimensional NMR study of Laurentian fulvic acid. *Environ. Sci. Technol.*, 2003; 37: 3935-3944.
57. Bruccoleri A., Pant BC, Sharma DK, Langford CH. Evaluation of primary photoproduct quantum yields in fulvic acids. *Environ. Sci. Technol.*, 1993; 27: 889-894.

Chapter 8

FACTORS AFFECTING NOM REACTIVITY WITH CHLORINE

David A. Reckhow, Paula L. Rees, Gladys Makdissy and Darleen Bryan

Department of Civil & Environmental Engineering, University of Massachusetts, Amherst, MA 01003, USA

8.1. INTRODUCTION

Natural organic matter (NOM) reacts quickly with aqueous chlorine to form an extensive mixture of halogenated and non-halogenated byproducts. These reactions occur in most US drinking water treatment plants where chlorine is still the disinfectant of choice. Toxicologists and epidemiologists have slowly amassed a convincing body of evidence that chlorination byproducts of NOM cause cancer and other serious health problems in humans [1].

Efforts by water treatment scientists and engineers to minimize disinfection byproduct (DBP) formation have been partly successful. Advances in process technology and process control have reduced DBP formation by 50% or more in treatment facilities across the country. As these efforts begin to reach their practical limit, the field is now looking to new approaches. Among these is the use of watershed management aimed at reducing organic DBP precursors in the source waters.

The origin of DBP precursors in watersheds is necessarily linked to the genesis of the larger body of aquatic NOM. While many researchers have studied dissolved organic carbon (DOC) origins, export and transformations, very few have conducted watershed investigations specifically aimed at DBP precursors. In this paper, we will focus on our current state of knowledge regarding the fraction of DOC that gives rise to disinfection byproducts. Data will be drawn from UMass studies as well as from the literature. These data will be used to make inferences on the processes that affect DBP precursors in the natural environment. Natural variations in NOM as regards chlorine reactivity will also be presented. Some underlying structural characteristics will be discussed within the context of these reactive properties.

The quantification of DBP precursors must be carried out by means of an empirical precursor test. A wide range of conditions have been employed for precursor tests since their earliest use in the mid 1970s. All involve some bench-scale treatment of the sample with chlorine, a reaction period, and then reduction of the residual chlorine followed by analysis of the disinfection byproducts, usually by gas chromatography. Researchers have not been in agreement about the exact conditions to be used (chlorine dose, pH, reaction time, temperature, etc.). Some of this disagreement occurs because of legitimate differences in the objectives of measuring precursor levels.

Most researchers would agree that the ideal precursor test should be capable of accurately measuring the total amount of precursor sites that are potentially accessible to reaction with the oxidant (usually free chlorine) under conditions typical of drinking water treatment. These tests should be completely independent of the inorganic matrix and they should be reproducible and robust. Since organic matter in natural waters is quite dilute, it is unlikely that there would be a substantial amount of reaction between organic molecules. Consequently, each organic molecule will react independently of the others and in accordance with the kinetic laws that govern their elementary reactions. These kinetics are primarily second-order (first-order in precursor organic, and first-order in chlorine species), or first-order overall (in precursor organic). In reactions with model compounds (known, simple organic molecules of high purity), we see a complex web of reactions that drifts from second-order to first-order as rapid bimolecular halogenation reactions subside and slower pathways involving some other rate-limiting step (e.g., hydrolysis) predominate.

These reactions always show direct proportionality between initial model compound concentration and byproduct formation as long as all other conditions are held constant. Non-reactive solutes (e.g., Na^+, Ca^{+2}, NO_3^-) will naturally remain constant through a test of this sort, and pH can be held fixed with an appropriate buffer. Reactive inorganic species are more difficult to control. Ideally, one would like to remove all such species (except, of course free chlorine), so that their concentrations always remain at zero. This is not really practical, so it is important to measure their initial concentrations (e.g., bromide, iodide, ammonia, sulfide). Ideally, one would also like to have a "chlorine buffer" or means of holding the chlorine residual constant instead of allowing it to become depleted upon reaction. At low chlorine levels, this requires too much intervention and potential loss of volatiles or introduction of contaminants to be practical. At very high chlorine levels (e.g., >10 times the demand), the relative depletion of chlorine is insignificant. However, such high levels are likely to access pathways that are not normally seen in typical drinking water systems. Our approach has been to use a compromise value of 20 mg/L for waters with 3-day demands of 10mg/L or less. This keeps the chlorine residual between 10 and 20 mg/L at all times, and thereby maintains a relatively constant chemical driving force that is not strongly influenced by the organic matter.

In contrast, the practice of using a fixed chlorine to carbon ratio typically results in a high initial residual which drops to some very low level at the end of the incubation time. Waters with a higher DOC and higher chlorine demand will require that the chlorine dose be higher. This is ideal when assessing the tendency of a particular water, with its unique DOC level, to produce DBPs, but it is not best if one is interested in the inherent reactivity of the organic compounds making up the dissolved organic matter (DOM). The latter is of greatest importance when considering the transport and inevitable dilution and degradation of NOM in watersheds.

8.2. MATERIALS AND METHODS

8.2.1. Study Site

This work is part of an ongoing set of studies that have so far focused on the Wachusett Reservoir watershed, which is a major source for Boston's unfiltered water supply. Eight major tributaries were monitored during dry and wet weather events. In addition, the Stillwater River, one of the two largest tributaries, was sampled intensively at various points along its length.

8.2.2. Sample Collection

Samples were collected in amber glass bottles that had been previously acid-washed. During collection, all bottles were stored in coolers with ice packs. All samples were returned to the UMass laboratory within 10 hours of collection, where they were stored under refrigeration until treatment or analysis.

8.2.3. Analytical Methods

Total organic carbon (TOC) was measured by the high-temperature combustion method [2] using a Shimadzu 5000 analyzer. Analyses were conducted within 2 days. Dissolved organic carbon (DOC) was measured following filtration with a pre-cleaned glass fiber filter (0.45 μm).

Samples intended for precursor analysis were treated with chlorine in the UMass laboratory within 2 days of collection. Samples were buffered at pH 7 and then treated with a chlorine dose of 20 mg/L (sufficient to leave a substantial residual at the end of the incubation period). Reactions were conducted in headspace-free glass vessels (chlorine demand free) in the dark at a fixed temperature (20°C). Controls were prepared by chlorinating high-purity water as described above.

Trihalomethanes and other neutral extractables (haloacetonitriles, haloketones, chloropicrin, etc.) were measured on all disinfected samples and controls. We used the standard micro-extraction method with GC and electron capture detection (ECD) [2].

The full suite of haloacetic acids were measured along with the THMs whenever samples were treated with chlorine. Haloacetic acids were measured by the micro-extraction method with methylation and separation/detection by GC with ECD. More specifically, we used the acidic methanol derivatization (US EPA method 552.2), which avoids the use of highly-toxic reagents required for the diazomethane method.

8.3. RESULTS AND DISCUSSION

8.3.1. Field Studies in the Wachusett Watershed

Figure 8.1 shows a typical dataset from a single dry weather sampling event. Note that there is substantial variability in the level of DBPs formed from the major tributaries as well as in the channel locations and small feeder streams to the Stillwater. Much of this variability is quite clearly due to differences in the TOC level. However, analysis of specific DBP precursor concentrations (precursor levels in µg/L divided by the TOC in mg/L) indicates that there is still a substantial variability that cannot be attributed to TOC alone.

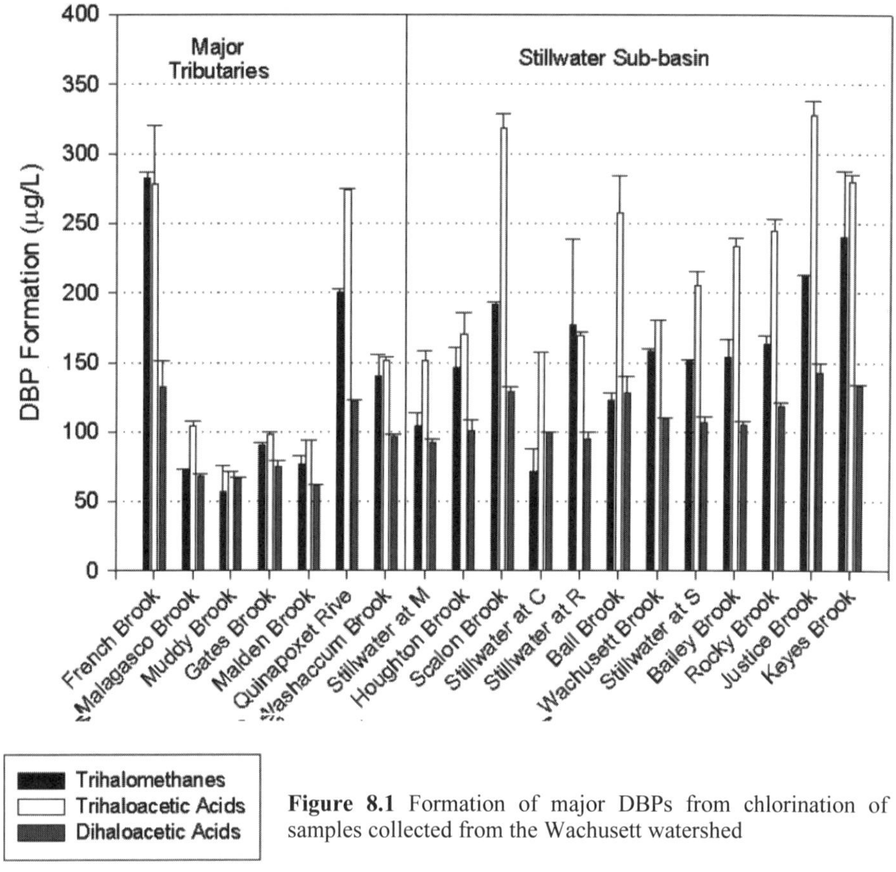

Figure 8.1 Formation of major DBPs from chlorination of samples collected from the Wachusett watershed

Figure 8.2 presents the specific THM precursor levels versus TOC over a half-dozen sampling campaigns during 2001-2002. First, the values run from as low as 20 µg/mg-C to almost 100 µg/mg-C. Second, there is a distinct decreasing trend with increasing TOC. This was not expected based on both fundamental NOM characterizations and on empirical observations. Highly colored waters have been shown to be dominated by humic substances, and these compounds are typically quite high in trihalomethane formation potential (Sp-THMPF) [3]. In contrast, groundwaters and many lake waters are higher in the hydrophilic fractions that are relatively low in reactivity. Furthermore, Chapra and colleagues [4] presented a different type of supportive evidence. They compiled 5 large datasets including 133 separate measurements, the sum of which indicated a trend toward higher specific precursor levels for waters of higher TOC [4].

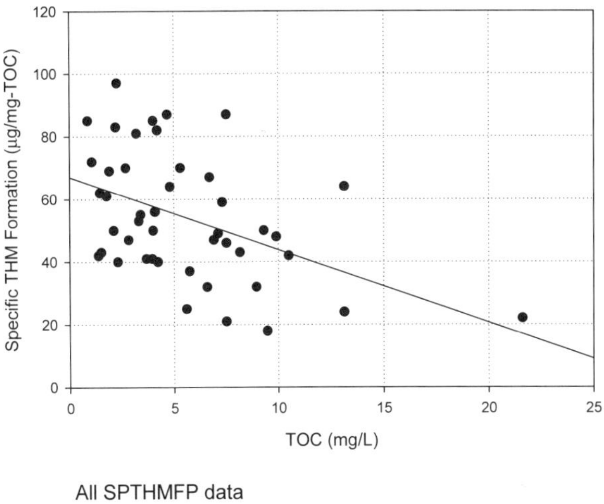

Figure 8.2 Relationship between specific THM precursor levels and TOC in Wachusett Reservoir tributaries

8.3.2. Evidence from the Literature

Cambridge: Another Intensive Regional Study. Work by Waldron and Bent [5] of the USGS supports the previously noted inverse relationship. This group studied the Cambridge (MA) watershed over a period of one year, measuring TOC, THM precursors and other DBP related parameters. Figure 8.3 shows the median values from each of 11 locations sampled during the study year. While the range is not as great as noted for the Wachusett watershed (perhaps because the median values were used), the negative correlation is nonetheless apparent.

New York: Long-term Monitoring of 2 Supplies. Edzwald and coworkers[6] examined raw and treated waters from the Canton and Oneida, NY water treatment plants over a period of more than 2 years. In both cases they found strong linear correlations between THMFP and TOC (Eqs. 8.1 and 8.2), and in both cases these had a positive intercept. The intercept from the Oneida (Glenmore Reservoir) data was quite substantial. The positive intercept indicates that low TOC waters have higher specific THMFPs than higher TOC waters.

$$THMFP = 75.8(TOC) + 13.6 \quad \text{(Grasse River)} \qquad (8.1)$$

$$THMFP = 45.1(TOC) + 113.4 \quad \text{(Glenmore Reservoir)} \qquad (8.2)$$

Comparison across different geographic regions. As previously noted, Chapra and colleagues [4] compiled 133 data points from 5 literature sources and concluded that the THMFP of natural waters increases with TOC raised to the power of about 1.25 (Eq. 8.3).

$$THMFP = 43.78(TOC)^{1.248} \tag{8.3}$$

which can be rearranged to:

$$\frac{THMFP}{TOC} = 43.78(TOC)^{0.248} \tag{8.4}$$

This suggests that the specific THMFP increases with increasing TOC. However, these data were not corrected in any way for differences in chlorination testing conditions. Of the five databases used by these authors, four were readily available in the refereed literature [6-9]. Figure 8.4 shows the data from these four, and a linear correlation of the log transform gives a model (Eq. 8.5) that is quite close to that reported by Chapra et al. [4] (Eq. 8.1), despite the missing fifth dataset.

$$THMFP = 42.01(TOC)^{1.236} \tag{8.5}$$

Direct comparison of DBP formation potential data across research groups is risky when there are multiple sets of chlorination conditions used in the compiled dataset. In addition, raw DBPFP data can be heavily influenced by the TOC of the water being tested, when chlorine doses are positively correlated with TOC. This is a problem even when comparing data obtained from the same DBPFP testing protocol. To a lesser extent varying raw water bromide levels can create apparent differences in the reactivity of organic matter. For this reason, it is helpful to normalize data based on treatment conditions, TOC and perhaps bromide levels. As a first approximation, we selected a well-calibrated power function model [7] to achieve some level of normalization (Eq. 8.6). This particular model was selected because of its large and well-documented database, and because many of the coefficients derived for this model are quite close to those determined by other researchers based on other datasets. The normalized conditions adopted for this analysis were arbitrarily selected as follows: 20 mg/L chlorine dose, 72 hour reaction time, pH 7, no bromide and temperature 20°C. These values as well as the model coefficients and offsets are represented in Eq. 8.6.

$$Std-THMFP = THMFP\left(\frac{20}{Cl_2dose}\right)^{0.409}\left(\frac{72}{time}\right)^{0.265}\left(\frac{7-2.6}{pH-2.6}\right)^{0.715}\left(\frac{1}{Br+1}\right)^{0.036}\left(\frac{20}{temp}\right)^{1.06} \tag{8.6}$$

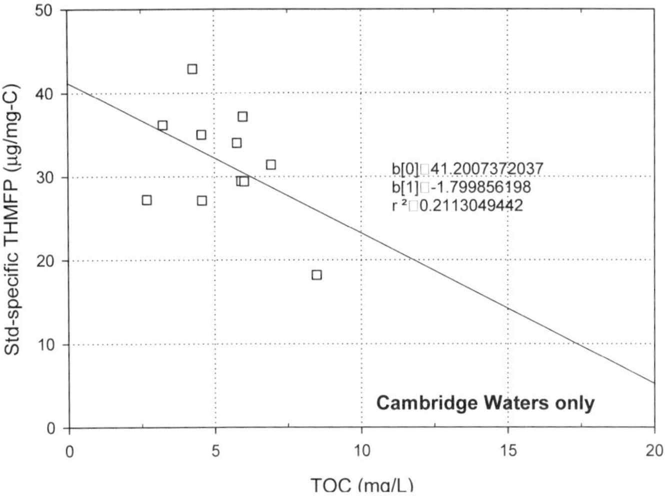

Figure 8.3 Relationship between specific THM precursor levels and TOC in Cambridge (MA) watershed

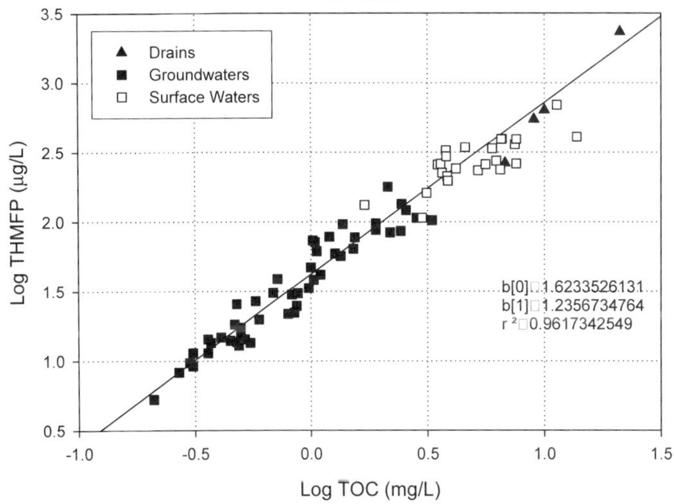

Figure 8.4 Re-testing of four databases used by Chapra et al. [4]

Figure 8.5 shows some of these data after manipulation to bring them all to the same set of formation potential conditions. Note that the THMFPs for the higher TOC waters were readjusted downward more than the others. This is because of the higher chlorine doses necessary for testing these waters, and therefore the artificially high reactive environment experienced by the organics in these waters. As a result, the slope of the overall regression line has dropped to 1.09.

Since all of the earlier datasets were exclusively populated with surface waters, the most appropriate comparison with the data in Figure 8.5 should also involve just the surface waters. Excluding all groundwaters and agricultural drains, the regression slope drops to 0.68. When plotted as specific THMFP on a linear scale, the surface water data resemble much more closely those correlations developed for the two Massachusetts watersheds. Again, the groundwaters and agricultural drains do not fit this relationship. Possible reasons for this will be discussed below.

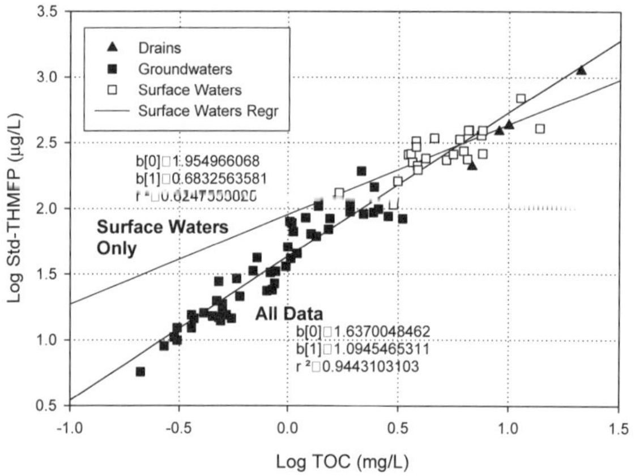

Figure 8.5 Re-evaluation of data from Figure 8.4 based on a common chlorination protocol

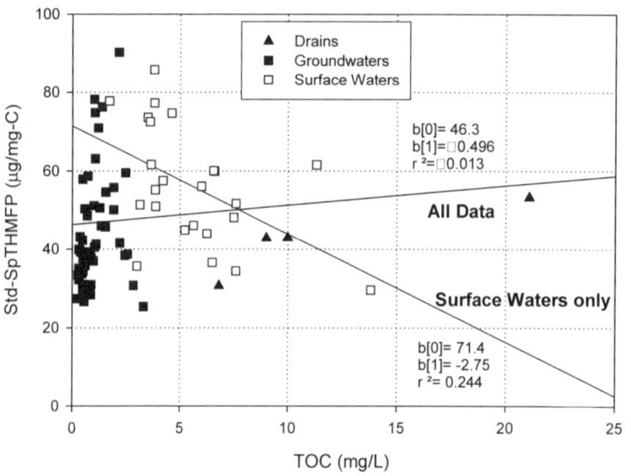

Figure 8.6 Standardized, specific THMFP versus TOC for the four datasets from Chapra et al. [4]

There are many other good databases that can be incorporated into this analysis. Figure 8.7 shows the results of a compilation of 27 datasets that are accessible in the open literature [5-26]. As before, the regression based on these data indicates that specific THMFP drops with increasing TOC. Figure 8.8 shows only the surface water data for clarity.

8.4. LEACHING EXPERIMENTS

In summary, all data sets comprised of surface waters show the same behavior. The specific THMFP drops as the TOC increases. As mentioned previously, we did not expect to see this behavior. However, once it was recognized we proposed that this could be attributed to ageing of NOM and selective degradation of non-precursor organics. This came partly from the observation that DOC levels dropped as residence time in the small and medium sized streams increased [27]. The proposal also came from realization that the most precursor-rich components (e.g., lignin residues, tannins) are also among the most slowly biodegraded.

To test this hypothesis, a series of leaching and degradation experiments were conducted in the laboratory. Briefly, leaves and needles were gathered in the Wachusett watershed in late winter (March) from the top of the leaf pack. For each leaf type, five bottles (four amber with and without biocide and one clear) were prepared containing a fixed mass of leaf matter in distilled water, seeded with a small amount of reservoir water (system intake). These bottles were left for seven days in the dark, during which time substantial amounts of organic matter were released into solution.

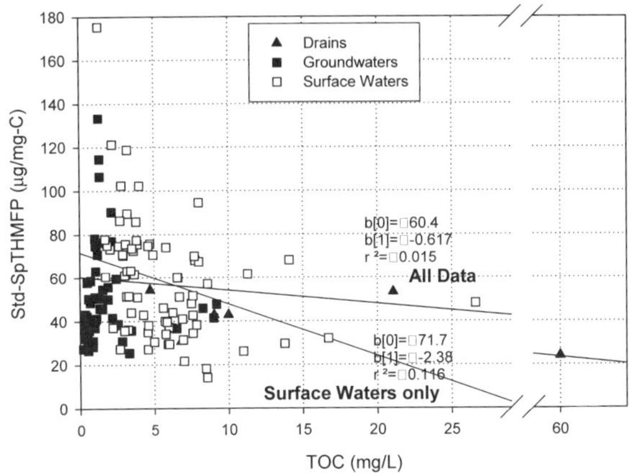

Figure 8.7 Standardized, specific THMFP versus TOC for 27 datasets from the literature

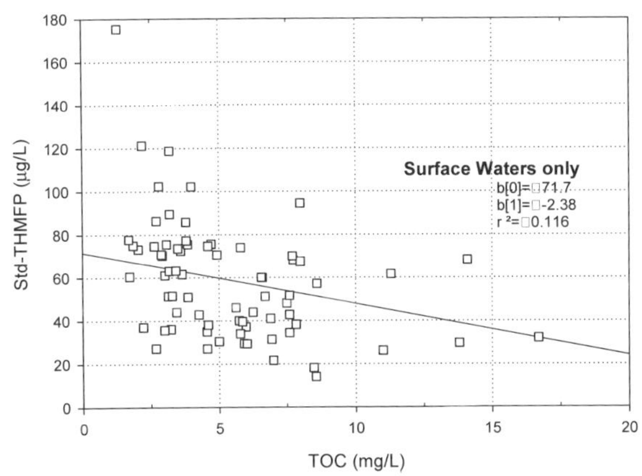

Figure 8.8 Standardized, specific THMFP versus TOC for surface waters based on 27 datasets from the literature

When subject to chlorination tests, the leached organic matter proved to be a moderately rich source of THM and haloacetic acid (HAA) precursors (Figure 8.9, black bars). Although there was some species variability, all three fell into the range of 30-50 μg/mg-C. After the 7-day leaching period, all particulate matter was removed from the bottles and the leached solutions were held under aerobic conditions for 80 days (20°C, bubbling with zero-grade air). During this period a substantial loss in DOC was observed as a result of biodegradation. At the end of the 80 days, samples were again analyzed for THM precursors, and these data are shown in Figure 8.9 as the gray bars. Biodegradation caused a substantial increase in the specific precursor level, thereby supporting the original hypothesis.

As an illustration of how this might occur, a fictitious set of degradation rates were selected for 8 classes of plant products that might be found in water (Table 8.1). These are not intended to represent actual environmental degradation rates, but merely to approximately represent the expected rank order. For each of these classes, a class-specific THMFP was included. These are roughly based on typical structures and known THM yields from model compound studies. The various classes of plant products were allowed to decay based on the fictitious rates and a first-order rate law. Along with these decay processes, the associated THMFP decayed as well (Figure 8.10). Also shown in Figure 8.10 is the modeled ratio of THMFP to residual TOC (i.e., the Sp-THMFP). This value climbs with time, due primarily to the rapid loss of the relatively non-reactive carbohydrates and the persistence of the highly reactive tannins.

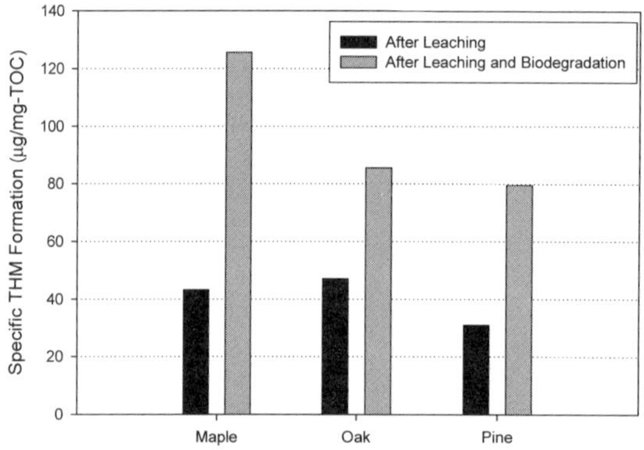

Figure 8.9 Specific THM precursors before and after biodegradation for three leaf types

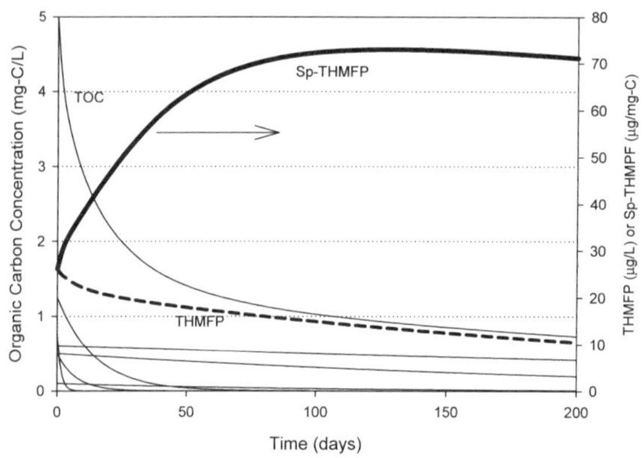

Figure 8.10 Fictitious example of selective degradation of NOM constituents and resulting specific THMFP (individual constituents are shown at bottom)

Table 8.1 Model parameters used to create Figure 8.10

Constituent	Initial abundance (% of TOC)	Half-life (d)	Sp-THMFP (µg/mg-C)
Fats and Waxes	2	60	0
Terpenoids	12	400	10
Tannins	4	200	400
Lignins	10	150	40
Cellulose	22	20	3
Hemicellulose	25	10	3
Protein	10	5	30
Sugars and Starches	15	1	3

The above analysis is most applicable to young NOM, particularly the organic matter in flowing surface waters. This is also a gross simplification of the natural processes, which do not result in direct mineralization of the NOM classes. Biodegradation rates certainly will depend on the immediate environment, the standing biomass, the presence of light, availability of terminal electron acceptors, and so on.

Groundwaters will likely be much older on average than surface waters. They are also depleted in some of the slowly degraded, but reactive aromatic fractions (e.g., tannins, lignins) due to adsorption to mineral surfaces. Agricultural drains and dystrophic lakes will also likely have organic matter that is old, yet present at high concentrations. These systems are relatively unproductive and experience low rates of biodegradation [28].

CONCLUSIONS

The trihalomethane precursor content of natural waters can vary sharply with the source type, location, and time of year. Although much of this variability can be attributed to differences in TOC, some cannot. By calculating the specific trihalomethane formation potential (Sp-THMFP), one can explore the nature of the variability in this intensive property of NOM.

When considering surface waters only, there is a consistent trend of increasing Sp-THMFP with decreasing TOC. This is as evident in the literature as it is in recently collected data from the Wachusett watershed in central Massachusetts. Evidence from laboratory leaching and degradation studies suggest that the observed relationship between Sp-THMPF and TOC is due to selective biodegradation of freshly solubilized NOM. This interpretation also fits in well with current understanding of NOM types, relative biodegradability and reactivity with chlorine.

ACKNOWLEDGEMENTS

The authors wish to thank the Massachusetts Department of Conservation and Recreation and the American Waterworks Research Foundation for their support in conducting these research projects.

REFERENCES

1. Xie YF; *Disinfection byproducts in drinking water: Formation, analysis and control*. Baca Raton: CRC Press; 2003.
2. APHA, AWWA, WEF; *Standard methods for the examination of water and wastewater*. 20th ed. Washington, D.C.: American Public Health Association; 1999.
3. Reckhow DA, Singer PC, Malcolm RL. Chlorination of humic materials. By-product formation and chemical interpretations. *Environ. Sci. Technol.*, 1990; 24:1655-1664.

4. Chapra SC, Canale RP, Amy GL. Empirical models for disinfection by-products in lakes and reservoirs. *J. Environ. Engineer.-ASCE,* 1997; 123:714-715.
5. Waldron MC, Bent GC. Northborough, MA: U.S. Geological Survey; 2001.
6. Edzwald JK, Becker WC, Wattier KL. Surrogate parameters for monitoring organic matter and THM precursors. *J. Amer. Waterworks Assoc.,* 1985; 77:122-132.
7. Amy GL, Chadik PA, Chowdhury ZK. Developing models for predicting trihalomethane formation potential and kinetics. *J. Amer. Waterworks Assoc.,* 1987; 79: 89-97.
8. Amy GL, Thompson JM, Tan L, Davis MK, Krasner SW. Evaluation of THM precursor contributions from agricultural drains. *J. Amer. Waterworks Assoc.,* 1990; 82: 57-64.
9. Miller RE, Randtke SJ, Hathaway LR, Denne JE. Organic carbon and THM formation potential in Kansas groundwaters. *J. Amer. Waterworks Assoc.,* 1990; 82: 49-62 .
10. Amy GL, Minear RA, Cooper WJ. Testing and validation of a multiple nonlinear-regression model for predicting trihalomethane formation potential. *Water Research.* 1987; 21:649-659.
11. Amy GL, Sierka RA, Bedessem J, Price D, Tan L. Molecular-size distributions of dissolved organic-matter. *J. Amer. Waterworks Assoc.,* 1992; 84: 67-75.
12. Chadik PA, Amy GL. Removing trihalomethane precursors from various natural waters by metal coagulants. *J. Amer. Waterworks Assoc.,* 1983; 75: 532-536.
13. Chang SD, Singer PC. The impact of ozonation on particle stability and the removal of TOC and THM precursors. *J. Amer. Waterworks Assoc.,* 1991; 83: 71-79.
14. Childress AE, Vrijenhoek EM, Elimelech M, Tanaka TS, Beuhler MD. Particulate and THM precursor removal with ferric chloride. *J. Environ. Engineer.-ASCE,* 1999; 125: 1054-1061.
15. Fujii R, Ranalli AJ, Aiken GR, Bergamaschi BA. Sacramento, CA: U.S. Geological Survey; 1998.
16. Garvey EA, Tobiason JE. Assessment of natural organic matter in Quabbin Reservoir. *J. Water Supply Res. Technol.-Aqua.* 2003; 52: 19-36.
17. Glaze WH, Wallace JL. Control of trihalomethane precursors in drinking-water: granular activated carbon with and without preozonation. *J. Amer. Waterworks Assoc.,* 1984; 76: 68-75.
18. McClellan JN, Reckhow DA, Tobiason JE, Edzwald JK, Smith DB. A comprehensive kinetic model for chlorine decay and chlorination by-product formation. Natural organic matter and disinfection by-products: Characterization and control in drinking water: Anaheim, CA. Washington, DC: ACS; 2000: 223-246.
19. Oliver BG, Lawrence J. Haloforms in drinking water: A study of precursors and precursor removal. *J. Amer. Waterworks Assoc.,* 1979; 71: 161-163.
20. Reckhow DA, Singer PC. Chlorination by-products in drinking waters: From formation potentials to finished water concentrations. *J. Amer. Waterworks Assoc.,* 1990; 82: 173-180.
21. Shukairy HM, Miltner RJ, Summers RS. Bromide's effect on DBP formation, speciation, and control: Part 1, Ozonation. *J. Amer. Waterworks Assoc.,* 1994; 86: 72-87.

22. Siddiqui M, Amy GL, Ozekin K, Westerhoff P. Empirically and theoretically-based models for predicting brominated ozonated by-products. *Ozone: Sci. Engineer.,* 1994; 16: 157-178.
23. Sinsabaugh RLI, Hoehn RC, Knocke WRLAE. Precursor size and organic halide formation rates in raw and coagulated surface waters. *J. Environ. Engineer.-ASCE,* 1986; 112: 139-153.
24. Symons JM, Speitel GEJr, Diehl AC, Sorenson HWJr. Precursor control in waters containing bromide. *J. Amer. Waterworks Assoc.,* 1994; 86: 48-60.
25. Taylor JK; *Quality assurance of chemical measurements*. Chelsea, MI: Lewis Publishers, 1987.
26. Thompson JM, Tan L, Amy GL, Davis MK. Agricultural drains in the Sacramento River delta: Trihalomethane precursor contributions to California State Project Water. *Annual Conference Proceedings; American Waterworks Association*; Orlando, FL. 1988: 935.
27. Reckhow DA, Rees PL, Bryan D. Watershed sources of disinfection by-product precursors. Submitted to: *Water Quality and Treatment*, 2004.
28. Wetzel RG. *Limnology: Lake and river ecosystems*. 3rd ed. San Diego: Academic Press; 2001.

Chapter 9

FLUORESCENCE AND ENERGY TRANSFER PROCESSES OF HUMIC SUBSTANCES AND RELATED MODEL COMPOUNDS IN TERBIUM COMPLEXES

Michael U. Kumke and Sascha Eidner

Institute of Chemistry, University of Potsdam,
Karl-Liebknecht-Str. 24-25, 14476 Potsdam-Golm, Germany

9.1. INTRODUCTION

Quenching of the intrinsic fluorescence of humic substances (HSs) upon complexation with metal ions has been employed to calculate the binding constants β [1-4]. Synchronous fluorescence, excitation-emission matrices and lanthanide ion probe spectroscopy also have been applied to elucidate the binding of metal ions by HSs [5-13]. Main drawbacks of these approaches are the unknown stoichiometry of the metal-HSs complex and the unknown source of the intrinsic HSs fluorescence, both of which are due to lack of information on HSs structures [14-17].

However, fluorescence techniques for the determination of β are very attractive because 1) no separation step is involved, 2) they are applicable at environmentally relevant concentrations, and 3) the system investigated experiences only a minor disturbance by the excitation light. Major limitations in the application of fluorescence techniques for the determination of β arise from lack of understanding of the complex intra- and intermolecular processes involved in HSs fluorescence. To shed more light on these processes, we approached the problem using simple substituted benzoic acids as model compounds for HSs in order to elucidate the fundamental fluorescence processes of HSs. The complexation of lanthanide ions by HSs and model compounds was investigated using steady-state and time-resolved luminescence techniques. In the luminescence measurements, the fluorescence of HSs and substituted benzoic acids as model compounds for HSs chromophores was recorded together with the intrinsic luminescence of Tb^{3+}. Selection of the model compounds was driven by 1) growing experimental evidence that HSs molecules are smaller than thought previously compared to the suggestions made earlier [18], 2) the presence of model-compound-related structures in precursor materials like lignins, and 3) analytical data obtained from degradation experiments like enzymatic hydrolysis [19]. Based on these facts we chose substituted benzoic acids such as salicylic acid and 3,5-dihydroxybenzoic acid as model compounds.

The first part of this paper compares the fluorescent properties of the model compounds with those of HSs. Special emphasis was given to the time-resolved fluorescence and how the fluorescence decay kinetics were changed upon Tb^{3+} complexation.

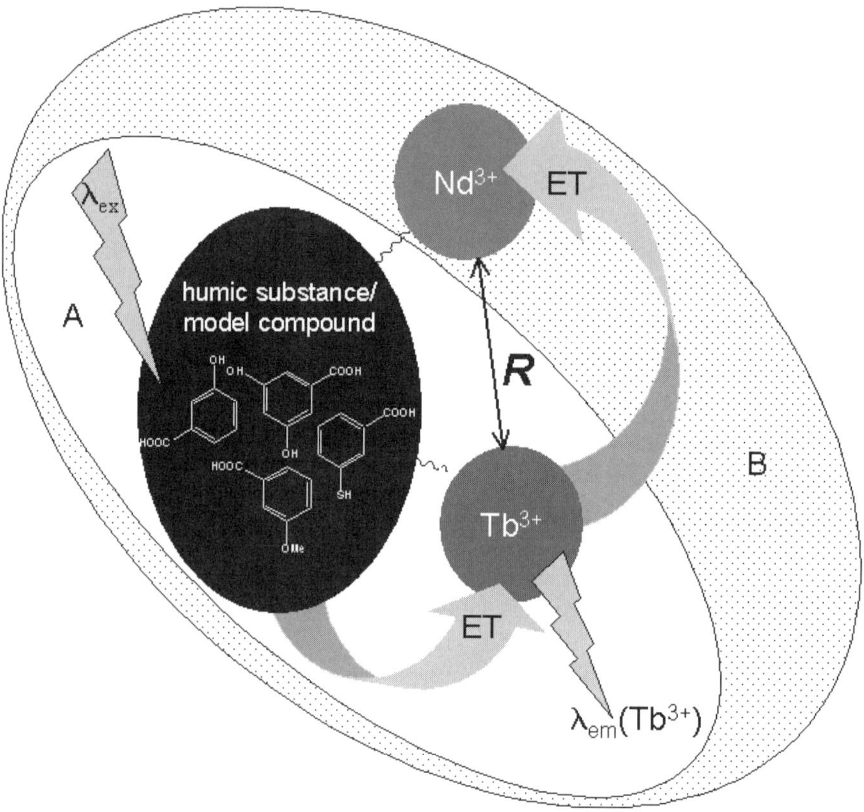

Scheme 9.1 Investigated energy transfer processes for A) the interaction between HSs (or model compound) and Tb^{3+} and B) the interlanthanide energy transfer between Tb^{3+} and Nd^{3+} for the determination of *R*

In the second part we investigated energy transfer from HSs and from model compounds to complexed Tb^{3+} and measured the quenching of the intrinsic fluorescence of the ligands as well as the luminescence enhancement of complexed Tb^{3+} (see Scheme 9.1). Furthermore, the inter-lanthanide energy transfer between Tb^{3+} and Nd^{3+} (both bound to HSs) was employed to calculate the (average) distance *R* between Tb^{3+} and Nd^{3+} in HSs (see Scheme 9.1). The distance calculations were based on the well-established concept of resonance energy transfer introduced by Förster [20-22].

9.2. MATERIALS AND METHODS

9.2.1. Materials

The substituted benzoic acids (benzoic acid, 3-methoxybenzoic acid, 3,5-dihydroxybenzoic acid, salicylic acid, mercaptosalicylic acid and gallic acid) were purchased from Aldrich and used without further purification. The concentration of the substituted aromatic benzoic acid was set to 10^{-4} M in all the experiments.

Several HSs samples were employed in the study, with a focus on Suwannee River fulvic acid (SRFA). Commercially-available humic substances were purchased from Aldrich and from the International Humic Substances Society (IHSS: Peat HA, Leonardite HA, Soil HA, Suwannee River HA (SRHA)). HSs were also isolated from soil seepage water (BS1 HA) and from a brown water lake (HO13 HA) [23]. The HSs samples were investigated at pH 5 and constant ionic strength (I = 0.02 M NaCl). The SRFA sample was also investigated at pH 3, 4, 5, and 6 at constant ionic strength (I = 0.02 M NaCl). The pH of solutions was adjusted with HCl and NaOH. The lanthanide ions were added using 0.001 M stock solutions of Tb^{3+}, Nd^{3+}, and La^{3+} nitrates. All solutions were stored in the dark until used. In the investigation of the inter-lanthanide energy transfer, typically 3 mL of a HSs solution (10 mg L^{-1}, pH and I const) and 30 µL of Tb^{3+} (same pH and I) were mixed and equilibrated for 5 min to give $[Tb^{3+}] \sim 10^{-5}$ M. The Tb^{3+} luminescence was recorded and subsequently in steps of 5 µL a solution containing a second lanthanide ion (Nd^{3+} or La^{3+} with $[Ln] = 10^{-3}$ M) was added, equilibrated and then measured up to final overall metal concentrations of $\sim 5 \times 10^{-5}$ M.

9.2.2. Methods

Steady-state fluorescence data were acquired with a FluoroMax III spectrometer (Jobin-Yvon, Grasbrunn, Germany) equipped with a R928P photomultiplier tube. The slits were set to a 2 nm spectral bandpass in excitation and 4 nm spectral bandpass in emission. In the steady-state fluorescence experiments with HSs, an excitation wavelength of λ_{ex} = 325 nm was applied and the HSs fluorescence was measured between 375 nm $< \lambda_{em} <$ 640 nm. The spectra were evaluated at a Tb^{3+}-luminescence peak (λ_{em} = 544 nm). In the case of the substituted benzoic acids, the excitation wavelength was chosen based on the absorption maximum of the compound (located in the longest wavelength range). Depending on the particular compound, an excitation wavelength λ_{ex} between 300 nm $< \lambda_{ex} <$ 335 nm was employed.

Time-resolved fluorescence experiments were performed with a FL920 lifetime spectrometer (Edinburgh Instruments, UK). The instrument was equipped with a R955 photomultiplier operated in the single photon counting mode. A N_2 laser (LTB, Berlin, Germany) with λ_{ex} = 337.1 nm was used as the excitation source with a repetition rate of 20 Hz. The terbium luminescence was detected at λ_{em} = 544 nm ($^5D_4 \rightarrow {}^7F_5$ transition). The data were analyzed by standard least-squares algorithms of a commercial software package (Edinburgh Instruments, UK). All fluorescence measurements were carried out at 293 K.

Time-resolved emission spectra (TRES) were measured with a streak camera system. A mode-locked Ti:sapphire laser (Tsunami, Spectra Physics, Germany) was used for excitation of the samples. If not otherwise noted, the samples were excited after third harmonic generation at λ_{ex} = 280 nm. The FWHM of the laser pulse was

100 fs. Fluorescence of the samples was detected with a streak camera (C5680, Hamamatsu, Germany) equipped with an imaging spectrograph (250is, Chromex). The streak camera was operated in the sync mode at 80.2 MHz. Time-resolved measurements were performed using two time windows of 3495 ps and 1998 ps, respectively. The TRES were recorded in the single photon counting mode. For each TRES, 10000 images (304 ms integration time each) were accumulated. The fluorescence decay times were calculated from the TRES by evaluating slices of $\Delta\lambda_{em}$ = 10 nm using a lifetime fitting module (TA-Fit, Hamamatsu).

9.3. RESULTS

9.3.1. Intrinsic Fluorescence of Simple Organic Compounds

Fluorescence techniques are powerful tools for the investigation of interactions like metal complexation. However, for HSs the connection between structure and possible contributions from intramolecular processes is still not fully understood [24, 25]. In order to elucidate the contributions of various possible intramolecular processes (*vide infra*), we chose substituted benzoic acids as model compounds for HSs and investigated their fluorescence characteristics.

The model compounds and HSs had similar fluorescence characteristics and dependences on solution parameters, e.g., the spectral range of emission, and broad, featureless emission spectra with a strong dependence on pH (Figure 9.1) [26-38].

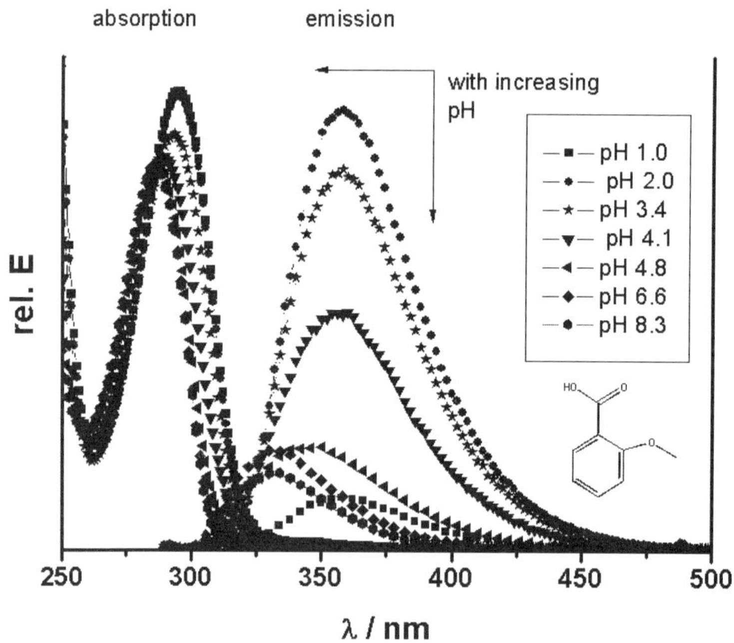

Figure 9.1 Absorption and fluorescence spectra of 3-methoxybenzoic acid (c = 10^{-4} M) at different pH. The excitation wavelength λ_{ex} was set to 285 nm in the fluorescence measurements

As an example, the pH dependence of the absorption and fluorescence of 3-methoxybenzoic acid is shown in Figure 9.1. The fluorescence efficiency and location of the emission maximum changed with pH. It shifted toward shorter wavelengths and the fluorescence intensity decreased except at pH 1, where the smallest fluorescence intensity was observed. The blue shift was also observed for the longest wavelength absorption band.

The time-resolved emission spectra (TRES) of the model compounds were also measured. As an example, the TRES of 3,5-dihydroxybenzoic acid is shown in Figure 9.2 at pH 2 (top) and pH 5 (bottom). The fluorescence spectra (x-axis) are presented as a function of time (y-axis) after excitation on a picosecond time scale. A gray scale represents the corresponding fluorescence intensities.

In Figure 9.2, the steady-state fluorescence spectrum is shown on top of each TRES. The TRES were markedly influenced by pH. At pH < 2 and pH > 10, the observed fluorescence decay of the model compounds (e.g., 3,5-dihydroxybenzoic acid and 3-methoxybenzoic acid) followed a mono-exponential decay law, indicating that only one species was involved in the fluorescence process (see top of Figure 9.2 for examples of fluorescence decay curves). This is further supported by the shape of the fluorescence spectrum during the excited state lifetime. At low and high pH, no shift of the emission maximum during the excited state lifetime was observed in the TRES. However, in the pH range between $2 < \text{pH} < 10$ a complex fluorescence decay was found as well as a shift of the emission maximum. As an example, the results of analysis of the fluorescence decays of 3-methoxybenzoic acid at different pH are shown in Figure 9.3. The fluorescence decays were analysed by a mono- or bi-exponential decay law depending on the pH of the sample. The fitted decay times are plotted as a function of pH. It can be seen that in the medium pH range the fluorescence decay of 3-methoxybenzoic acid follows a complex, bi-exponential decay law.

Moreover, for the medium pH range the spectral shape and location of the fluorescence maximum observed in TRES of the model compounds changed during the excited state lifetime. Comparison of the fluorescence decay at the blue and red edges of the TRES of 3,5-dihydroxybenzoic acid (Figure 9.2, bottom) emphasizes the spectral shift of the fluorescence maximum during the fluorescence lifetime. On the blue edge (left decay in Figure 9.2, bottom), the fluorescence decay is relatively short; it becomes much longer at longer wavelengths (right decay in Figure 9.2, bottom), indicating a shift in the emission maximum with time.

The TRES of 3,5-dihydroxybenzoic acid at pH 5 in the presence of Tb^{3+} ions is shown in Figure 9.4 and the results of analysis of the fluorescence decays at 360 nm $< \lambda_{em} <$ 370 nm and at 422 nm $< \lambda_{em} <$ 432 nm are compared in Table 9.1 with corresponding lifetime data in the absence of Tb^{3+} (see also Figure 9.2, bottom). It is striking that the temporal and spectral evolution of the TRES is altered by complexation of lanthanide ions. While a shift in the emission maximum and a complex, emission wavelength-dependent fluorescence decay was found in the absence of Tb^{3+} at pH 5, these effects were almost eliminated on addition of Tb^{3+}. This can also be seen from the calculated fluorescence decay times for the different spectral ranges. In the absence of Tb^{3+}, τ_2 differs almost by a factor of two while in the presence of Tb^{3+} equal values for τ_2 were found (see Table 9.1). It is interesting to note that no influence of Tb^{3+} on the fluorescence decay was observed with 3-methoxybenzoic acid (see also Table 9.1). This might indicate the importance of hydroxyl groups in the interaction with metal ions and on subsequent fluorescence processes.

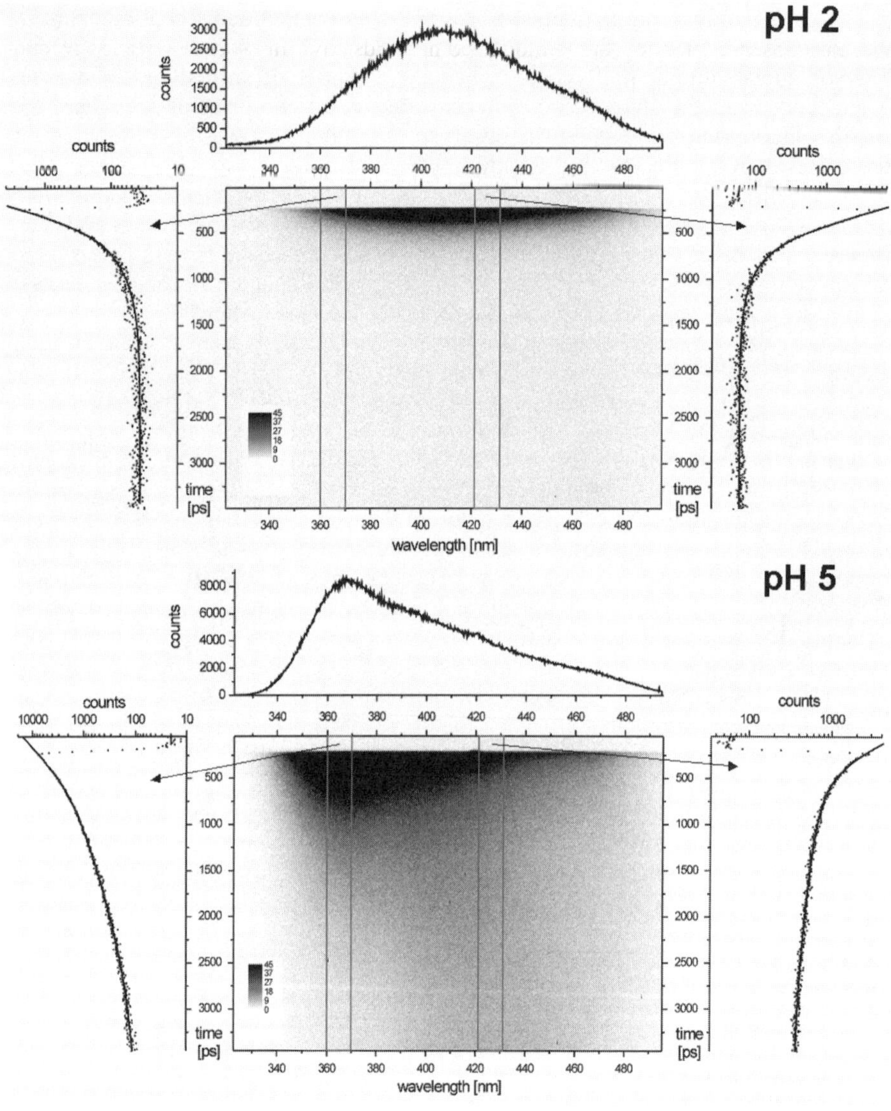

Figure 9.2 The TRES of 3,5-dihydroxy-benzoic acid (c = 10^{-4} M) at pH 2 (top) and pH 5 (bottom) were obtained with an excitation wavelength λ_{ex} = 280 nm. The FWHM of the excitation pulse was 100 fs. The steady state fluorescence spectrum calculated from the TRES spectrum is shown on top of the TRES. Fluorescence decay curves for the wavelength range 360 nm < λ_{em} < 370 nm and 422 nm < λ_{em} < 432 nm are presented on the left and right hand side

9.3.2. Energy Transfer from Ligands to Lanthanide ions

The effect of Tb^{3+} addition on the intrinsic fluorescence of model compounds and HSs was investigated.

Fluorescence spectra of gallic acid (10^{-4} M) with increasing concentration of Tb^{3+} are shown in Figure 9.5. Only a minor decrease in the fluorescence intensity of gallic acid was observed on addition of Tb^{3+}. Inefficient fluorescence quenching was also observed for the other model compounds investigated, in contrast to the observed quenching of HSs fluorescence on Tb^{3+} addition. The effect of Tb^{3+} on the fluorescence intensity of salicylic acid and of HSs of different origins are compared in Figure 9.6. The normalized fluorescence intensity ratios I/I_0 are shown. The fluorescence of salicylic acid was not quenched, which is obvious from the independence of the ratios I/I_0 on the concentration of Tb^{3+}. The data points are randomly distributed around a ratio of one (Figure 9.6). On the other hand, the fluorescence of HSs was quenched on addition of Tb^{3+} ions, which can be seen from the decreasing ratio I/I_0. The extent of fluorescence quenching was highly dependent on the HSs sample. The fluorescence of HSs isolated from soil was quenched more effectively than the fluorescence of HSs isolated from aquatic sources.

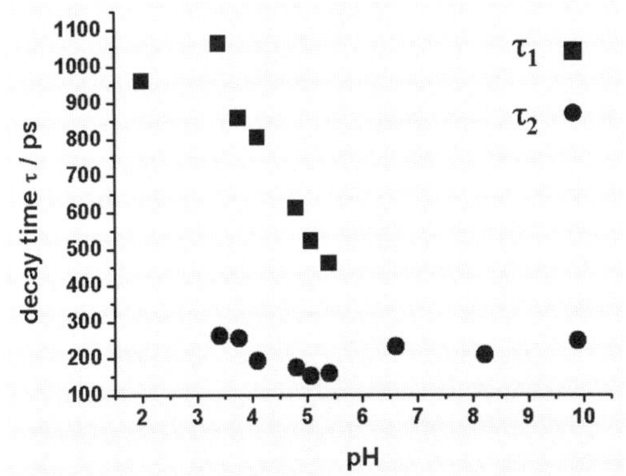

Figure 9.3 Fluorescence decay times of 3-methoxybenzoic acid ($c = 10^{-4}$ M) at different pH. The fluorescence decay curves were extracted from the TRES ($\lambda_{ex} = 280$ nm, 365 nm $< \lambda_{em} <$ 385 nm)

Table 9.1 Comparison of the fluorescence decay times of 3,5-dihydroxybenzoic acid and 3-methoxybenzoic acid (10^{-4} M) in the absence and presence of Tb^{3+} (10^{-4} M) at pH 5.

			addition of Tb^{3+}	
	τ_1 / ps	τ_2 / ps	τ_1 / ps	τ_2 / ps
3,5-dihydroxybenzoic acid				
350-360nm	202	736	183	607
422-432nm	162	1297	98	592
3-methoxybenzoic acid				
345-365nm	143	1022	152	1036
390-420nm	208	4122	226	4022

A second luminescence signal originating from Tb^{3+} was observed in the fluorescence quenching experiments (see Figure 9.5) [39-41]. While the fluorescence of the model compounds and HSs stayed unchanged or were quenched, respectively, the Tb^{3+} luminescence increased with increasing concentration of Tb^{3+}. It is interesting to note that in control experiments in pure water in the absence of organic ligands under identical experimental conditions (Tb^{3+} concentration, excitation wavelength, slit widths, pH, ionic strength), no luminescence signal was observed for Tb^{3+}. Only at high Tb^{3+} concentrations (> 100 μM) a small luminescence signal from Tb^{3+} was detected. Moreover, the luminescence sensitisation of Tb^{3+} in complexes with model compounds was also investigated at pH 2, where no Tb^{3+} luminescence was observed. This indicates that only the luminescence of complexed Tb^{3+} is detected.

In the presence of model compounds as well as HSs even at very low Tb^{3+} concentrations (< μM), a Tb^{3+}-sensitized luminescence signal was detected at pH 4, 5, and 6. No higher pH values were tested because of the formation of Tb^{3+} hydroxides. The Tb^{3+} luminescence could be easily identified in the experimental emission spectra because of the characteristic sharp luminescence peaks at λ_{em} = 489 nm and 544 nm (Figure 9.5).

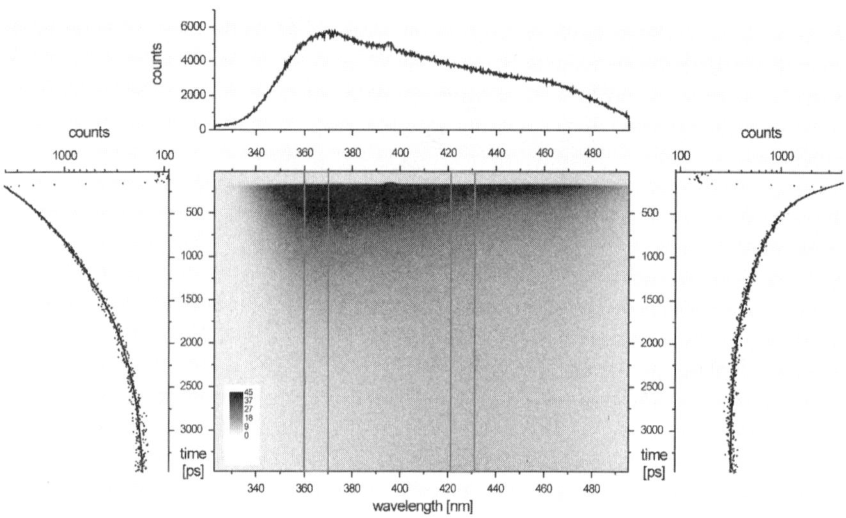

Figure 9.4 TRES of 3,5-dihydroxybenzoic acid (c = 10^{-4} M) in the presence of Tb^{3+} at pH 5 at a molar ratio of 1:1 (λ_{ex} = 265 nm). Also shown are the fluorescence decay curves (on the left and right for the wavelength ranges 360 nm < λ_{em} < 370 nm and 422 nm < λ_{em} < 432 nm, respectively) and the steady-state fluorescence spectrum (top, the small peak at ~398 nm is a minor contribution from scattered light of the second harmonics of the Ti/sapphire laser). See Table 9.1 for the corresponding fluorescence decay time

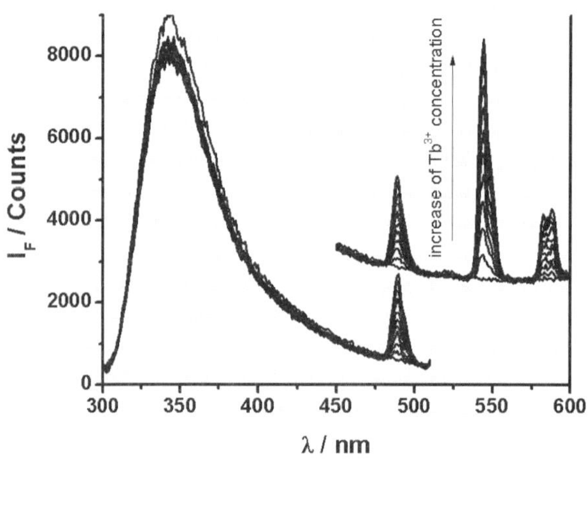

Figure 9.5 Steady-state fluorescence spectra of gallic acid (c = 10^{-4} M) at pH 4 with increasing Tb^{3+} concentration. ($10^4 c_{Tb}$ = 0.4, 0.8, 1.2, 1.6, 2, 2.3, 2.7, 3.1, 3.5, 3.8). Also shown is the luminescence of Tb^{3+}. The fluorescence was excited at λ_{ex} = 295 nm)

The observed increase of the Tb^{3+} luminescence intensity depended on the type of model ligand and on the HSs origin. HSs samples of aquatic origin showed a strong Tb^{3+} luminescence enhancement, while for soil derived HS the luminescence enhancement was considerably smaller. In order to compare the observed increase for different model compounds and for HSs, the luminescence intensity at $\lambda_{em} = 544$ nm was normalized. Curves with different slopes at low Tb^{3+} concentrations were obtained for the model compounds and HSs. Well defined limiting plateaus were found for HSs at higher Tb^{3+} concentrations. This effect was less obvious for the model compounds, and an almost linear relationship was observed for gallic acid and 3,5-dihydroxybenzoic acid (see Figures 9.7 and 9.8).

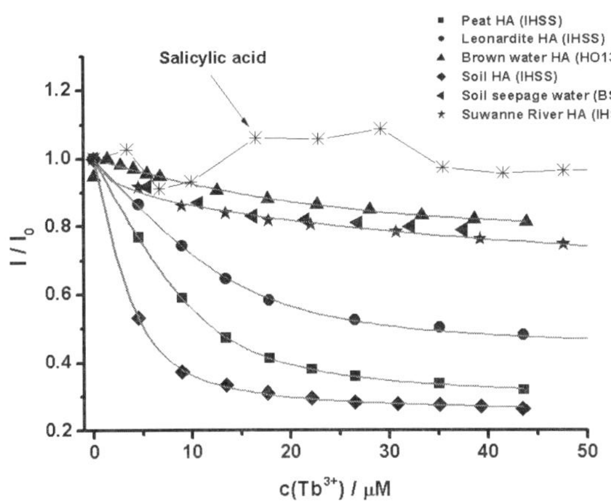

Figure 9.6 Quenching of the intrinsic fluorescence of HSs of different origin on addition of Tb^{3+}. Also shown is the effect of Tb^{3+} on the fluorescence of salicylic acid. For HSs the concentration of all samples was 10 mg/L and the pH was adjusted to 5 ($\lambda_{ex} = 325$ nm, 425 nm $< \lambda_{em} <$ 475 nm)

The luminescence decay of Tb^{3+} was investigated in the presence of model compounds or HSs. A luminescence decay time of $\tau_{Tb} = 400$ µs was measured for Tb^{3+}, which agrees well with literature values [42-45]. The fluorescence decay times of Tb^{3+} in the presence of model compounds were also mono-exponential and depended on the molar ratio of model compound:Tb^{3+}, which is a consequence of the increased binding of Tb^{3+} with increasing ratio. It is interesting to note that the luminescence decay times of Tb^{3+} complexed with model compounds were shorter compared to Tb_{aq}^{3+} (aq). For the samples containing Tb^{3+} and HSs, the experimental decays were no longer mono-exponential (Figure 9.10, left). A complex luminescence decay observed for Tb^{3+} bound to HSs has to be attributed to the complexity of the HSs sample and the combination of energy transfer and backtransfer processes in the HSs-Tb^{3+} complexes.

9.3.3. Inter-lanthanide Energy Transfer in Lanthanide-HSs Complexes

Because of the unknown structure of HSs, the analysis of data from metal complexation experiments is difficult, mainly because the stoichiometry of the complexes formed is unknown. This allows one only to calculate conditional binding constants (e.g., assuming a 1:1 stoichiometry). To circumvent this problem and gain more insight into structural characteristics of HSs that influence metal binding, energy transfer between Tb^{3+} and Nd^{3+} (both bound to HSs) was investigated.

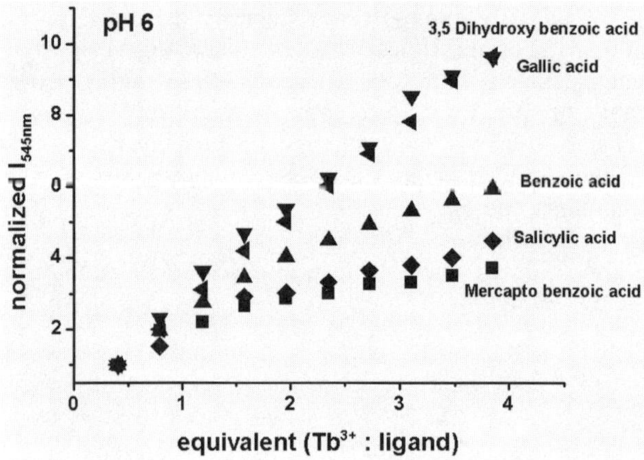

Figure 9.7 Luminescence enhancement of Tb^{3+} (λ_{em} = 544 nm) on addition of model compounds at pH 4. Depending on the model compound investigated, the excitation wavelength was chosen to be 275 nm < λ_{ex} < 325 nm. The signals were normalized to the Tb^{3+} luminescence signal at the lowest concentration ($c_{Tb} = 4 \times 10^{-5}$ M) in order to compare the relative luminescence enhancement

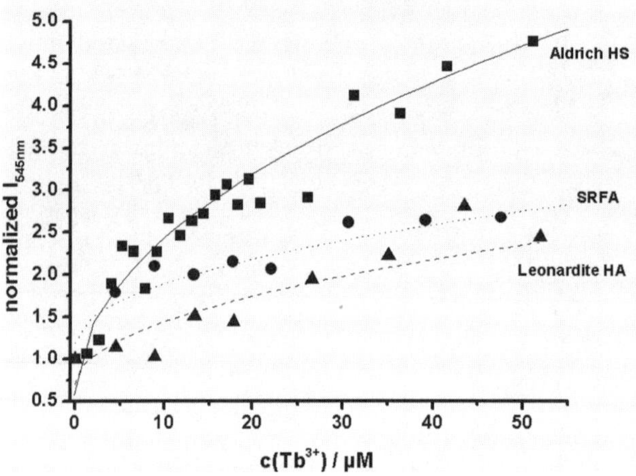

Figure 9.8 Luminescence enhancement of Tb^{3+} in the presence of HSs (10 mg L^{-1}, pH 5) at λ_{em} = 544 nm normalized to the lowest Tb^{3+} concentration

It is well established that the luminescence of Tb^{3+} is quenched by Nd^{3+}. The luminescence quenching is based on a dipole-dipole interaction and can be understood as a resonance energy transfer process between an excited donor (Tb^{3+}) and an acceptor (Nd^{3+}). Horrocks and co-workers used inter-lanthanide energy transfer for the investigation of Ca^{2+}-binding sites in a number of enzymes [46-50]. The theory is well established and is based on a concept originally developed by Förster [20-22]. The basic idea is that the rate constant of energy transfer k_{ET} (and consequently the energy transfer efficiency *E,* where I_{DA} and I_D are the fluorescence intensities or luminescence lifetimes of the donor in the presence and absence of the acceptor, respectively) depends on $1/R^6$ where R is the distance between donor and acceptor.

$$k_{ET} = \frac{\Gamma \cdot \kappa^2 \cdot \Phi_D}{R_{DA}^6} J \tag{9.1}$$

$$E = 1 - \frac{I_{DA}}{I_D} \text{ and } E = 1 - \frac{\tau_{DA}}{\tau_D} \tag{9.2}$$

$$R = R_0 \left(\frac{1}{E} - 1 \right)^{\frac{1}{6}} \quad (9.3)$$

R_0 is the so-called Förster radius and is a characteristic of the donor-acceptor pair. The value of R_0 is defined by the spectral overlap integral J, which contains the fluorescence spectrum ($I_F(\nu)$) of the donor and the absorption spectrum ($\varepsilon(\nu)$) of the acceptor, and the relative orientation of the transition dipole moments between donor und acceptor (described by the orientation factor κ). Due to the energy transfer, the luminescence intensity and the luminescence lifetime of the donor (Tb^{3+}) are quenched. From measurements of the donor luminescence intensity (and/or the donor luminescence lifetime) in the absence and presence of acceptor (Nd^{3+}), the distance R between both can be calculated if R_0 is known, as for Tb^{3+}/Nd^{3+} ($R_0 = 9$ Å) [46-48].

Titration of HSs samples containing 10 μM Tb^{3+} with Nd^{3+} was performed and the luminescence intensity of Tb^{3+} at 544 nm was monitored. The concentration of Nd^{3+} was stepwise increased up to 100 μM. Under the experimental conditions applied, no luminescence of directly excited Tb^{3+} was observed and only Tb^{3+} bound to HS was detected (see Section 3.2). It was found that the luminescence intensity of Tb^{3+} decreased with increasing Nd^{3+} concentration, indicating either an energy transfer from Tb^{3+} to Nd^{3+} (both bound to HSs) or replacement of HS-bound Tb^{3+} ions by Nd^{3+} ions.

Control experiments were performed to distinguish between those two processes. La^{3+} was added to check the contribution of Tb^{3+} release from HSs due to replacement with other lanthanide ions. Complexation capabilities of the lanthanide ions should be very similar and La^{3+} and Nd^{3+} should have the same replacement capabilities for HSs-bound Tb^{3+}. However, the spectroscopic properties of La^{3+} are very different from Nd^{3+}. La^{3+} has no energy acceptor capabilities and, therefore, an energy transfer from Tb^{3+} to La^{3+} is not possible [46-51]. It was found that the contribution due to release of Tb^{3+} from the HSs is small in the μM concentration range. Thus, the observed decrease in luminescence intensity of Tb^{3+} on addition of Nd^{3+} can be attributed mainly to an energy transfer between both ions bound to HSs.

The luminescence lifetime of Tb^{3+} in the presence and absence of Nd^{3+} was determined (Figure 9.9). The lifetime of Tb^{3+} decreased with increasing Nd^{3+} concentration. No such decrease was observed for Tb^{3+}/Nd^{3+} solutions in water in the absence of HSs: the luminescence decay time of Tb^{3+} was unchanged in the presence of Nd^{3+} (Figure 9.9).

The average distance R between Tb^{3+} and Nd^{3+} was determined for HSs of different origin and for SRFA at different pH. The dependence of R on the loading with Nd^{3+} for HSs of different origin at pH 5 is shown in Figure 9.10. The average distance R between Tb^{3+}/Nd^{3+} bound to HSs decreased with increasing Nd^{3+} concentration. Depending on the Nd^{3+} concentration and the origin of the HSs, R values between 1.2 nm and 0.85 nm were calculated. It is interesting to note that for each HSs the average distance R approached a minimum distance R_{min} that seems to be specific for the HSs investigated.

The dependence of R on the loading of SRFA with Nd^{3+} at different pH is compared in Figure 9.11. The average distance between Tb^{3+} and Nd^{3+} approached a limiting distance R_{min} at each pH. The distance R was clearly dependent on pH. The average distance R was larger at lower pH. For example, for $[Nd^{3+}] = 20$ μM we found an average distance of 0.95 nm at pH 4, while an average distance R of 0.8

nm was calculated at pH 6.

9.4. DISCUSSION

The fluorescence emission spectra of HSs are broad, with no vibronic structure. They are located in the spectral range 350 nm < λ_{em} < 600 nm with a maximum around 470 nm ± 15 nm [24-27,29,34,35]. The fluorescence efficiency and location of the fluorescence maximum depend on solution parameters of which pH is the most important [26,31,33,36,37]. In addition, the fluorescence decay kinetics of HS are very complex [52-55]. Possible reasons are 1) the heterogeneity of HSs themselves, 2) various intra- and intermolecular interaction processes like energy transfer between different chromophores of HSs, and 3) processes that involve excited state reactions like proton transfer (excited state intramolecular proton transfer, ESIPT) or conformational motions (e.g., twisted intramolecular charge transfer, TICT). The steady-state and time-resolved fluorescence properties of the model compounds investigated resemble many of the HS fluorescence characteristics. In the medium pH range (3 < pH < 9), the complexity of HSs fluorescence decay as well as the shift of the fluorescence maximum in the TRES are also found for the model compounds. For the model compounds the observed complexity of the fluorescence can be attributed to excited state reactions that involve the transfer of protons (for example, ESIPT and interaction with water) and can be characterized by the so-called Förster cycle [20,21]. The importance of proton transfer reactions in the excited state is further supported by the observation that the temporal and spectral fluorescence characteristics are changed on addition of Tb^{3+}. This was previously reported for HSs and Eu^{3+} and the possible contribution of ESIPT reactions has been suggested [55]. The observed alterations in the temporal and spectral fluorescence characteristics of model compounds add further evidence for this conclusion.

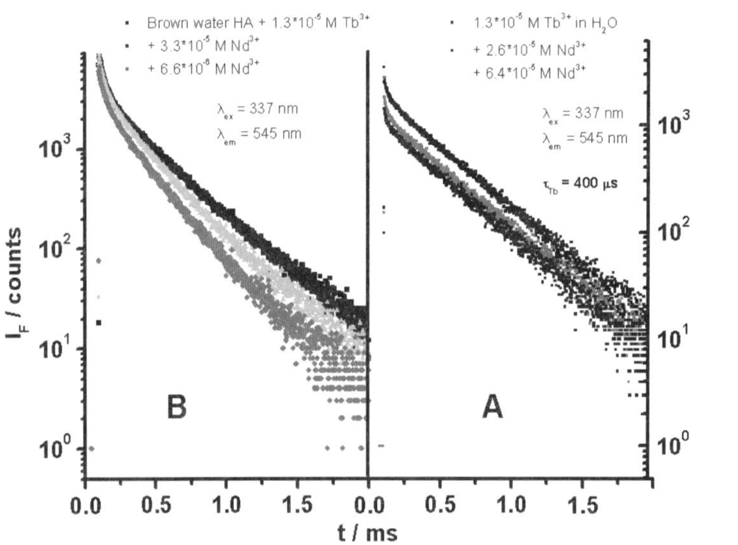

Figure 9.9 Dependence of Tb^{3+} luminescence decay on the presence of Nd^{3+} in a) pure water and b) bound to brown water HA (HO13 HA, 10 mg L^{-1}, pH 5)

While the temporal and spectral characteristics of the model compounds were in good agreement with the HSs fluorescence, a major difference was observed in the fluorescence quenching studies with Tb^{3+}. The intrinsic fluorescence of HSs was quenched on Tb^{3+} addition, while for the majority of the model compounds no (or only very minor) fluorescence quenching was found. The latter is expected, since for f-elements only a small spin-orbit coupling should be operative in complexes because the f-electrons are shielded by the closed s- and p-shells (in Tb^{3+} the 4f-electrons are shielded by 5s and 5p orbitals). It is more likely that other processes cause the HSs fluorescence quenching upon addition of Tb^{3+} (and in general by lanthanide ions): 1) the competition between H^+ and Tb^{3+} for carboxylate and hydroxyl groups and consequently the alteration of photo-induced processes connected to these groups like ESIPT and 2) the influence on the molecular flexibility caused by intra- or intermolecular bridging (e.g., comparable to the formation of H-bonds), which limits the rotational freedom of parts of the molecule. This could change the fluorescence of the molecule when photo-induced geometry changes are involved in the HSs fluorescence process. Twisted intramolecular charge transfer (TICT) processes are an example of such intramolecular rotations.

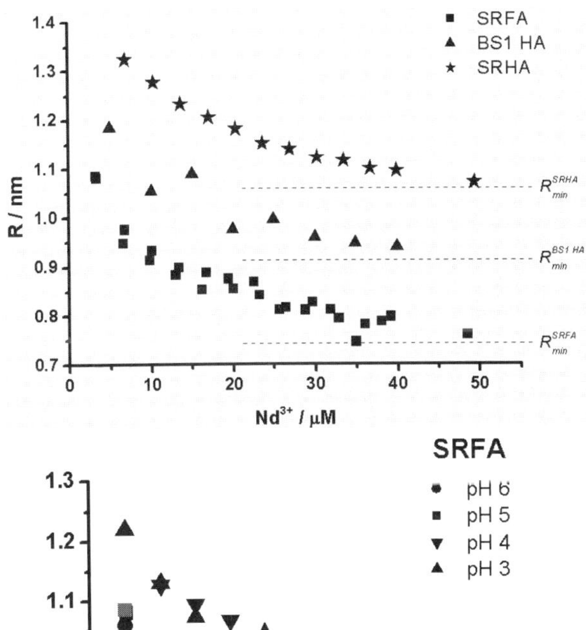

Figure 9.10 Dependence of the average distance R between Tb^{3+} (c_{Tb} = 10 µM) and Nd^{3+} bound to HSs of different origin (pH 5, c_{HS} = 10 mg L^{-1}) on the concentration of Nd^{3+}

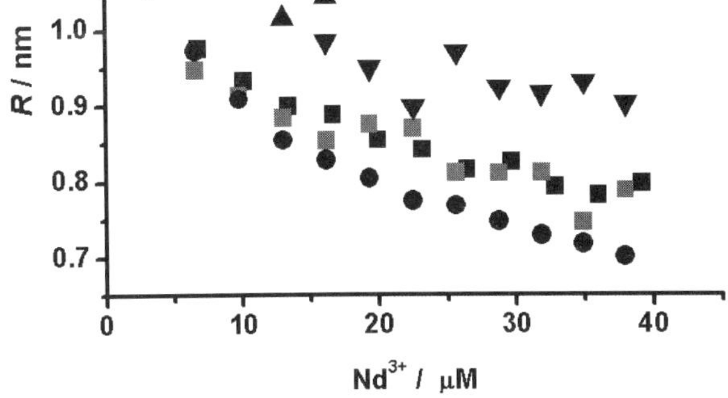

Figure 9.11 Dependence of the average distance R between Tb^{3+} (c_{Tb} = 10 µM) and Nd^{3+} bound to SRFA (pH 5, c_{HS} = 10 mg L^{-1}) on the concentration of Nd^{3+}. Also shown is the effect of pH on R

The fact that Tb^{3+} does not quench the fluorescence of the model compounds makes it attractive to assume that intramolecular (and possibly also intermolecular if association is considered as well) reorientations play a major part in the HSs fluorescence process [54,55]. Metal ions bound to HS can interfere with these processes by forming complexes with HSs and hence limiting the rotational freedom of molecular moieties. The observed effect would then be fluorescence quenching because the fluorescing form of the HSs can no longer be formed in the photoinduced process.

The observed sensitization of Tb^{3+} luminescence by organic ligands has been observed earlier and was attributed to energy transfer from the organic ligand to the Tb^{3+} ion [56-59]. The triplet state of the organic ligand has been identified as the energy-transferring state, which also explains why no fluorescence quenching was observed with model ligands. It also supports the assumption that the limitation of conformational motions is a major reason for the observed HSs fluorescence quenching by Tb^{3+}. The energies of the triplet state of the organic ligand and the emitting $^5D_4$-state of Tb^{3+} determine the efficiency of energy transfer and the possible radiationless decay due to an energy back transfer process from Tb^{3+} to the ligand [59]. The observed luminescence quantum efficiency of the complex is in many cases smaller than the luminescence quantum efficiency of the Tb_{aq}^{3+} ion itself due to an effective energy back transfer. However, because of the low molar absorptivity of Tb^{3+} ($< 1\ M^{-1}cm^{-1}$) the so-called “antenna effect” leads to an increase of the observable luminescence intensity of the organic complex (under identical excitation and emission conditions) [58,59]. This is related to the much larger absorptivities of the ligands ($> 1000\ M^{-1}cm^{-1}$).

For the model compounds, the normalized intensity of the sensitized Tb^{3+} luminescence gave curves with different slopes and in some cases almost a linear relationship. Well-defined plateaus observed for HSs were not found in the investigated concentration range. For HSs, the height of the plateaus formed depended on the HSs origin. A direct comparison between model compounds and HSs is difficult since the number of binding sites and the calculation of a molar ratio between HS and Tb^{3+} is not possible like it is for the model compounds. The fact that the observed luminescence sensitization of Tb^{3+} upon binding to HSs depends on the origin of the HS reflects differences of structure in terms of the presence of chromophores with 1) suitable triplet energies and 2) of $RCOO^-$- and RO^-- groups connected to those chromophores. For the soil derived HSs samples, the observed luminescence sensitization was considerably smaller than for HSs of aquatic origin. This indicates that in those samples the triplet energy may be too close to the $^5D_4$ energy of Tb^{3+}, allowing efficient energy back- transfer from Tb^{3+} to the HSs triplet state.

The measurement is very sensitive and specific because of 1) the sharp, well-defined Tb^{3+} luminescence peaks, and 2) only sensitized luminescence is observed, hence only the luminescence of the bound Tb^{3+} contributes to the detected luminescence signal. The measurement of fluorescence enhancement is easier and more sensitive compared to luminescence quenching techniques since the detection of photons against a low background is superior to the detection of a (minor) decrease of a luminescence signal. The Tb^{3+} luminescence is well separated from the HSs fluorescence and can easily be detected, especially with a time-gated detection scheme: Tb^{3+} luminescence is observed on a μs timescale while the HSs fluorescence decays on a nanosecond timescale. The limitation of the detection of sensitized luminescence is the still unknown connection between metal binding sites and HS chromophores that have suitable triplet state energies. The exact mechanism of the energy transfer between HS and Tb^{3+} is not known and is the subject of ongoing research.

Two basic mechanisms with different dependences on the distance between donor and acceptor have to be distinguished: 1) the dipole-dipole mechanism with its $1/R^6$-relationship and 2) the exchange mechanism, in which the rate constant of the energy transfer step decreases exponentially with the distance between donor and acceptor. A first rough calculation of a hypothetical R_0 between HSs and Tb^{3+} based on the fluorescence spectrum of Suwannee River fulvic acid and the absorption spectrum of Tb^{3+} yielded a value of approximately 4 Å, which is quite small and suggests that under the assumption made the exchange mechanism is the more probable. That would mean that the fraction of chromophores with metal binding capabilities in close proximity and with suitable triplet energies could be probed. Further research is needed to shed more light on the connection between HSs-metal binding sites and chromophores.

The limitations arising from the unknown HSs structure can be circumvented with the introduction of Nd^{3+} and the use of inter-lanthanide energy transfer as a molecular ruler. The spectroscopic properties of the Tb^{3+}/Nd^{3+} donor-acceptor pair are well characterized in terms of the operative dipole-dipole energy transfer mechanism and subsequently of the Förster distance R_0, which is 9.3 Å [46-50]. Here, only the relative distance of the bound Tb^{3+} and bound Nd^{3+} determines the energy transfer efficiency. Unbound Tb^{3+} or unbound Nd^{3+} ions do not interfere with the measurement as shown by control experiments. Although information on the absolute structure of HSs is still not accessible, relative information on HSs conformation and how the conformation of HSs is altered by solution parameters (e.g., pH, ionic strength, addition of metal ions) can be monitored – in solution and under environmentaly relevant concentrations of a few mg L^{-1}. The R_{min} seems to depend on the source of HSs and reflects structural (binding) properties of HSs as shown for HSs of different origin.

9.5. CONCLUSIONS

Excited state reactions play an important role in the overall fluorescence process of HSs. Proton transfer reactions and conformational reorientations after excitation are the major processes involved. Both are strongly influenced by metal ion complexation because of competition for binding sites between metal ions and protons and because of an inter- and intramolecular "bridging effect" of metal ions that limits the rotational freedom of the HSs molecules. Experiments with other model compounds and biopolymers (e.g., lignin and chitin) are planned and will further improve our understanding of those processes in HSs. Our results underline limitations of the HSs fluorescence quenching approach for the determination of conditional binding constants with metals. The contributions of real fluorescence quenching and spectral shifts in the fluorescence spectrum need to be elucidated, which is very difficult with standard 2D-fluorescence spectra. A solution might be the measurement of excitation-emission matrices in order to get the complete picture of the alterations in the HSs fluorescence upon metal complexation. With the application of Tb^{3+} and Eu^{3+} as luminescence probes, the metal complexation can be monitored from a different angle. Only bound Tb^{3+} is detected after indirect excitation via an intramolecular energy transfer step. The combination of HSs fluorescence quenching and luminescence enhancement via energy transfer could be helpful for the determination of better defined stability constants from fluorescence experiments with HSs. The energy of the triplet state of the chromophore is the most important parameter for energy transfer. This limits the luminescence enhancement of Tb^{3+} to a certain group of chromophores. The relative abundance of chromophores with suitable triplet states is an interesting parameter for the structural characterization of HSs.

The limitations of the unknown structure of HSs can be overcome with the application of inter-lanthanide energy transfer measurements. Here, only the distance between Tb^{3+} and Nd^{3+} is important when bound to HS. The measurements are very sensitive and can be performed at the μM-concentration level. That is, samples in realistic concentration ranges can be investigated. Effects of pH, ionic strength or the addition of metal ions (not necessarily lanthanides) on the conformation of HS can be monitored. Currently we are exploiting the inter-lanthanide energy transfer approach for the investigation of inter- and intramolecular association reactions of HSs.

ACKNOWLEDGEMENTS

The authors wish to thank Melanie Hans and Sascha Prenzel for their help with the preparation of samples. They also greatly appreciate the support of Prof. H.-G. Löhmannsröben and are thankful for access to instrumental equipment. The authors also thank Prof. F.H. Frimmel for the supply of the humic substances HO13 HA and BS1 HA.

REFERENCES

1. Dobbs JC, Susetyo W, Carreira LA, Azarraga LV. Competitive binding of protons and metal ions in humic substances by lanthanide ion probe spectroscopy. *Anal. Chem.*, 1989; 61: 1519-1524.
2. Ryan DK, Weber JH. Fluorescence quenching titration for determination of complexing capacities and stability constants of fulvic acid. *Anal. Chem.*, 1982; 54: 986-990.
3. Saar RA, Weber JH. Comparison of spectrofluorometry and ion-selective electrodes potentiometry for determination of complexes between fulvic acid and heavy metal ions. *Anal. Chem.*, 1980; 52: 2095-2100.
4. Gamble DS, Underdown AW, Langford CH. Copper(II) titration of fulvic acid ligand sites with theoretical, potentiometric, and spectrophotometric analysis. *Anal. Chem.*, 1980; 52: 1901-1908.
5. Wu SL, Horrocks WdeW. General method for the determination of stability constants of lanthanide ion chelates by ligand-ligand competition: laser-excited Eu^{3+} luminescence excitation spectroscopy. *Anal. Chem.*, 1996, 68: 394-401.
6. Smith DS, Kramer JR. Fluorescence analysis for multi-site aluminium binding to natural organic matter. *Environ. Internat.*, 1999; 25: 295-306.
7. Smith DS, Kramer JR. Multi-site aluminium speciation with natural organic matter using multiresponse fluorescence data. *Anal. Chim. Acta*, 1998; 363: 21-29.
8. Bidoglio G, Ferrari D, Selli E, Sena F, Tamborini G. Humic acid binding of trivalent Tl and Cr studied by synchronous and time-resolved fluorescence.

Environ. Sci. Technol., 1997, 31: 3536-3543.
9. Da Silva JCGE, Ferreira MA, Machado AASC, Rey F. Classification of binding sites of Al(III) in fulvic acid extracted from leaf litters and soils by synchronous fluorescence spectroscopy and multidimensional chemometric analysis. *Anal. Chim. Acta,* 1996; 333: 71-82.
10. Thomason JW, Susetyo W, Carreira LA. Fluorescence studies of metal-humic complexes with the use of lanthanide ion probe spectroscopy. *Appl. Spectrosc.* 1996; 50: 401-408.
11. Machado AASC, Da Silva JCGE, Maia JAC. Multi-wavelength analysis of synchronous fluorescence spectra of the complexes between a soil fulvic acid and Cu(II). *Anal. Chim. Acta*, 1994; 292: 121-132.
12. Cabaniss SE. Synchronous fluorescence spectra of metal-fulvic acid complexes. *Environ. Sci. Technol.*, 1992; 26: 1133-1139.
13. Shuman, MS. Dissociation pathways and species distribution of aluminium bound to an aquatic fulvic acid. *Environ. Sci. Technol.*, 1992; 26: 593-598.
14. Seritti A, Morelli E, Nannicini L, Giambelluca A, Scarano G. Fluorescence emission characteristics of naturally occuring organic matter in relation to metal complexation studies. *Sci. Total Environ.*, 1994; 148: 79-81.
15. Susetyo W, Carreira LA, Azarraga LV, Grimm DM. Fluorescence techniques for metal-humic interactions. *Fresenius J. Anal. Chem.*, 1991; 339: 624-635.
16. Cabaniss SE, Shuman MS. Fluorescence quenching measurements of copper-fulvic acid binding. *Anal. Chem.*, 1988; 60: 2418-2421.
17. Ryan DK, Ventry LS, Cabaniss SE, Shuman MS. Exchange of comments on fluorescence quenching measurements of copper-fulvic acid binding. *Anal. Chem.*, 1990; 62: 1523-1526.
18. Stenson AC, Marshall AG, Cooper WT. Exact masses and chemical formulas of individual Suwannee River fulvic acids from ultrahigh resolution electrospray ionization Fourier transform ion cyclotron resonance mass spectra. *Anal. Chem.*, 2003; 75: 1275-1284.
19. Kumke MU, Specht CH, Brinkmann T, Frimmel FH. Alkaline hydrolysis of humic substances – Spectroscopic and chromatographic investigations. *Chemosphere,* 2001; 45: 1023-1031.
20. Förster T. *Fluoreszenz organischer Verbindungen*. Göttingen: Vandenhoeck & Ruprecht, 1951.
21. Valeur B. *Molecular Fluorescence*. Weinheim: Wiley-VCH, 2002.
22. Clegg RM. Fluorescence resonance energy transfer. In: Winefordner JD, Wang XF, Herman B eds. *Chemical Analysis: Fluorescence imaging spectroscopy and microscopy*. New York: Wiley, Volume 137, 1996: 179-254.
23. Abbt-Braun G, Frimmel FH. The relevance of reference materials – isolation and general characterization. In: Frimmel FH, Abbt-Braun G, Heumann KG, Hock B, Lüdemann HD, Spiteller M eds. *Refactory organic substances in the environment*. Weinheim: Wiley-VCH, 2002: 1-54.
24. Kumke MU, Frimmel FH. Stationary and time-resolved fluorescence for humic substances characterization. In: Frimmel FH, Abbt-Braun G, Heumann KG, Hock B, Lüdemann HD, Spiteller M eds. *Refactory organic substances in the environment*. Weinheim: Wiley-VCH, 2002: 215-231.
25. Illenseer C, Löhmannsröben HG, Skrivanek T, Zimmermann U. Laser spectrosopy of humic substances. In: Ghabbour EA, Davies G eds. *Understanding humic substances – Advanced methods, properties and applications*. Cambridge: Royal Society of Chemistry, 1999: 129-145.
26. Provenzano MR, Miano TM, Senesi N. Concentration and pH effects on the

fluorescence spectra of humic acid-like soil fungal polymers. *Sci. Total Environ.*, 1989; 81/82: 129-136.

27. Senesi N, Miano TM, Provenzano MR, Brunetti G. Spectroscopic and compositional comparative characterization of IHSS reference and standard fulvic and humic acids of various origin. *Sci. Total Environ.*, 1989; 81/82: 143-156
28. Schmiedel U, Frimmel FH. Fluorescence of chlorinated fulvic acids. *Vom Wasser*, 1991; 77: 333-348.
29. Senesi N, Miano TM, Provenzano MR. Fluorescence spectroscopy as a means of distinguishing fulvic and humic acids from dissolved and sedimentrary aquatic sources and terrestrial sources. In: Bhattacharji S, Friedman GM, Neugebauer HJ, Seilacher A. Allard B, Boren H, Grimwall A eds. *Lecture notes in earth sciences: Humic substances in the aquatic and terrestrial environment*. Berlin: Springer-Verlag, Volume 33, 1991: 63-73.
30. Patterson HH, Cronan CS, Lakshman S, Plankey BJ, Taylor TA. Comparison of soil fulvic acids using synchronous scan fluorescence spectroscopy, FTIR, titration and metal complexation kinetics. *Sci. Total Environ.,* 1992; 113: 179-196.
31. Pullin MJ, Cabaniss SE. Rank analysis of the pH-dependent synchronous fluorescence spectra of six standard humic substances. *Environ. Sci. Technol.*, 1995; 29: 1460-1467.
32. Ahmad SR, Reynolds DM. Synchronous fluorescence spectroscopy of wastewater and some potential constituents. *Water Res.*, 1995; 29: 1599-1602.
33. Casassas E, Marques I, Rauler R. Study of acid-base properties of fulvic acids using fluorescence spectrometry and multivariate curve resolution methods. *Anal. Chim. Acta*, 1995; 310: 473-484.
34. Mobed JJ, Hemmingsen SL, Autry JL, McGown LB. Fluorescence characterization of IHSS humic substances: total luminescence spectra with absorbance correction. *Environ. Sci. Technol.*, 1996; 30: 3061-3065.
35. Matthews BJH, Jones AC, Theodorou NK, Tudhope AW. Excitation-emission-matrix fluorescence spectroscopy applied to humic acid bands in coral reefs. *Marine Chem.*, 1996; 55: 317-332.
36. Da Silva JCGE, Machado AASC, Silva CSPCO. Simultaneous use of evolving factor analysis of fluorescence spectral data and analysis of pH titration data for comparison of the acid-base properties of fulvic acids. *Anal. Chim. Acta*, 1996; 318: 365-372.
37. Da Silva JCGE, Machado AASC. Procedure for the classification of fulvic acids and similar substances based on the variation with pH of their synchronous fluorescence spectra. *Analyst*, 1997; 122: 1299-1305.
38. Kalbitz K, Geyer S, Geyer W. A comparative characterization of dissolved organic matter by means of original aqueous samples and isolated humic substances. *Chemosphere*, 2000; 40: 1305-1312.
39. Panak P, Klenze R, Kim JI, Wimmer H. A study of intramolecular energy transfer in Cm(III) complexes with aromatic ligands by time-resolved laser fluorescence spectroscopy. *J. Alloys Compounds*, 1995; 225: 261-266.
40. Kim JI, Rhee DS, Wimmer H, Buckau G, Klenze R. Complexation of trivalent actinide ions (Am^{3+}, Cm^{3+}) with humic acid: a comparison of different experimental methods. *Radiochim. Acta*, 1993; 62: 35-43.
41. Wimmer H, Kim JI, Klenze R. A direct speciation of Cm(III) in natural aquatic systems by time-resolved laser-induced fluorescence spectroscopy (TRLFS). *Radiochim. Acta*, 1992; 58/59: 165-171.

42. Horrocks WDeW, Albin M. Lanthanide ion luminescence in coordination chemistry and biochemistry. In: Lippard SJ ed. *Progress in inorganic chemistry*. New York: Wiley, Volume 31, 1984: 1-104.
43. Horrocks WdeW, Sudnick DR. Lanthanide ion luminescence probes of the structure of biological macromolecules. *Accounts Chem. Res.*, 1981; 14: 384-392.
44. Sabbatini N, Guardigli M, Lehn, JM. Luminescent lanthanide complexes as photochemical supramolecular devices. *Coord. Chem. Rev.*, 1993; 123: 201-223.
45. Selvin PR. Principles and biophysical applications of lanthanide-based probes. *Ann. Rev. Biophys. Biomol. Structure*, 2002; 31: 275-302.
46. Horrocks WdeW, Rhee MJ, Snyder AP, Sudnick DR. Laser-induced metal ion luminescence: interlanthanide ion energy transfer distance measurements in the calcium-binding proteins, Parvalbumin and Thermolysin. Metalloprotein models address a photophysical problem. *J. Amer. Chem. Soc.*, 1980; 102: 3650-3652.
47. Snyder AP, Sudnick DR, Arkle VK, Horrocks WdeW. Lanthanide ion luminescence probes. Characterization of metal ion binding sites and intermetal energy transfer distance measurements in calcium-binding proteins. 2. Thermolysin. *Biochem.*, 1981; 20: 3334-3339.
48. Horrocks WDeW, Collier WE. Lanthanide ion luminescence probes. Measurement of distance between intrinsic protein fluorophores and bound metal ions: Quantification of energy transfer between Tryptophan and terbium(III) or europium(III) in the calcium-binding protein parvalbumin. *J. Amer. Chem. Soc.,* 1981; 103: 2856-2862.
49. Breen PJ, Hild EK, Horrocks WDeW. Spectroscopic studies of metal ion binding to Tryptophan-containing parvalbumin. *Biochem.*, 1985; 24: 4991-4997.
50. Mulqueen P, Tingey JM, Horrocks WDeW. Characterization of lanthanide(III) ion binding to calmodulin using luminescence spectroscopy. *Biochem.*, 1985; 24: 6639-6645.
51. Lis S, Elbanowski M, Makowska B, Hnatejko Z. Energy transfer in solution of lanthanide complexes. *J. Photochem. Photobiol. A: Chemistry*, 2002; 150: 233-247.
52. Kumke MU, Abbt-Braun G, Frimmel FH. Time-resolved fluorescence measurements of aquatic natural organic matter. *Acta Hydrochim. Hydrobiol.a*, 1998; 26: 73–81.
53. Kumke MU, Tiseanu C, Abbt-Braun G, Frimmel FH. Fluorescence decay of natural organic matter (NOM) – influence of fractionation, oxidation, and metal ion complexation. *J. Fluoresc.*, 1998; 8: 309–318.
54. Frimmel FH, Kumke MU. Fluorescence decay of humic substances (HS) – A comparative study. In: Davies G, Ghabbour EA eds. *Humic Substances: structure, properties, and uses*. Cambridge: Royal Society of Chemistry, 1998: 113 – 122.
55. Tiseanu C, Kumke MU, Frimmel FH, Klenze R, Kim JI. Time-resolved spectroscopy of fulvic acid and fulvic acid complexed with Eu^{3+} - A Comparative Study. *J. Photochem. Photobiol. A*, 1998; 117: 175–184.
56. Buono-Core GE, Li H, Marciniak B. Quenching of excited states by lanthanide ions and chelates in solution. *Coord. Chem. Rev.*, 1990; 99: 55-87.
57. Wu P, Brand L. Resonance energy transfer: methods and applications. *Anal. Biochem.* 1994; 218: 1-13.

58. Selvin PR. Lanthanide-based resonance energy transfer. *IEEE J. Selected Topics Quantum Electronics*, 1996; 2: 1077-1087.
59. Alpha B, Ballardini R, Balzani V, Lehn JM, Perathoner S, Sabbatini N. Antenna effect in luminescent lanthanide cryptates: a photophysical study. *Photochem. Photobiol.*, 1990; 52: 299-306.

Part 3

METAL BINDING AND MOBILITY: DATA AND THEORY

Chapter 10

PHYSICO-CHEMICAL INTERPRETATION OF COMPLEXATION/BINDING CAPACITY OR EFFECTIVE HUMATE LIGAND CONCENTRATION FOR An(III) AND An(VI) HUMATE COMPLEXATION

Gunnar Buckau

Institut für Nukleare Entsorgung, Forschungszentrum Karlsruhe, P.O. Box 3640, 76021 Karlsruhe, Germany

10.1. INTRODUCTION

Humic acid shows metal ion complexation behavior frequently reported to deviate from that of "normal" organic carboxylic acids. The metal ion complexation of simple organic acids can in general be easily described by the mass action law with the organic acid ligand concentration equal to the molecular concentration of the organic acid and an associated stability constant. The simplest case is where one metal ion reacting with one organic acid molecule results in a reaction of 1:1 stoichiometry, i.e. (1) a metal ion reacts with one reactive molecule, and (2) if a molecule complexes a metal ion this molecule is not available for further reactions. As a result, the total ligand concentration is equal to the total organic acid molecular concentration and the free ligand concentration is equal to the total minus the concentration of complex. In the case of protonation resulting in inaccessibility for metal ion complexation, the degree of protonation is applied for correction of the total and free organic acid concentration. Such a simple approach is not applicable to humic and fulvic acid metal ion complexation. Problems involved are lack of understanding of the protonation function and previous misinterpretation of the mass distribution with far too high molar masses assumed. Furthermore, despite marked progress through access to spectroscopic data in recent years, the binding mode of metal ions and the response of individual humic or fulvic acid molecules to metal ion complexation is not well understood. One important observation, however, is that no significant differences in the metal ion complexes is found for different physico-chemical conditions and different degrees of humate ligand saturated with the respective metal ions.

Humic and fulvic acids are not substances with molecules of a sole composition and molar mass but are a mixture of molecules with a distribution of mass and functionalities. Nevertheless, the metal ion complexation of humic and fulvic acids shows a surprisingly consistent and systematic behavior. The functional behavior towards metal ion complexation of these substances is relatively well defined and falls in a narrow range. For strongly complexing metal ions, the complexation may be described as follows.

- If a humic acid is titrated with the metal ion, progressive complexation occurs. Approaching a certain complex concentration/amount, the complex formation approaches a plateau value. The complexation may be described by a simple interaction isotherm with one interaction constant and one ligand saturation value. Thereby, the concentration of reactive humic acid entities decreases by one unit per complex formed. The observation can be summarized as a relatively simple complexation behavior with a surprisingly narrow distribution in the complexation properties.
- As the plateau value is approached, humic acid flocculates, whereas fulvic acid shows strong increase in light scattering. This behavior is directly comparable with protonation. In both cases the complex shows less hydrophilic properties and can be isolated by sorption on XAD-8 resin (used as a standard isolation method). Sorption on other surfaces is also observed. The observation shows that the hydrophilic character and the charge repulsion of humic and fulvic acids is diminished by protonation as well as by metal complexation.
- With increasing protonation or ionic strength, the plateau value decreases. The problem encountered is that the plateau value corresponding to the total humate ligand concentration does not correlate with the total concentration of proton exchanging groups in a simple fashion. Similarly, the decrease in the plateau value does not strictly follow the decrease in the degree of ionization of humic and fulvic acid proton exchanging functional groups. This is not surprising. Strictly speaking, straightforward use of the degree of ionization for proton exchanging functional groups is only allowed where this corresponds to one complex per ionized functional group and one functional group per reactive entity/molecule. In the case of humic and fulvic acids, there are more than one functional groups per molecule/reactive entity. Combination of the observation that the complexation follows a simple 1:1 interaction isotherm and the decrease in the concentration of reactive entities/molecules with protonation (and to a lesser extent ionic strength) shows that the key questions are:

(i) how does the number concentration of reactive entities/molecules decrease with the progress of protonation (or, what is the concentration of reactive entities at a given progress of protonation (including full ionization)), and

(ii) how does this relate to the decrease in reactive entities by one unit per complex formed?

These questions are the basis for the present paper, where the interaction is discussed of one humic acid (Gohy-573 HA) in one ionic medium (0.1 M $NaClO_4$) with the trivalent Am^{3+} and Cm^{3+} actinide ions, and the hexavalent uranyl ion (UO_2^{2+}). The latter has a comparable charge in the equatorial plane as the homogeneous charge of the Am^{3+} and Cm^{3+} ions and very comparable complexation behavior is observed.

The paper does not go into questions related to the potential for a weaker type of interaction after flocculation of humic acid, nor the potential for a stronger interaction at very low metal ion concentrations. It also does not make strong reference to the observed comparably slow kinetics of the metal ion complexation, especially rearrangements shifting the inventory of complexed metal ions to a slower dissociation mode, and the reason for the comparably slow dissociation with at least two kinetic components. The fundamental and surprisingly simple and systematic complexation properties deduced, however, open up further discussion and specifically tailored investigations.

The main aim of the paper is to show that very complicated approaches frequently proposed, published and used are mainly based on false interpretation of the ligand function (specifically using the concentration of ionized groups instead of the concentration of reactive entities/molecules). This feeds a view of humic and fulvic acids as being an excessively complicated system or even beyond the reach of a systematic description. The key point about this paper is a different and very straightforward, simple interpretation of existing published data. Most of the observations discussed are common knowledge in the scientific community. Consequently, there is a deliberate very restricted use of references. The present paper also reflects a progressive understanding of the mass distribution of humic and fulvic acids applying modern mass spectroscopic techniques in recent years, changing the entire understanding of humic and fulvic acid systems and functions. Last but not least, the present paper is the outcome of discussion with a large number of colleagues in the field. This is reflected in the acknowledgements.

10.2. ELEMENTAL COMPOSITION AND OXYGEN FUNCTIONAL GROUP CONTENT

A brief discussion of elemental composition and functional group content is given. The discussion is focused on numbers and properties important for the purpose of this paper. For more detailed information on distribution of the elemental compositions, including heteroatoms and different origins as well as different functional groups, the literature needs to be consulted.

The elemental composition of humic substances varies within limits. An overall elemental composition is well described by approximately $CO_{0.5}H$, omitting minor contributions of sulfur and nitrogen. Interestingly enough, this overall formula is also obtained by removing one half water molecule per carbon in carbohydrates. The carbon has a formal oxidation state of around zero.

Examples of the content of the most abundant oxygen containing proton exchanging groups are given in Table 10.1 [1,2]. The amounts of proton exchanging functional groups obtained by different operational methods cannot be allocated to specific groups *per se*. Nevertheless, certain types of functional groups are frequently reported by the use of these operational methods. The Gohy-573 HA at the bottom of Table 10.1 is the one for which complexation data are discussed in this paper. The numbers should be compared also to the "total proton exchange capacity" by potentiometric pH titration of 6.40 meq/g for Gohy-573 HA (cf. Figure 10.1). The latter refers to the assumption that no further deprotonation occurs on increasing pH beyond about 10.25. There is an excellent agreement between the Ba exchange capacity at high pH and the total proton exchange capacity by potentiometric pH titration. The latter number is only about 3 % lower than the former one. This is on the order of experimental uncertainty. The numbers show that the total proton exchange capacity can be correctly evaluated, whereas allocation of parts of this inventory to different types of groups is questionable.

Table 10.1 Proton exchange capacities by different operational methods with groups frequently reported to be associated with these methods: "total" (Ba exchange capacity under alkaline conditions), and "carboxylic group content" by "Ca-acetate" (Ca exchange capacity in acetate buffer medium), "radiometric" (attachment of ^{14}C labeled methyl groups via diazomethane), and "potentiometric pH titration" (potentiometric pH titration, where the proton exchanging groups are quantified as those protonated at and below titration curve turning point slightly below pH 8)

Humic Acid	Ba-Exchange "Total"	Ca-Exchange "Carboxylic"	Radiometric "Carboxylic"	Pot. pH Titration "Carboxylic"
Aldrich HA [2]				
meq/g	7.12	4.41	3.9	5.06
% of "total"	(100)	62	55	71
Aldrich HA [3]				
meq/g	7.06	4.80	-	5.43
% of "total"	(100)	68	-	77
Gohy-573 HA [3]				
meq/g	6.61	4.75	-	5.38
% of "total"	(100)	72		81
	Average (%)	67 ± 5	55	76 ± 5

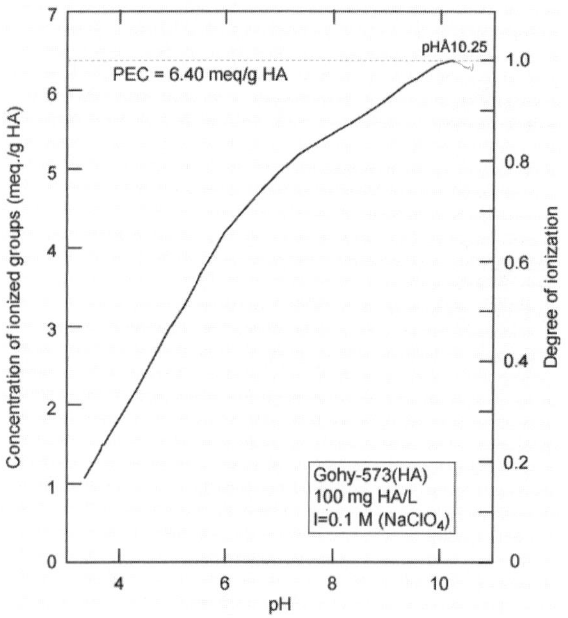

Figure 10.1 Potentiometric pH titration of the Gohy-573 humic acid

The total exchange capacities of two batches of Aldrich humic acid (Aldrich HA) independently purified by two different groups are basically indistinguishable.

The total proton exchange capacity is about 7 % higher than for Gohy-573 HA. This shows that careful handling of humic acid leads to high reproducibility for the functional group content and that there are differences between humic acids from different sources. More important is that the presumed contents of carboxylic groups deduced from different operational methods deviate from each other. This is obvious from Figure 10.1, showing that one cannot identify a sharp distinction between different proton exchanging groups along with progressive protonation. This should be kept in mind when the proton exchange behavior is represented by a distinct number of "acidic groups" from fitting the curve in Figure 10.1. It is especially surprising when pK_a values are allocated to a limited number of "proton exchanging groups", including in the lower pH range. Potentiometric titration does not allow going much further into the acidic range due to lowering of the electrode response to proton ion concentration/activity and thus the protonation endpoint cannot be determined.

Further important information is the number of additional oxygen functional groups because of their relevance both for the hydrophilic character of humic and fulvic acids as well as their key role in complexing metal ions. The distribution of oxygen can be estimated from the data in Table 10.1. About two thirds of the proton exchanging groups can be estimated to be of carboxylic group character and about one third of phenolic character. Using direct titration data for quantification of carboxylic groups, the fraction of the carboxylic groups becomes slightly higher. With the radiometric method, the carboxylic fraction is evaluated to be somewhat lower. With the average elemental composition given above, omitting heteroatoms (sulfur and nitrogen) and the very light hydrogen, there are 50 mmol $CO_{0.5}$ units per gram of humic acid (50 mmol C and 25 mmol O per gram). With 4.7 mmol/g humic acid carboxylic functional groups and 2.3 mmol/g humic acid phenolic groups (2/3 carboxylic and 1/3 phenolic of 7 meq/g total (cf. Table 10.1)), there is 11.7 mmol per gram humic acid oxygen bound in these functional groups. This leaves somewhat more than half of the oxygen bound in other forms than these proton exchanging groups (13.3 mmol/g). With one carbon atom per proton exchanging functional group, there is 43 mmol C per gram in other structures. This means that there is about one oxygen atom per three carbon atoms that are not bound in proton exchanging groups.

In summary, the hydrophilic character of humic and fulvic acid is due to the high oxygen functional group content. There is around 7 mmol/g proton exchanging groups. Protonation of these functional groups leads to flocculation of humic acid and an increase in light scattering of fulvic acid. Because of the high content of oxygen in non-proton exchanging groups, however, complete loss of hydrophilic character does not occur even at full protonation of the proton exchanging groups.

10.3. MASS DISTRIBUTION

The mass distribution of Aldrich humic acid and humic and fulvic acids from the groundwater Gohy-573 is shown in Table 10.2. The concentration of molecules in solution if no association occurs is given by the number average distribution maximum (M_n). It should be noted that the mass distribution as given in Table 10.2 represents numbers that are slowly becoming generally accepted. This is important because of the previous perception of humic and fulvic acid molecules as having much higher mass distribution numbers from misinterpretation of size measurements by gel permeation/size exclusion chromatography and field flow fractionation methods [3]. An important number for metal ion complexation that can be deduced is the average number of proton exchanging functional groups per average molecule. For Gohy-573 HA, with 6.61 meq/g of proton exchanging functional groups (cf. Table 10.1) and an number average molar mass of 521 Da (cf. Table 10.2), the average number of proton exchanging functional groups per average molecule is 3.44.

For a distribution of around one third of phenolic character and two thirds of carboxylic character, an average of 2.3 carboxylic and 1.2 phenolic groups per molecule is obtained. Under the assumption that the proton exchanging functional group content is equally distributed with mass for molecules of different mass, about one proton exchanging group per molecule is found at the lower end of the mass distribution. At the upper end of the mass distribution, the corresponding number is less than 20. Because the O/C ratio decreases with increasing molar mass [4], the number of proton exchanging functional groups on the molecules with higher mass may be expected to be lower. Given the relatively narrow mass distribution and the above discussed uncertainty in the allocation of different nature of the proton exchanging groups, the most important number to carry over for further discussion is the around 3.4 proton exchanging groups per average molecule.

Table 10.2 Calculated parameters of the distributions M_{conc} *vs.* mass (for details, see Wolf et al. [3])

Humic substance	Mass range (Da)	M_n (Da)	M_w (Da)	M_w/M_n
Gohy-573 FA	150 – 3000 [5000]	428 [428]	609 [613]	1.42 [1.43]
Gohy-573 HA	150 – 3000 [5000]	521 [523]	752 [759]	1.44 [1.45]
Aldrich HA	150 – 3000 [5000]	445 [446]	626 [627]	1.40 [1.41]

10.4. BINDING CAPACITY AND EFFECTIVE HUMATE LIGAND CONCENTRATION (EHLC)

The plateau/saturation value for complexation is frequently called the binding capacity or complexation capacity. It represents the effective humate ligand concentration (EHLC) under given physico-chemical conditions. As mentioned above, application of this EHLC results in a simple and consistent description of the complexation reaction. In this section the relation between the EHLC and the nature of humic and fulvic acids is discussed.

10.4.1. Impact of pH and Ionic Strength

As mentioned in the Introduction, titration of humic or fulvic acid with a complexing metal ion results in a progressive increase in the amount of metal humate complex until a plateau value is reached. Approaching this plateau/saturation value, flocculation of humic acid takes place. One example is shown in Figure 10.2, where the dissolved and flocculated Am(III) humic acid complex are shown together as AmHA [5]. In Figure 10.3, these plateau/saturation values are shown for the UO_2^{2+} and Am^{3+} complexation with Gohy-573 HA as a function of pH in 0.1 M $NaClO_4$ (the same ionic medium as used for potentiometric pH titration in Figure 10.1) [6-8]. The plateau/saturation value decreases with increasing protonation of the humic acid. At pH 3 this value is only about 11 % of what is found at pH 6. At higher pH values the Am^{3+} ion shows hydrolysis and thus direct evaluation is not possible.

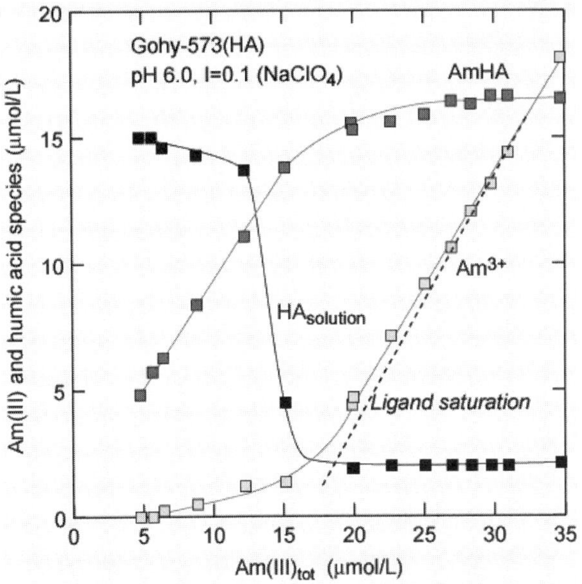

Figure 10.2 Titration of the humic acid Gohy-573 HA with the Am^{3+} ion at pH 6.0 in 0.1 M $NaClO_4$. The humate ligand saturation is given by AmHA approaching a constant value with a corresponding increase in the non-complexed Am^{3+} ion and an increase in the total Am(III) concentration. Approach to the humate ligand saturation is preceded by the flocculation of AmHA (decrease in $HA_{solution}$)

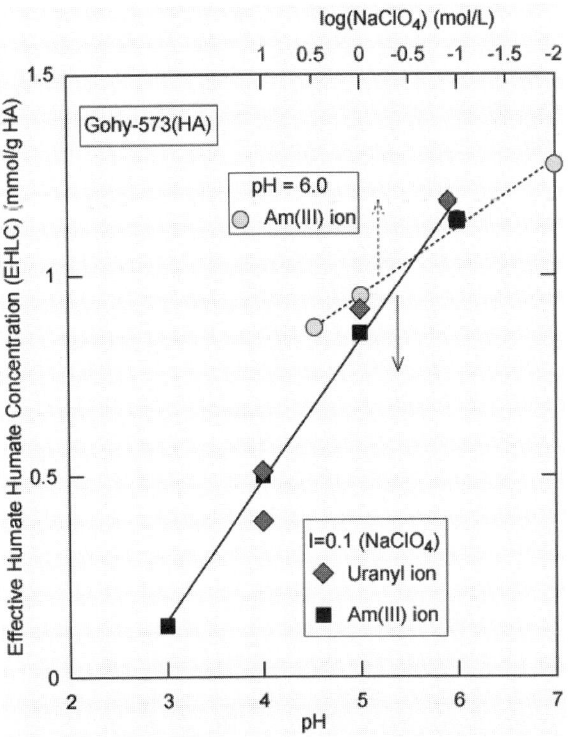

Figure 10.3 Effective humate ligand concentration (EHLC) of the uranyl (UO_2^{2+}) and Am^{3+} ions with the humic acid Gohy-573 HA as a function of pH (in 0.1 M $NaClO_4$) (lower scale), and as a function of ionic strength (at pH = 6.0) (upper scale). The EHLC values are given by the total amount of Am humic acid complex at humate ligand saturation (cf. Figure 10.2) under these different physico-chemical conditions

The data for the uranyl ion are directly measured at pH 4. At higher pH values hydrolysis occurs. Data above pH 4 are deduced from interaction data below the saturation value by making use of an invariable interaction constant and accounting for the impact of hydrolysis. The straight line in the figure may not be understood as a functional relation, especially as the plateau/saturation value then would fall below zero around pH 2.6. It may, however, be used for description of the data in the concerned pH range (see below).

In Figure 10.3, the impact of ionic strength on the plateau/saturation value is shown for a pH value of 6.0. For comparison with the impact of pH, a logarithmic scale is used (upper scale of Figure 10.3). The principal impact of ionic strength is the same as for pH. However, the magnitude of the impact is lower. This shows that the accessibility of humic acid (that is, the concentration of reactive entities) decreases both with increasing protonation and also with increasing cation concentration of the ionic medium. These results lead to the question of how protonation correlates with the effective humic acid ligand concentration (the plateau/ saturation value).

10.4.2. Relation between EHLC and pH

The relation between the EHLC with Am^{3+} and UO_2^{2+} and the protonation function of the humic acid Gohy-573 HA in 0.1 M $NaClO_4$ is shown in Figure 10.4. Without the intention of allocating functional relationships between pH and EHLC as well as the concentration of ionized groups, the respective data are represented by straight lines. The surprising finding is a ratio between the concentration of ionized groups and EHLC of 3.41 (meq/mol). This should be compared with the number of 3.44 proton exchanging functional group per average molecule (see above). The outcome is that the EHLC decreases by one unit/molecule for the protonation of the same amount of proton exchanging groups as found per average molecule. As a consequence, the EHLC can be written as given in Eq. 10.1, where PEC is the total proton exchange capacity, n is the average number of proton exchanging groups per molecule and α is the degree of ionization.

$$EHLC = PEC/n\ \alpha \qquad (10.1)$$

In order to visualize this function, for simplicity a number of three functional groups per molecule is used. Furthermore, the weaker impact of ionization is ignored. The simplest approach is starting with three molecules, each with three ionized groups. With three-protonations occurring on one molecule, there are two reactive molecules left over for metal ion complexation (each with three ionized groups per molecule). This is in agreement with the experimental findings. The second interpretation would require a random protonation where the ligand concentration is a function of the probability of encountering an ionized group on a specific molecule. As the most probable random distribution case, three protonation takes place with one protonation per molecule. In this case the interaction probability with each molecule is 2/3 multiplied by 3 (molecules), resulting in the same EHLC (namely 2). The problem with the latter approach is that the EHLC is a number measured via the total concentration of metal ion complexes formed at a specific pH (and ionic strength), and the free ligand concentration is decreased by one unit for each metal ion complex formed. This follows from the simple reaction description discussed above (see also Eq. 10.4).

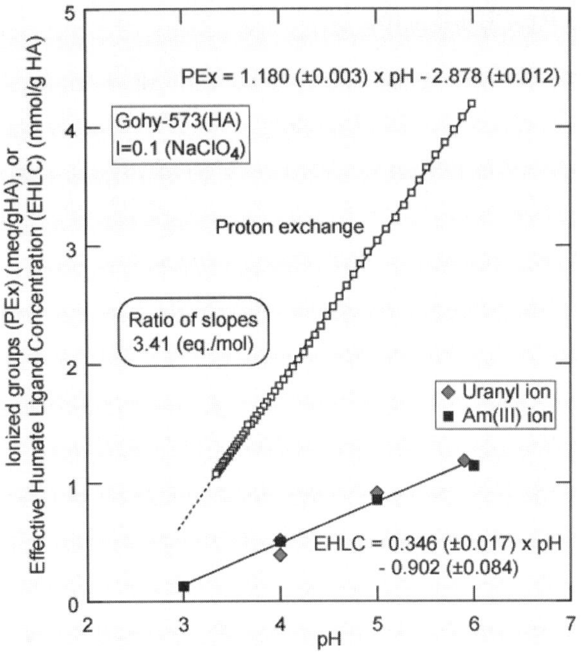

Figure 10.4 Variation with pH of experimental data and representation of these data by linear regression for the concentration of ionized groups and EHLC for the UO_2^{2+} and Am^{3+} ions with the humic acid Gohy-573 HA in 0.1 M $NaClO_4$. The ratio between the slopes of linear regression is 3.41 (meq/mol).

$$[HA]_{free} = EHLC - MHA \qquad (10.2)$$

Combination of Eqs. 10.1 and 10.2 leads to Eq. 10.3.

$$[HA]_{free} = PEC/n\,\alpha - MHA \qquad (10.3)$$

This implies that for each complex formed, the number of reactive molecules decreases by one unit. If, however, there are three molecules each with an interaction probability of 2/3, formation of a complex with one molecule would leave two molecules each with 2/3 metal ion interaction/complex formation probability. Consequently, the free humic acid ligand concentration after three protonations and one metal ion complexation would be 4/3 and not 1 as experimentally found and also correctly postulated from the first approach with non-random protonation. In the non-random protonation case the three protonations would lead to a decrease in one ligand unit, and the metal ion complexation with one of the remaining molecules each with three ionized groups would lead to a decrease in the free humic acid concentration by another ligand unit. This is in agreement with Eq. 10.3 and reflects the experimental findings.

A final conclusion on the possibility to use a refined random protonation approach requires further data evaluation. The simple approach of preferred consecutive protonation following the initial random one is very tempting, reflects experimental findings and would be supported by neighboring impacts from the first protonation changing the overall functional groups behavior of the residual ionized ones, making them more prone to protonation. Direct evidence for such an approach, however, is still pending. Furthermore, the impact of variation of ionic strength is still not well understood.

10.4.3. Extrapolation to Full Ionization

In the range of metal ion hydrolysis, the ligand saturation plateau values cannot be directly measured. Therefore, numbers for the EHLC of Am^{3+} beyond pH 6 need to be indirectly deduced. In Figure 10.5, the relationship between the EHLC and the protonation behavior is extrapolated to full ionization, i.e. to a pH of 10.25. For this extrapolation the constant ratio between EHLC and the protonation of 3.41 meq/mol is used (cf. Figure 10.4). This extrapolation is not based on direct experimental evidence, but it leads to a fully consistent description of the system and the complexation process. The reason is that the EHLC at full ionization becomes 1.88 mmol/g humic acid. The inverse value of this number is 533 g/mol humic acid. This is basically identical with the number averaged molar mass of individual molecules. This means that the EHLC at full ionization is simply the number concentration of humic acid molecules. Starting from this point, the process of protonation and complexation is then described by Eq. 10.3, namely the decrease in humic acid ligand concentration by one unit (inactivation of one molecule) for consecutive protonation of its ionized groups, followed by a decrease by one ligand unit per complex formation (inactivation of one molecule as a consequence of one complex formation). In this description, the weaker impact of ionic strength is omitted.

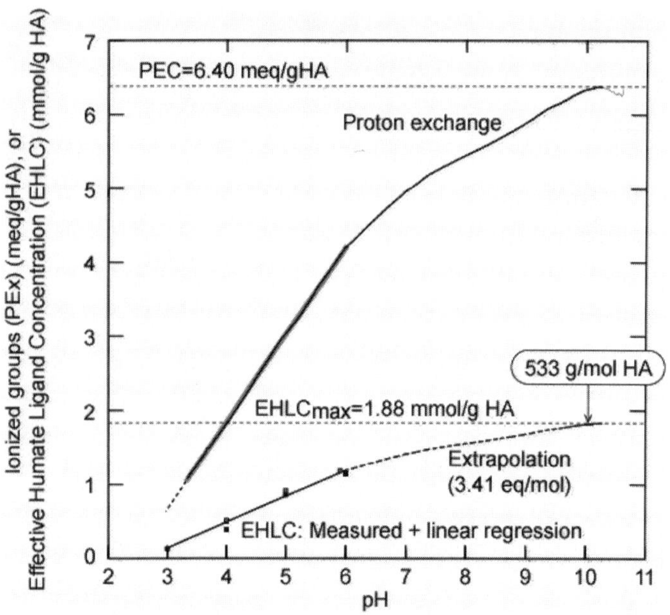

Figure 10.5 Extrapolation of the concentration of ionized groups and EHLC for the UO_2^{2+} and Am^{3+} ions for pH between 6 and 10.25 by a constant ratio of 3.41 meq/mol (cf. Figure 10.4). The maximum value for EHLC of 1.88 mmol/g HA is achieved at pH 10.25. This corresponds to a molar mass of the humate ligand of 533 g/mol (inverse value of $EHLC_{max}$) and is basically identical with the number averaged mass distribution value from mass spectroscopy by TOF-SIMS (cf. Table 10.2)

10.4.4. Interaction Constant

The interaction process is described by Eq. 10.4 with $[HA]_{free}$ given by Eqs. 10.2 or 10.3.

$$ß = [MHA]/[M_{free}][HA]_{free} \qquad (10.4)$$

Equation 10.4 can be rearranged making use of the degree of saturation (X) as [MHA]/EHLC. Plotting log(X/(1-X)) against log([M_{free}]) results in a straight line with unit slope and one specific interaction constant [9]. In Figure 10.6, this is done for a large number of complexation data for Am^{3+} and the analogous Cm^{3+} ion with various samples of Gohy-573 humic acid. The ionic strength varies from 0.001 to 5 mol/L ($NaClO_4$) at a constant pH of 6.0. The EHLC is experimentally determined for each condition and system. The data fall on one single straight line with the slope of unity. The same is found for variation of pH using the experimentally determined EHLC in Eq. 10.2. This shows that the interaction process for an individual complex formation is basically invariant with respect to the ionic strength and pH. Furthermore, the complex formation strength is invariant with respect to the degree of humic acid ligand saturation (X) for values between about 0.01 and basically full saturation ([MHA] > EHLC) (right hand scale in Figure 10.6). This is also reflected in the invariant shape of absorption spectra of the complexed Am^{3+} ion in the upper approximately 90 % of ligand saturation.

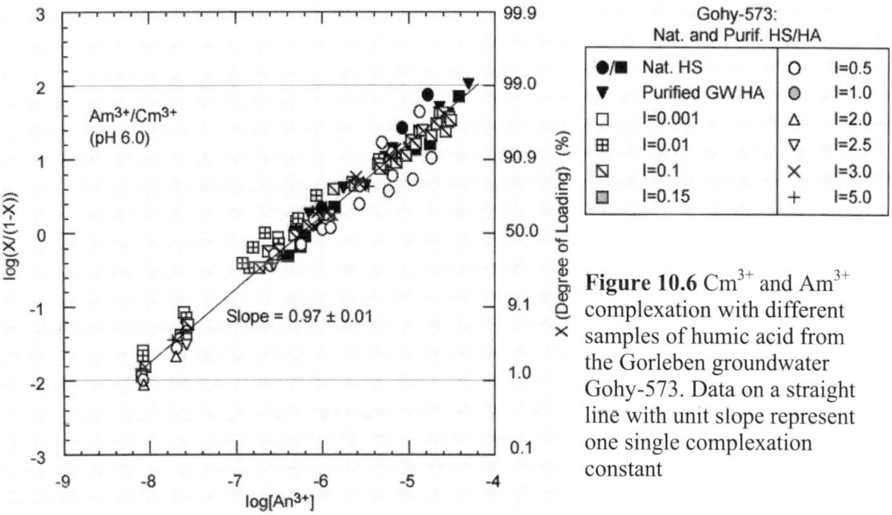

Figure 10.6 Cm^{3+} and Am^{3+} complexation with different samples of humic acid from the Gorleben groundwater Gohy-573. Data on a straight line with unit slope represent one single complexation constant

10.5. SUMMARY, CONCLUSIONS AND OUTLOOK

The metal ion complexation of humic and fulvic acid is discussed. The paper restricts itself to the strongly complexing trivalent Am and Cm actinide ions and equally strongly complexing hexavalent uranyl ion (UO_2^{2+}). The reason for selecting these three ions is that a large number of complexation data are available based especially on spectroscopic investigations with one and the same batch of a well characterized humic acid. It is shown that, contrary to frequently published "models" relying on a large number of fitting parameters, the complexation can be described by an effective humate ligand concentration varying with pH and ionic strength and one single complexation constant. This is not a new concept and it has been used by some groups for a long time. The trust in this simple approach, however, has been hampered by the lack in understanding of the experimentally determined effective humate ligand concentration.

The present paper shows that the variation in the ligand behavior of humic acid can be easily related to the proton exchange behavior and the molar mass distribution. This provides trust in the simple approach and also verifies the very narrow distribution in complexation functional properties of humic and fulvic acids.

ACKNOWLEDGEMENTS

The present paper is based on a general development of mass spectroscopic analytical techniques showing that humic and fulvic acids have much lower mass distribution numbers than previously thought. Towards the end of the nineties it became obvious that humic and fulvic acid molecules have their mass distribution maximum in the range of around 500 mass units. The community, however, has only partly recognized this in very recent years. The author would like to acknowledge the many researchers that have worked against the accepted incorrect views of the mass distribution and provided this important key information, especially M. Wolf and W. Szyamczak (GSF Research Center, Germany). The basis for the present paper started at a meeting 10-11 January 2002, at the home institution of the author with the following participants (in addition to the author): Melissa Denecke (same institution as the author), Nick Bryan (University of Manchester), Cooper Langford (University of Calgary), Willem van Riemsdijk (Wageningen University) and Ulf Wahlgren (Albanova, Stockholm). Throughout the past years the issue has also been discussed with I. Pashalidis (University of Cyprus), providing experimental data and ideas on the complexation behavior in the upper pH range. J.I. Kim (same institution as the author) has provided continuous support with discussions, infrastructure resources and coworkers required for this work for more than two decades. The author would like to acknowledge the input of all these persons as a prerequisite for the present paper. Finally, the European Commission is acknowledged for support of this research since about the mid-eighties, including the most recent project "Humic substances in performance assessment of nuclear waste disposal: Actinide and iodine migration in the far-field" (HUPA, EC Contract No. FIKW-CT-2001-00128).

REFERENCES

1. Schmeide K, Zänker H, Heise KH, Nitsche H. Isolation and characterization of aquatic humic substances from the bog "Kleiner Kranichsee". In: Buckau G ed. *Effects of humic substances on the migration of radionuclides: Complexation and transport of actinides, first technical progress report.* Report FZKA 6124, 1998: 161.
2 Kim JI, Buckau G, Li GH, Duschner H, Psarros N. Characterization of humic and fulvic acids from Gorleben groundwater. *Fresenius J. Anal. Chem.*, 1990; 338: 245.
3. Wolf M, Buckau G, Geckeis H, Ngo Man T, Hoque E, Szymczak W, Kim JI. Aspects of measurement of the hydrodynamic size and molecular mass distribution of humic and fulvic acids. In: Ghabbour EA, Davies G eds. *Humic*

substances: Structures, models and functions. Cambridge: Royal Society of Chemistry, 2001: 51

4. Plancque G, Amekratz B, Moulin V, Toulhoat P, Moulin Ch. Molecular structure of fulvic acids by electrospray with quadrupole time-of-flight mass spectrometry. *Rapid Commun. Mass Spectrom.*, 2001; 15: 827.
5. Kim JI, Buckau G, Bryant E, Klenze R. Complexation of americium(III) with humic acid. *Radiochim. Acta*, 1989; 48: 135.
6. Czerwinski KR, Kim JI, Rhee DS, Buckau G. Complexation of trivalent actinide ions (Am^{3+}, Cm^{3+}) with humic acid: The effect of ionic strength. *Radiochim. Acta*, 1996; 72: 179.
7. Montavon G, Mansel A, Seibert A, Keller H, Kratz JV, Trautmann N. Complexation studies of UO_2^{2+} with humic acid at low metal ion concentrations by indirect speciation methods, *Radiochim. Acta*, 2000; 88: 17.
8. Higgo JJW, Kinniburgh D, Smith B, Tipping E. Complexation of Co^{2+}, Ni^{2+}, UO_2^{2+} and Ca^{2+} by humic substances in groundwaters. *Radiochim. Acta*, 1993; 61: 91.
9. Buckau G, Kim JI, Klenze R, Rhee D-S, Wimmer H. A comparative study of the fulvate complexation of trivalent transuranic element ions. *Radiochim. Acta*, 1992; 57: 105.

Chapter 11

PHYSICAL AND CHEMICAL KINETICS IN HUMIC DISPERSIONS

Martina Klučáková and Miloslav Pekař

Institute of Physical and Applied Chemistry, Faculty of Chemistry, Brno University of Technology, Purkyňova 118, 612 00 Brno, Czech Republic

11.1. INTRODUCTION

Kinetics plays a very important role in nature and our life. Humic substances (HSs), as the indispensable element of the environment, are governed by kinetics in interesting ways. In coal and lignite, HSs are stored through many years and centuries, whereas soil HSs take part in the incessant "life cycle" of plant nutrition and humification. The functions of soil are impossible without organic matter. Organic substances that constitute living plants are synthesized and maintained by a vital force. They are unstable after the death of the plant [1] and can be transformed to HSs (humified). On the other hand, the birth and growth of living plants is not possible without organic matter in soil. Thus, HSs in soil are continuously renewed and the kinetic rate constants of these processes are relatively high. But in coal and lignite, HSs sleep. It is similar to waiting for their opportunity, as long as material can be utilized in appropriate ways (which could not be combustion in this case). Coal or lignite is a kind of reservoir. It is a container in which the kinetic rate is very low. It is surprising that the humic acids (HAs) isolated from a dynamic soil system and a stationary coal system are so similar. Nature is a mighty enchantress. It can keep required amounts of HSs in two different systems under very different conditions in various ways.

When the conditions around coal or lignitic HAs are changed and they can take their opportunity, they are able to behave very similarly to the behavior of soil HSs. Added to soil, lignitic HAs are able to play the same role in plant nutrition or immobilization of heavy metals. Although HSs structure is strongly influenced by conditions of their birth, the general properties of HAs are the same. It is promising for life because lignite is a very abundant source of HAs that can be utilized for soil revitalization. Now, let us consider a little more precisely the interaction of HAs with metal ions because one of the most important roles of HAs in soils is transport of nutritional elements and immobilization of toxic metals.

Our approach stresses the role of both chemical and physical kinetics in the functioning of HSs. Although kinetic studies are not ignored within humic research, they are not so numerous as various structural and spectroscopic investigations of equilibrium states. We believe that transport properties and rates of interactions, either covalent or non-covalent, of HSs are as important for evaluating and understanding their role in natural systems and human-driven applications as solving structural questions. Despite their mixture character, uncertainties in structure, ongoing discussion of polymeric or supramolecular-aggregate essence, HSs are definitely colloids. Thus, besides their molecular character, their particular colloidal state may significantly affect their behavior and properties. Indeed, in natural systems HSs may occur in different colloidal states.

In this work we compare kinetic behavior of lignitic HAs in three different colloidal states – sol, gel and suspension (of solid humics). Mobility of molecular entities decreases from the sol to the suspended state. Humic sols can be expected to occur preferably in water systems and to have all or most of their functional or reactive groups free for interactions. Humic gels might be found in water sediments, swelled peat, soil or coal. Some of their functionalities should be engaged in forming the gel structure, but the whole system including its interior is still accessible to various interacting particles, penetrable through the gel network. In suspensions, humic constituents are tightly bonded in the solid fraction and essentially only the outer surface (including the surface of pores) is accessible to intermolecular interactions. The suspended state can appear in unswollen soil or water streams. Copper(II) was selected as a model interacting metal in this study.

Many authors [2-8] determined copper as a metal with very high affinity for HSs. Its bivalent ions can form very stable and strong complexes with HA. EPR study showed that Cu^{2+} ions at low concentrations coordinate strongly on binding sites with three or four oxygen donor atoms. At higher ratios of copper to HA the weaker sites are gradually occupied and water remains the highly probable ligand in these complexes. Experimental results correspond with interaction energies computed for various types of Cu-HA complexes by quantum chemistry calculations [9].

11.2. THEORY

In this section, theoretical equations for description of either chemical or physical kinetics in all three colloidal systems are collected. Inspired by Occam's razor, we tried to use as simple a model as possible but still be able to describe peculiarities in the data obtained.

11.2.1. System A - Solid HA

Mathematical description of adsorption kinetics was derived on the basis of the well-known Langmuir isotherm, Eq. 11.1, where a, a_{max}, b and c are the adsorbed amount for a given concentration, the maximum adsorbed amount, the adsorption coefficient and the equilibrium concentration of cupric ions, respectively.

$$a = a_{\max} \frac{bc}{1 + bc} \tag{11.1}$$

Langmuir derived Eq. 11.1 on the basis of kinetic ideas: at equilibrium the adsorption rate r_{ads} is equal to the desorption rate r_{des} and the total rate r is equal to zero, Eq. 11.2, where k_{ads}

$$r = r_{ads} - r_{des} = k_{ads}\, c\left(1 - \frac{a}{a_{\max}}\right) - k_{des} \frac{a}{a_{\max}} \tag{11.2}$$

and k_{des} are the rate constant of adsorption and desorption, respectively. To solve Eq. 11.2, all concentrations have to be transformed to concentration of Cu^{2+} ions according to material balance. If the initial concentration of Cu-HA complex is equal to zero, its final form is given by Eq. 11.3, where t is time, and x_0 and y_0 are initial concentrations of metal ions and active

$$-\frac{dc}{k_{ads}\, dt} = c^2 + c\left(y_0 - x_0 + \frac{1}{b}\right) - \frac{x_0}{b} \tag{11.3}$$

centers in HA particles, respectively. The adsorption coefficient $b\left(=k_{ads}/k_{des}\right)$ can be determined from the Langmuir adsorption isotherm (Eq. 11.1). At equilibrium, Eq. 11.3 is quadratic with following coefficients:

$$\alpha = 1, \quad \beta = y_0 - x_0 + \frac{1}{b}, \quad \text{and} \quad \gamma = -\frac{x_0}{b}$$

If c_{eq} is the equilibrium concentration of cupric ions, Eq. 11.3 can be modified to linear form Eq. 11.4. From the slope of the time dependence of fraction ϕ_A we can determine k_{ads}. The value of k_{des} can be calculated from the equilibrium measurements (adsorption coefficient b) [10].

$$\phi_A = \ln \frac{c + c_{eq} + \frac{\beta}{\alpha}}{c - c_{eq}} = \left(k_{ads}\sqrt{b + 4\alpha\gamma}\right) t + const \tag{11.4}$$

Adsorption of cupric ions on HA is a complex process including three main stages:

- transport of Cu^{2+} ions to humic solid,
- diffusion of Cu^{2+} ions through humic particle to active centres,
- formation of bonds between Cu^{2+} ions and HA.

Transport from the liquid phase to solid particles is strongly influenced by stirring. In the majority of cases its rate is much higher than the rate of the two other processes and therefore can be usually neglected. The kinetics of diffusion and intrinsic interaction in copper-HA are analyzed in a later section.

11.2.2. System B – Humic Gel

Basic diffusion theory applied to this situation can be found in detail in our previous work [11,12]. A simple cylindrical arrangement corresponding to a one-dimensional semi-infinite system is used in this study and so the well known Fick's law for non-stationary diffusion is valid in this case (Eq. 11.5), where c_g is the concentration of Cu^{2+} ions at time t and diffusion distance x in a humic gel, and D_g is the diffusion coefficient of Cu^{2+} ions in the humic gel.

$$\frac{\partial c_g}{\partial t} = D_g \frac{\partial^2 c_g}{\partial x^2} \tag{11.5}$$

From Eq. 11.5 the following relation between the total diffusion flux (m) and concentration of diffusing ions at the interface gel/solution (c_s) is obtained:

$$m = 2c_s \sqrt{\frac{D_g t}{\pi}} \tag{11.6}$$

From the slope of the flux-time straight line, the diffusion coefficient D_g can be obtained. Similarly, Eq. 11.7 for the concentration profile of diffusing ions in a humic gel was derived [12].

$$\frac{c_g}{c_s} = erfc \frac{x}{\sqrt{4D_g t}} \tag{11.7}$$

Equations 11.5 and 11.6 are valid only for diffusion from solution with constant concentration (a continuously saturated solution was used in previous work [11,12]).

In the case of a solution with a changing concentration (due to the diffusion process), the system has to comply with initial and boundary conditions shown in Table 11.1.

Table 11.1 Initial and boundary conditions for diffusion from time-variable sources

$t = 0$	$x \in \langle 0; L \rangle$	$c_g = 0$
$t = 0$	$x < 0;\ x > L$	$c = c_0$
$t > 0$	$x = 0$	$D\ \partial c / \partial t = D_g\ \partial c_g / \partial t$
$t > 0$	$x = L/2$	$D_g\ \partial c_g / \partial t = 0$
$t \to \infty$	$x = 0$	$\varepsilon = c_g / c$

For diffusion from the time-variable solute concentration the following equations can be derived [13],

$$m = \frac{2\varepsilon c_0}{1+\varepsilon\sqrt{D_g / D}} \sqrt{\frac{D_g\, t}{\pi}} \tag{11.8}$$

and

$$c_g = \frac{\varepsilon c_0}{(1+\varepsilon\sqrt{D_g / D}}\, erfc \frac{x}{\sqrt{4 D_g\, t}} \tag{11.9}$$

where ε is the ratio of the ion concentration in the gel and in the solution at final equilibrium (that is, at the "end of diffusion"), c_0 is the initial solution concentration and D is the diffusion coefficient of cupric ions in solution.

The "chemical" interactions of cupric ions with humic gel were studied by potentiometry and conductometry. If chemical bonds between HA and cupric ions are realised by carboxylic and phenolic groups, system acidity and conductivity have to increase during the complexation. The summary reaction scheme can be written as Eq. 11.10,

$$Cu^{2+} \; + \; H_2R \; \rightarrow \; CuR \; + \; 2H^+ \tag{11.10}$$

where H_2R is a binding site for a bivalent ion in the HA structure. This simple scheme is a second order reaction in which the reverse reaction can be neglected due to the high stability of Cu-HA complexes. The following kinetic equation is valid for this reaction type, where k_2

$$\frac{dz}{dt} = k_2 \left(y_0 - z\right)\left(c_0 - z\right) \tag{11.11}$$

is the rate constant of the second order reaction and z is the concentration of CuR complexes. The integrated form of Eq. 11.11 is

$$\frac{1}{y_0 - c_0} \ln \frac{c_0 \left(y_0 - z\right)}{y_0 \left(c_0 - z\right)} = k_2\, t \tag{11.12}$$

For conductometric and potentiometric data, Eq. 11.12 was modified to the following forms [11,12], where G_0, G_t and G_{eq} are the initial conductivity, conductivity at time t and at equilibrium, respectively and where Δh is the increase of concentration of H^+ ions in a given time ($\Delta h = h_t - h_0$). According to Eq. 11.10, Δh is directly proportional to the amount of formed complexes CuR. The rate constant k_2 of the second order reaction was marked as k_B and k_C in modified Eqs. 11.13 and 11.14.

Mathematical description of diffusion with chemical reaction was derived in our previous work [11,12]. It was shown that the chemical kinetics of copper-HA interactions are complex and cannot be directly included in Fick's law (Eq. 11.5). The following simple equation 11.15 is valid under conditions where the immobilizing reaction proceeds very rapidly compared with the diffusion process [13]

$$k_{\mathrm{B}}t + const = \frac{1}{y_0 - c_0} \ln \frac{3G_0 - G_{\mathrm{t}}}{G_{\mathrm{eq}} - G_{\mathrm{t}}} = \phi_{\mathrm{B}} \qquad (11.13)$$

$$k_{\mathrm{C}}t = \frac{1}{y_0 - c_0} \ln \frac{y_0\left(c_0 - \frac{\Delta h}{2}\right)}{c_0\left(y_0 - \frac{\Delta h}{2}\right)} = \phi_{\mathrm{C}} \qquad (11.14)$$

$$\frac{\partial c_{\mathrm{g}}}{\partial t} = D_{\mathrm{ef}} \frac{\partial^2 c_{\mathrm{g}}}{\partial x^2} = \frac{D_{\mathrm{g}}}{\kappa + 1} \frac{\partial^2 c_{\mathrm{g}}}{\partial x^2} \qquad (11.15)$$

Equation 11.15 is formally similar to Fick's law (Eq. 11.5) and can be solved in the same way. Here, D_{ef} is the effective diffusion coefficient, which includes the influence of chemical reaction, and κ is the proportionality constant between free (diffusionable) ions and immobilized ones [11,12].

11.2.3. System C – Humic Sol

For more detailed study of the complexation mechanism, we decided to study interactions of cupric ions with HA in dissolved form – a humic sol. At first we assumed reaction 11.10 and appropriate kinetic equations for these interactions. But it was found [14] that our kinetic data do not correspond to this simple summary reaction. The minimum typical of a reaction intermediate in consecutive reactions was observed in the measured time dependence of pH during complexation. The intermediate is H^+ ion in this case. Therefore we devised other mechanism more appropriate to previous experimental results [11]:

$$\mathrm{Cu}^{2+} + \mathrm{HR} \leftrightarrow \mathrm{CuR}^+ + \mathrm{H}^+ \qquad (11.10)$$

$$2\mathrm{CuR}^+ \rightarrow \mathrm{CuR}_2 + \mathrm{Cu}^{2+} \qquad (11.16)$$

The mechanism in reaction 11.16 was proposed on the basis of the assumption that the affinity of HA for copper is very high and that the humic particles are less mobile than Cu^{2+} ions. Therefore, Cu^{2+} ions can very quickly occupy the majority of functional groups (HR), which causes liberation of H^+ ions in the first step. Because all Cu^{2+} ions need two functional groups, it is necessary to find them not too far from each other to form stable Cu-HA complexes. Thus, rearrangement of the structure of HA-Cu is necessary (second step). Functional groups that cannot be connected by Cu^{2+} ions can reconnect with H^+ ion (reverse reaction of the first step). The production of Cu^{2+} ions in the second step of reaction 11.16 is the reason for the conductivity increase in systems with constant pH (see also section 11.4.2). Equation 11.17 was derived for the second step of the mechanism where G_0 and G_t are conductivities at the beginning and at time *t*, respectively. Rate constant k_D is related to the second step of the mechanism in reaction 11.16. The unit of this constant is [$S^{-1}s^{-1}$] because conductivity is a measure of concentration in this case [14].

$$\phi_D = \frac{G_t - G_0}{G_0\, G_t} = k_D\, t + const \qquad (11.17)$$

Equation 11.16 was derived on the assumption that the total conductivity increase is caused mainly by the reconnected Cu^{2+} ions in the second step of the mechanism. The more exact form of the kinetic Eq. 11.18 was used in this study.

$$\phi_E = \frac{G_t - G_{12}}{G_{12}\left(2G_{12} - G_t\right)} = k_E\, t - G_{12} \qquad (11.18)$$

This equation operates with conductivity G_{12}, which is the value at the beginning of the second step of the proposed mechanism. The value of the rate constant k_E thus should not be influenced by the production of H^+ ions in the first reaction step. Its units are the same as in the case of Eq. 11.17 because both forms were derived for the second order reaction.

11.3. MATERIALS AND METHODS

11.3.1 Materials

Humic Acids. Solid HAs were obtained from South-Moravia lignite by means of alkaline extraction. The finely ground lignite was mixed with 0.25 M NaOH solution and heated for 180 min at 80 °C with a reflux condenser. The alkali-soluble portion was separated using a centrifuge. HA were precipitated from solution with HCl to pH = 1, centrifuged and washed repeatedly with deionized water until free of Cl^- ions. HAs prepared using this process were dried and finely milled. The yield of HA was about 45 % of the starting material. Elemental composition, total and carboxylic acidities and some optical properties of the resulting HA are listed in Table 11.2. The IR spectrum of prepared HA was published in previous work [10].

Humic Gel. HAs were dissolved in 0.5 M NaOH. This solution was acidified with concentrated HCl to pH = 1-2 and centrifuged. The supernatant was discarded. Humic gel was repeatedly washed with deionized water and centrifuged. The resulting humic gel contained about 90 % of the liquid phase.

Humic Sol. HAs were dissolved in 0.1 M NaOH. To neutralize NaOH the appropriate amount of 0.1 M HCl was added to this solution. The resulting humic sol had a concentration of 0.102 g L^{-1}, ionic strength 0.1 M and pH 6.85.

Other Materials. NaOH and HCl were used as received. Cupric chloride was used for the preparation of solutions for all equilibrium and kinetic measurements.

Table 11.2 Elemental composition, total and carboxylic acidity and some optical properties of HA[a]

C %	H %	N %	O %	S %	COOH meq/g	Total Acidity, meq/g	Ash Content, %[b]	E_4/E_6	Δlog K
40.2	41.0	1.53	16.8	0.39	4.36	5.49	9.43 %	5.10	0.71

[a]All values are normalized on dry, ash-free HA; [b] ash contents measured in water free HA sample

11.3.2. Methods

Characterization of Prepared HA. Elemental analysis was determined with a Carlo Erba CHNSO Microprobe Flash 1112 Analyzer. The content of carboxylic groups was determined by the standard Ca-acetate method [15]. Total acidity was obtained from conductometric and potentiometric titration: HA were suspended in deionized water in ratio 1 g : 50 mL, stirred 24 h to achieve equilibrium and titrated with standard NaOH solution [10]. Optical properties were obtained by the standard IHSS method.

Adsorption Experiment. Finely ground solid HAs were suspended in $CuCl_2$ solution in the ratio 1 g HA/ 50 mL and shaken at laboratory temperature. Solutions with initial concentrations in the range 0.05-0.80M were used for equilibrium measurements. The initial concentration of copper(II) in kinetic experiments was 0.05M. At equilibrium (after 24 h) or at given times the solid HAs were separated by centrifugation. The supernatant was analysed using UV/vis spectrophotometry (Hitachi U-3300), potentiometry (Sentron Titan K185-016) and conductometry (Hanna Instruments HI8820N). The quantity of the adsorbed ions at time *t* or at equilibrium was calculated on the basis of the decrease of absorbance at 810 nm (prior to measurements, the calibration curve for cupric ions was determined and from the increase of concentration of H^+ ions (two H^+ ions are liberated by binding of one Cu^{2+} ion).

Diffusion Experiment. Humic gel was gently pressed into a silicone rubber tube. The length of the tube was chosen to simulate a semi-infinite system. The tube with gel was placed in a solution of Cu^{2+} ions. At different times the humic gel was sliced and analysed. Concentrations of Cu^{2+} ions in separate slices were determined from the extract (concentrated HCl) using UV/vis spectrophotometry (Hitachi U-3300).

Study of Reaction Kinetics in a Humic Gel. The kinetics of chemical interactions of cupric ions with humic gel were studied using potentiometry (Sentron Titan K185-016) and conductometry (Hanna Instruments HI8820N). The solution was mixed with humic gel in ratio 1 g gel/ 50 mL and its pH and conductivity were measured under stirring until there was no further change. The total decrease of

Cu^{2+} concentration also was measured spectrophotometrically (Hitachi U-3300).

Study of Reaction Kinetics in a Humic Sol. The complexation kinetics of humic sols were studied using potentiometry (Sentron Titan K185-016) and conductometry (Hanna Instruments HI8820N). The humic sol was mixed with Cu^{2+} solution in ratio 1/1 (v/v) and conductivities were measured under stirring until stabilization. The total decrease of Cu^{2+} concentration also was measured spectrophotometrically (Hitachi U-3300).

11.4. RESULTS

11.4.1 Adsorption on Solid HA

Examples of adsorption isotherms for various initial pH values are shown in Figure 11.1. Experimental data measured using UV/vis spectrophotometry were fitted by the Langmuir isotherm (Eq. 11.1). The calculated parameters are listed in Table 11.3.

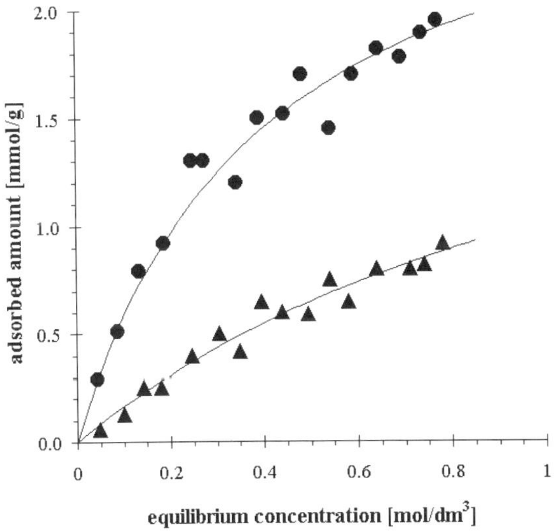

Figure 11.1 Langmuir isotherms measured spectrophotometrically for initial pH 2.0 (▲) and 4.5 (●)

Table 11.3 Parameters of Langmuir isotherm for various initial pH-values [10]

Initial pH	Maximum Adsorbed Amount, mmol/g	Adsorption Coefficient, mL/mol
2.0	2.44	0.72
2.8	2.59	1.52
3.6	2.63	2.03
4.5	2.93	2.47

Comparison of results obtained for equilibrium adsorption using UV/vis spectrophotometry and potentiometry is presented in Figure 11.2. Parameters of the Langmuir isotherm were computed for the adsorbed amount calculated from the difference between spetrophotometric and potentiometric measurements are listed in Table 11.4.

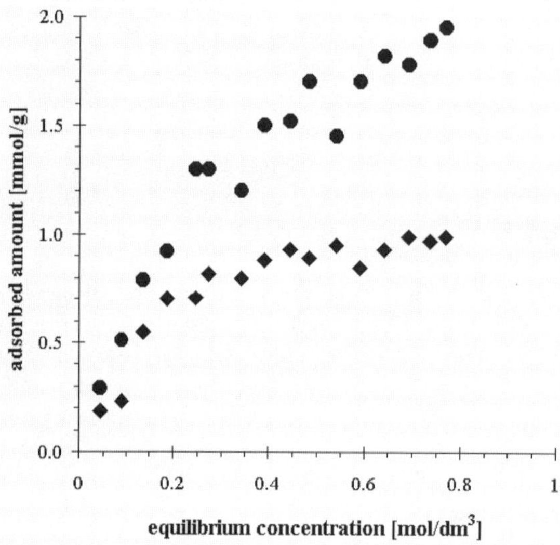

Figure 11.2 Comparison of experimental adsorption isotherms obtained using UV/vis spectrophotometry (●) and potentiometry (◆) for initial pH 4.5

Table 11.4 Parameters of Langmuir isotherm for adsorption data obtained from difference between spectrometric and potentiometric measurements

Initial pH	Maximum Adsorbed Amount, mmol/g	Adsorption Coefficient, mL/mol
2.0	2.44	0.72
2.8	2.52	0.70
3.6	2.59	0.71
4.5	2.68	0.69

Examples of kinetic data measured spectrophotometrically for various initial pH values are shown in Figure 11.3. In Figure 11.4 we can see the same data fitted by Eq. 11.4. Computed rate constants of adsorption and desorption are listed in Table 11.5. Comparisons of experimental kinetic data obtained using UV/vis spectrophotometry and potentiometry are presented in Figure 11.5. All kinetic measurements in this section were carried out with Cu^{2+} solutions with an initial concentration of 0.05 M.

11.4.2. Diffusion in Humic Gel

Because the HA content in humic gels ranged from 10 to 12 %, all concentration of Cu^{2+} ions in the humic gel were normalized to the unit weight of HA. Figure 11.6 shows the dependence of total diffusion flux on the square root of time for various initial concentrations of Cu^{2+} solution. Experimental data are fitted by Eq. 11.6 (saturated solution) and Eq. 11.8 (time variable Cu^{2+} concentration in solution).

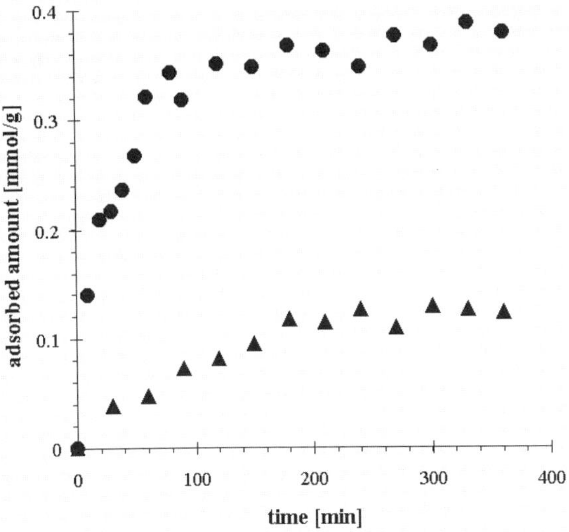

Figure 11.3 Kinetic data measured using UV/vis spectrophotometry for initial Cu^{2+} concentration 0.05 M and pH 2.0 (▲) and 4.5 (●)

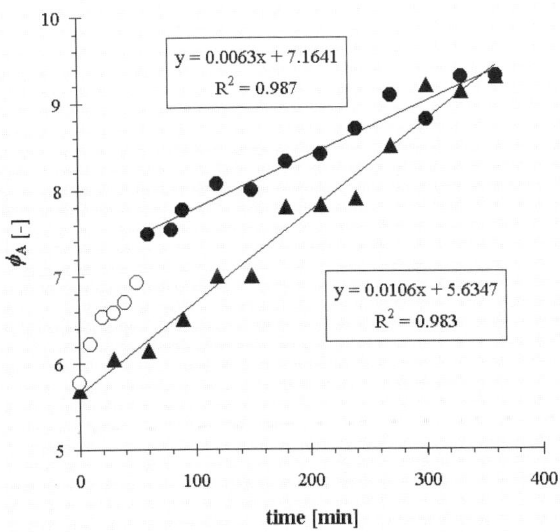

Figure 11.4 Spectrophotometric data fitting by Eq. 11.4 for initial Cu^{2+} concentration 0.05 M and pH 2.0 (▲) and 4.5 (●)

Table 11.5 Adsorption and desorption rate constants computed according to Eq. 11.4

Initial pH	Adsorption Rate Constant, L/mol s	Desorption Rate Constant, s^{-1}
2.0	2.34×10^{-1}	3.25×10^{-4}
2.8	2.26×10^{-1}	3.22×10^{-4}
3.6	2.39×10^{-1}	3.37×10^{-4}
4.5	2.46×10^{-1}	3.56×10^{-4}

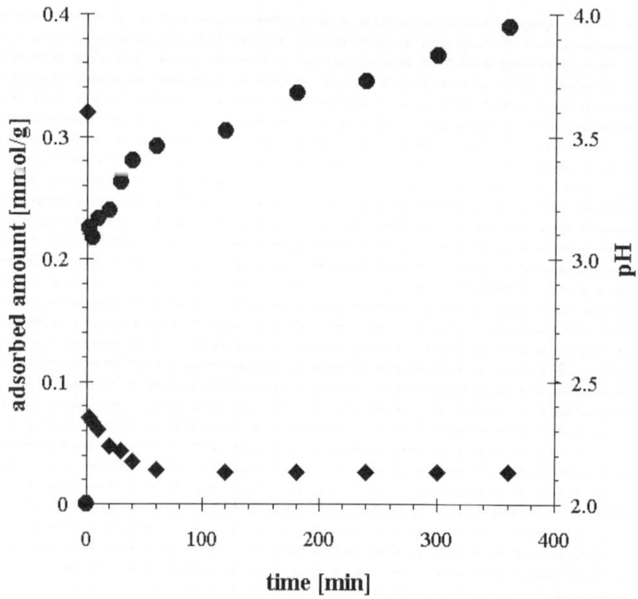

Figure 11.5 Comparison of kinetic data obtained using UV/vis spectrophotometry (●) and potentiometry (◆) for initial Cu^{2+} concentration 0.05 M and pH 3.6

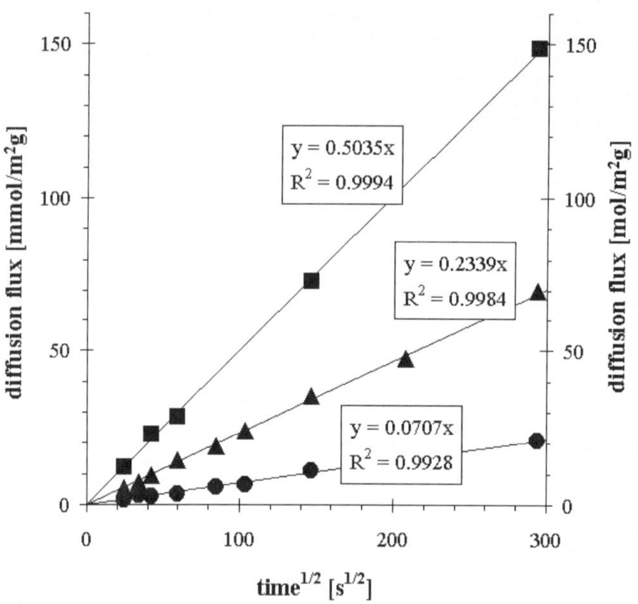

Figure 11.6 Experimentally determined values of total diffusion flux for initial Cu^{2+} concentrations 0.01M (●) and 0.40M (▲) in mmol/m²g fitted by Eq. 11.6 and saturated solution (■) in mol/m²g fitted by Eq. 11.8

Examples of concentration profiles for three initial concentrations of Cu^{2+} ions in solution are presented in Figure 11.7. Experimental data are valid up to 24 h. Time development of the concentration profile is shown in Figure 11.8. The initial concentration of Cu^{2+} ions in solution was 0.40 M. All experimental data in Figures 11.7 and 11.8 were fitted by Eq. 11.9.

Figure 11.9 presents the time development of concentration in the HA gel at a distance of 1 mm from the solution/gel interface for various initial concentrations of Cu^{2+} ions in solution. These curves were computed using Eq. 11.7 and the value of the effective diffusion coefficient was determined as in ref. 12.

Figure 11.10 compares experimental data with concentration profiles calculated using Eq. 11.7 with diffusion coefficients determined for humic gels [12] and published for diffusion in water [16].

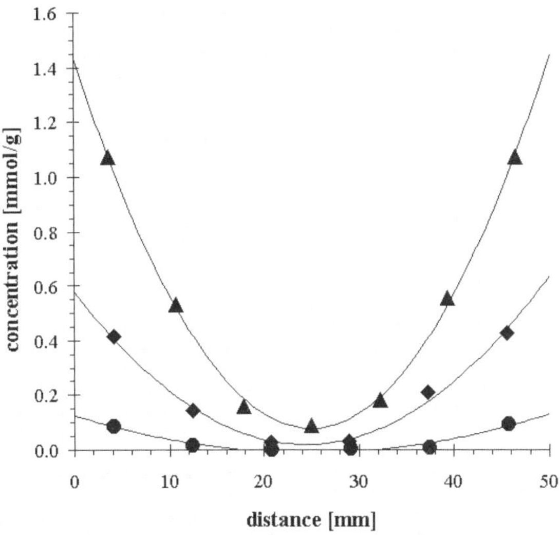

Figure 11.7 Concentration profiles after 24 h for three initial concentrations of Cu^{2+} ions: 0.02M (●), 0.20M (◆) and 0.40M (▲)

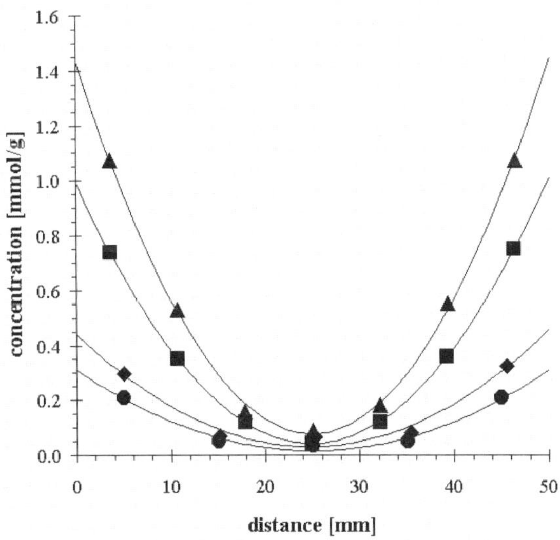

Figure 11.8 Time development of concentration profile for 0.40M solution: 30 min (●), 2 h (◆), 12 h (■) and 24 h (▲)

11.4.3. Complexation in Humic Gel

An example of kinetic data obtained using potentiometry and conductometry for a 0.10 M Cu^{2+} solution is shown in Figure 11.11. The amount of Cu-HA complexes formed in dependence on the initial Cu^{2+} concentration is presented in Figure 11.12. Apparent rate constants computed from Eqs. 11.13 and 11.14 are listed in Table 11.6.

11.4.4. Complexation in Humic Sol

Figures 11.13 and 11.14 show examples of complexation kinetic data in a humic sol for two initial concentrations of Cu^{2+} ions. Open points in these figures represent the control measurements: these values are valid for Cu^{2+} solution mixed with 0.1M NaCl solution in ratio 1:1 v/v and they are constant during the measurements. Experimental data obtained from conductometry fitted by Eqs. 11.17 and 11.18 are shown in Figure 11.15. The values of rate constants determined from Eqs. 11.17 and 11.18 are listed in Table 11.7. Apparent rate constants valid for simple summary reaction 11.10 are also in Table 11.7.

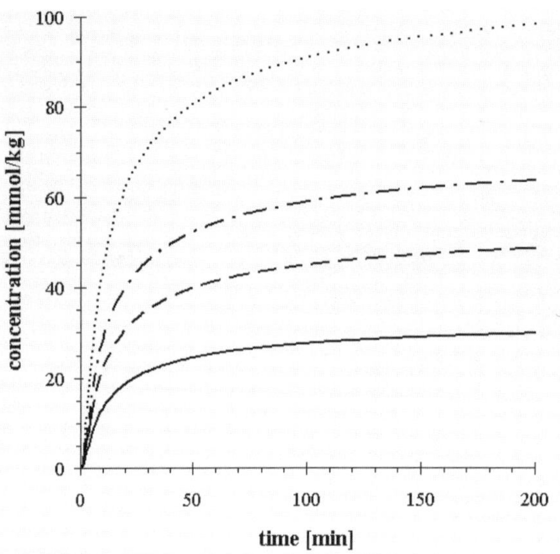

Figure 11.9 Time development of concentration in gel (distance 1 mm from gel/solution interface) for various initial Cu^{2+} concentrations in solution: 0.01M (full line), 0.10M (dashed line), 0.20M (dot-and-dash line) and 0.40M (dotted line)

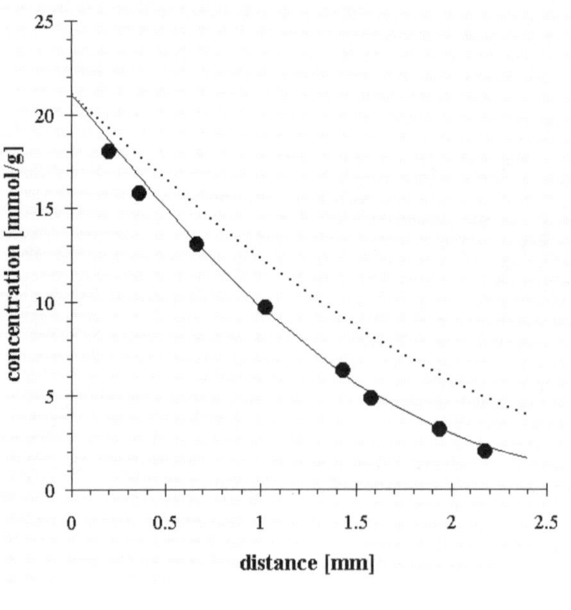

Figure 11.10 Comparison of experimental data for saturated solution (●) with concentration profiles computed by Eq. 11.7 using the diffusion coefficient determined for humic gel [12] (full line) and water [16] (dotted line)

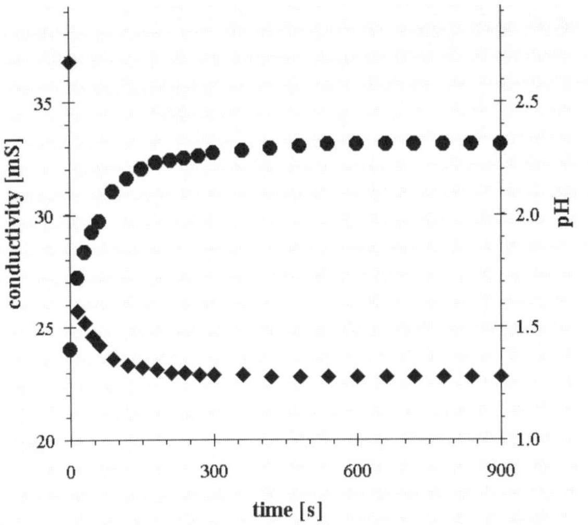

Figure 11.11 An example of kinetic data for complexation of Cu^{2+} ions (initial concentration 0.10M) with humic gel measured by conductometry (●) and potentiometry (◆)

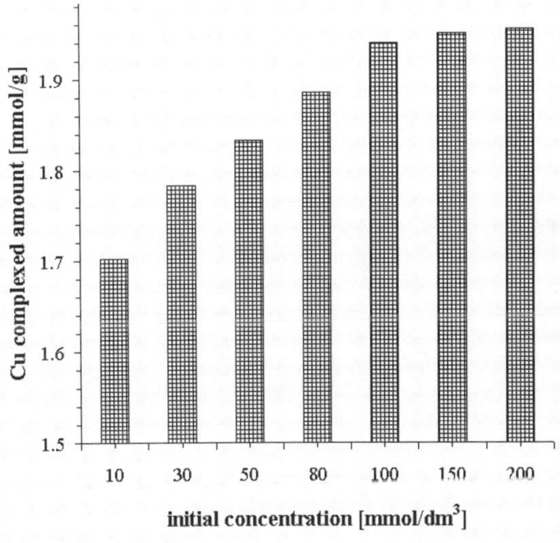

Figure 11.12 The dependence of formed Cu-HA complexes in humic gel on initial concentration of Cu^{2+} solution

Table 11.6 Apparent rate constants of complexation of Cu^{2+} ions with humic gel determined from conductometry k_B (Eq. 11.13) and potentiometry k_C (Eq. 11.14)

Initial Concentration, mmol/mL	Apparent Rate Constant k_B, L/mol s	Apparent Rate Constant k_C, L/mol s
10	3.21×10^{-3}	5.39×10^{-3}
30	3.79×10^{-4}	2.82×10^{-4}
50	1.26×10^{-4}	1.36×10^{-4}
80	5.89×10^{-5}	7.07×10^{-5}
100	3.51×10^{-5}	4.16×10^{-5}
150	2.37×10^{-5}	2.51×10^{-5}
200	1.65×10^{-5}	1.92×10^{-5}

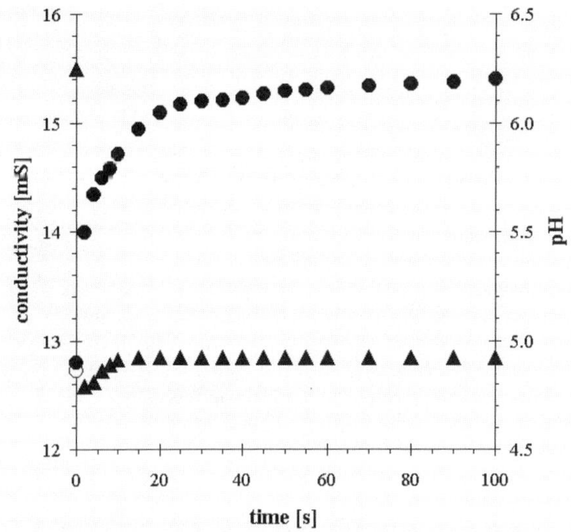

Figure 11.13 Kinetic data for complexation of Cu^{2+} ions (initial concentration 0.10M) with humic sol measured by conductometry (●) – control (○) - and potentiometry (▲)

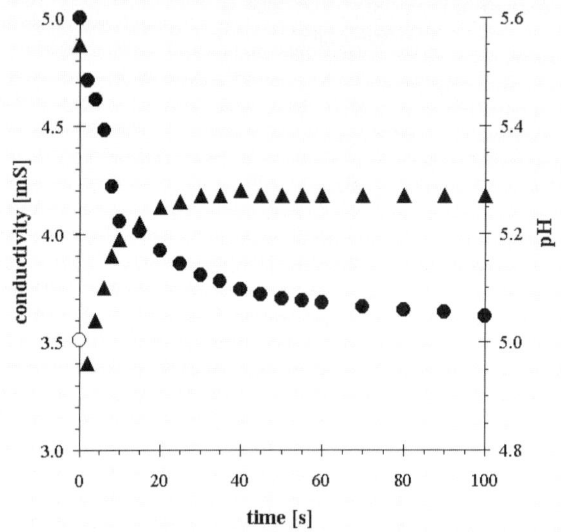

Figure 11.14 Kinetic data for complexation of Cu^{2+} ions (initial concentration 0.005M) with humic sol measured by conductometry (●) – control (○) - and potentiometry (▲)

Table 11.7 Apparent rate constants of complexation of Cu^{2+} ions with humic sol determined from various kinetic equations

Initial Concentration, mmol/mL	Rate Constant k_C (Eq. 11.13), mL/mol s	Rate Constant k_D (Eq. 11.17), 1/S s	Rate Constant k_E (Eq. 11.18), 1/S s
25	1.28×10^{0}	9.78×10^{-2}	3.94×10^{-2}
40	4.32×10^{-2}	8.47×10^{-2}	3.88×10^{-2}
50	2.18×10^{-2}	8.86×10^{-2}	3.87×10^{-2}
75	7.89×10^{-3}	9.34×10^{-2}	3.96×10^{-2}
100	1.56×10^{-3}	8.65×10^{-2}	3.89×10^{-2}

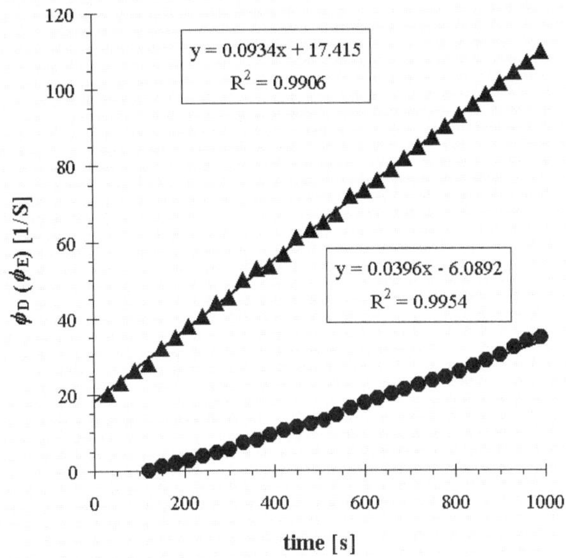

Figure 11.15 Experimental data for interaction of Cu^{2+} ions (initial concentration 0.75M) with humic sol fitted by Eqs. 11.17 (●) and 11.18 (▲)

11.5. DISCUSSION

11.5.1. Adsorption on Solid HA

Before thinking about adsorption kinetics let us refer to the equilibrium study. It was found [10] that experimental equilibrium data obtained for various initial pH values can be very well fitted by the Langmuir isotherm (see Figure 11.1). While the maximum adsorbed amount increases only slightly with pH, the adsorption coefficient is strongly influenced by the system acidity (see Table 11.3). From physical chemistry it is well known that the rate or equilibrium constant is independent of the concentrations of reactants and products. Therefore, the adsorption coefficient as the ratio of the rate constants of adsorption and desorption should be independent of the concentration of H^+ ions produced by the binding of metal ions with carboxylic and phenolic functional groups in HA. How can we explain this discrepancy? From Eq. 11.2, we find that the production of H^+ ions is not included in Langmuir theory. Therefore, this model cannot be complete for the description of adsorption of metal ions on HA. From comparison of spectrophotometric and potentiometric data (Figure 11.2) it follows that metal ions also can be adsorbed on other binding sites in the HA structure. Such sites are not able to split off H^+ ions. Further support for this hypothesis comes from equilibrium data: with decrease of pH only a slight decrease of a_{max} is observed in contrast to a noticeable decrease of parameter b, indicating weaker adsorption. Potentiometric adsorption data shows much earlier saturation than data from spectrophotometry (Figure 11.5). Thus, the difference between photometric and potentiometric data can be ascribed to adsorption on sites not occupied by protons. Treating only these data we can see that values of the adsorption coefficients for these other binding sites are not dependent on system acidity and the Langmuir model is valid (see Table 11.4). On the basis of these experimental results a new model for equilibrium adsorption was derived and published in ref. 7.

$$a_h = a_{max,h} \frac{b_h\, c}{h^2 + b_h\, c} \tag{11.19}$$

The adsorbed amount a_h represents metal ions that are bonded with HA through carboxylic or phenolic OH groups and can be determined from potentiometry. Parameters of Eq. 11.19 are the maximum amount bonded by these groups $a_{max,h}$ and the adsorption coefficient for these groups b_h. The equilibrium solution concentration c is the same as in the Langmuir isotherm (Eq. 11.1) and h is the equilibrium concentration of H^+ ions. The total adsorbed amount of metal ions on HA (a_{total}) is the sum of a_h (Eq. 11.19) and a (Eq. 11.1). The values of a_{total} and a_h can be determined from spetrophotometry and potentiometry, respectively. The amount of metal ions adsorbed on other binding sites is the difference between a_{total} and a_h [17].

Evaluating kinetic data given in Figure 11.3 with the Langmuir kinetic equations, the discrepancies at higher pH values are confirmed (see Figure 11.4). The whole dependence is not accounted for by a single line as predicted by Eq. 11.3 but by at least two lines. In accordance with previous deductions, only the last linear part, where no change in pH was detected, was fitted by the Langmuir kinetic equation. Estimated rate constants collected in Table 11.5 are not dependent on the pH value, as the true rate constant should be.

Summarizing the results for adsorption on solid humic acid, it can be stated that kinetic experiments confirmed what was indicated by equilibrium measurements. The adsorption mechanism is complex, does not agree with original Langmuir ideas and several types of active sites are involved. In strongly acidic systems phenolic and carboxylic groups cannot split off H^+ ions and copper is adsorbed mainly on other binding sites (straight line in Figure 11.4).

11.5.2. Diffusion in Humic Gel

Interactions with solids may be complicated by diffusion effects. Therefore, the data in the preceding section, particularly the kinetic data, may be affected by diffusive transport of interacting ions. In this section we tried to study diffusion of metal ions experimentally in a much more easily tractable colloidal system: a humic gel. It can be viewed not only as a pure model system but also as a model of humic gels, which can occur in nature where conditions for swelling appear (e.g. peats, water sediments).

The first experiments [11] were made from the solution in which saturation by copper ions was maintained. Then we added measurements from solutions with various initial concentrations below saturation (i.e. with changing Cu^{2+} concentration during the diffusion experiment). Selected data fitted by appropriate diffusion equations are shown in Figure 11.6 and good agreement between experiment and theory is seen. In previous work [12] the values of the diffusion coefficients of Cu^{2+}, Co^{2+} and Ni^{2+} ions were determined from slopes of the lines for saturated solutions (see Eq. 11.6).

As expected, the lower the initial concentration of diffusing solution the lower the metal concentration within the gel, see Figure 11.7. The profiles are steep even after 24 hours and the concentration in the middle of the diffusing gel medium increases very slowly (see also Figure 11.8). Diffusion transport is inherently slow even in the studied gels with high water content and the diffusion coefficients are not significantly lower than in water [16]. Figure 11.9 shows that achieving a final concentration even as close as 1 mm to the diffusion boundary is a matter of several hours with little effect of the initial concentration of the diffusing solution.

Due to the complexation ability of HA, diffusion is inseparably accompanied by chemical reaction. Concentration profiles reported here from the diffusion studies thus inevitably contain also the effect of complexation. From experimental diffusion data, a single diffusion effect cannot be separated. A very simple model of diffusion coupled with chemical reaction, Eq. 11.15, indicates that chemical reaction diminishes the value of the diffusion coefficient, resulting in the observed quantity. Figure 11.10 shows the decelerating effect of chemical reaction on cupric ion transport through the humic gel. The dotted line was calculated with the diffusion coefficient tabulated for water [16] (remember that the gel used in experiments contained about 90% of water), whereas the solid line was obtained with our effective value [12]. Chemical reaction immobilizes diffusing ions, extracts them from the diffusing zone, decreases the concentration of diffusing particles and probably also lowers the concentration gradient, which is the driving force of diffusion. Some information on reaction kinetics with solid humics was obtained in a previous section. However, interaction kinetics with the gel form may be quite different. Therefore, kinetic studies with humic gel were made using the same methodology as with the solid humics.

11.5.3. Complexation in Humic Gel

Figure 11.11 shows, as expected, that reaction of copper ions from solution with humic gel is much more rapid than combined diffusion and reaction with the gel. The equilibrium state is achieved within 5-10 minutes and equilibrium concentrations are of course dependent on the initial amount of metal ions (Figure 11.12). This Figure also shows some adsorption isotherm-like behavior, i.e. probably saturation of the humic gel surface upon interaction with copper ion solutions of increasing concentration.

It is also interesting to compare results of interactions with humic gel and solid. Although Figures 11.1 and 11.12 do not refer to the same experimental conditions, it is clear from their comparison that specific “sorption capacity” is much higher for humic gel than for solid HA. In fact, the sorption capacity of the humic gel obtained in experiments included in Figure 11.12 is about 1.5 times higher than the maximum capacities of solid HAs determined for various pH values and given in Table 11.3. Clearly, the “semi-solid” gel state possesses many more binding sites than the solid state.

Rate parameters determined for the gel state show strong concentration dependence (Table 11.6) and are therefore only effective or apparent quantities. This indicates a more complex mechanism than supposed by reaction 11.16.

Comparing values of forward rate constants estimated for suspension (Table 11.5) and gel (Table 11.6), an apparently higher rate for the former system might be found. Raw kinetic data show something different (see Figures 11.3 and 11.11). Indeed, the reaction rate is not determined by the value of the rate constant alone but also by the concentration of reacting species. The concentration of reactive sites in the gel-containing system is thus much higher and even with a low rate constant a high reaction rate is achieved. As the concentrations of copper ions in both systems comparable, this means that the amounts of (accessible) humic active sites are quite different. Comparison of rate constants listed in Tables 11.5 and 11.6 can also be misleading because Eqs. 11.13 and 11.14 were derived only for the forward reaction (see Eq. 11.10) and the dissociation of formed Cu-HA complexes was neglected.

Kinetic experiments with humic gel thus confirmed the role of chemical reaction in this colloidal system. Of course, the hypothesis underlying Eq. 11.15 could not be directly tested in this way. As the equilibration reaction of humic gel with aqueous cupric ions was attained within a couple of minutes, this hypothesis perhaps could be accepted as a rough approximation.

11.5.4. Complexation in Humic Sol

To complete knowledge of kinetics of chemical interactions between copper ions and HAs, experiments with the HA in sol form were made. The main difference to similar experiments with the gel form, discussed in the preceding section, was occurrence of an extreme on the pH-time profile (Figures 11.11 and 11.13). This minimum was explained [14] with a multistep but still simple mechanism (Eq. 11.16). Estimated rate constants for various kinetic models are collected in Table 11.7. The simpler model (Eq. 11.10) shows strong dependence of rate constants on initial concentration and should not be considered as adequate. Kinetic Eq. 11.17 based on our model (Eq. 11.16) gives much better results in this regard but still shows some scatter of estimated rate constants with initial concentration (Table 11.7). Reconsidering the chemical background of the mechanism in Eq. 11.16, a more appropriate kinetic Eq. 11.18 was derived. With this equation, rate constants are effectively independent of initial Cu^{2+} concentration (Table 11.7). It is noticeable that from the goodness of fit shown in Figure 11.15 this seemingly subtle difference between model Eqs. 11.17 and 11.18 cannot be detected.

Comparing apparent rate constants k_C for the sol (Table 11.7) and the gel (Table 11.6), it seems that interaction in humic sols are slower, but experimental kinetic data in Figures 11.11, 11.13 and 11.14 show a higher complexation rate in the sol. In this case the Cu:HA concentrations ratio also influences the rate of complexation. Because apparent rate constants depend on the copper concentration, it is very probable that their values also are influenced by HA concentration. The content of HA is much lower in the sol and could be the reason of this discrepancy.

Kinetic experiments in sol systems revealed another interesting point. In Figure 11.13, the conductivity increases during the whole run and the initial control conductivity (i.e. measured with the same solutions but without addition of HAs) corresponds to the initial conductivity of the reacting system. Figure 11.14, obtained for lower initial concentration of cupric ions, shows something different. Conductivity decreases during the whole reaction and the initial control conductivity is obviously lower than in the reacting system. The latter points to some very rapid interaction between copper and humics immediately after mixing the two solutions. As pH increases, the conductivity decrease is much more understandable than its increase in Figure 11.13. Clearly, interactions and their mechanism are affected by concentration if there are interacting particles.

Figures 11.13 and 11.14 show the different conductivity curves with similar time dependences on pH. The conductivity increase was observed also in systems with lower initial concentrations of Cu^{2+} ions (0.025-0.1M). When the initial copper concentration was equal to 0.015M, the conductivity was constant and corresponded with a blank experiment (mixing of Cu^{2+} and NaCl solutions without HA). This concentration is practically the same as the initial concentration of HA in the system (0.0153M). These results show that conductivity increases in systems with a ratio Cu:HA > 1 and decreases if Cu:HA < 1.

In humic gel some functional groups of humic molecules are involved in forming the gel structure by physical interactions. Although physical bonds are weaker than chemical bonds, they can, once formed, protect particular groups from chemical interactions. Furthermore, humic molecules cannot change their conformations upon interaction with copper ions as they are anchored in the gel structure. Therefore, there are no extremes of pH in kinetic curves in systems in which the humic gel was used.

11.6. CONCLUSIONS

This study leads to two principal conclusions. Equilibrium (adsorption) data sometimes show certain discrepancies, which are fully revealed only by kinetic experimentation. Detailed kinetic studies clearly show that properties of the same material depend on its colloidal state. Compositional complexity of HAs is manifested in non-simple kinetic behavior of their interactions with metal ions. Moreover, this complexity is multiplied by the ability of humic molecules or macromolecules to appear in different colloidal states. Unraveling structure and chemical composition of HAs is thus not sufficient to describe and understand their behavior and function, particularly in natural environments. Investigation of their actual colloidal state and its properties should be also tackled. Functional groups of humic molecules can be engaged also in physical interactions, participating, for example, in forming bonded colloidal systems. These interactions can prevent the parts of humic molecules from participating in chemical reactions. Besides the various interactions, for example, metal ions-humic molecules, re-conformations of humic molecules should be considered. They are manifested only in kinetic, dynamic studies. Different mechanisms operate in interactions of the same kind in different HA colloidal states.

REFERENCES

1. Kumada K. *Chemistry of soil organic matter*. Tokyo: Japan Scientific Societies Press, 1987.
2. Bryan ND, Robinson VJ, Livens FR, Hesketh N, Jones MN, Lead JR. Metal-humic interactions: A random structural modeling approach. *Geochim. Cosmochim. Acta*, 1997; 61: 805-820.
3. Lubal P, Široký D, Fetsch D, Havel J. The acidobasic and complexation properties of humic acids. Study of complexation of Czech humic acids with metal ions. *Talanta*, 1998; 47: 401-412.
4. Davies G, Ghabbour EA, Cherkasskiy A, Fataftah A. Tight metal binding by solid phase peat and soil humic acids. In: Clapp CE, Hayes MHB, Senesi N, Bloom PR, Jardine PM eds.*Humic substances and chemical contaminants*. Anaheim: Soil Science Society of America, 2001: 371-395.
5. Baruah MK, Upreti MC. Preferential uptake of Fe(III) by humic acid extracted from lignite, *Fuel*, 1994; 73: 273-275.
6. Jin X, Bailey GW, Yu YS, Lynch, AT. Kinetics of single and multiple metal

ion sorption processes on humic substances. *Soil Sci.*, 1996; 161: 509-520.

7. Manunza B, Deiana S, Maddau V, Gessa C, Seeber R. Stability constants of metal-humate complexes: titration data analyzed by bimodal Gaussian distribution. *Soil Sci. Soc. Amer. J.*, 1995, 59: 1570-1574.
8. Martyniuk H, Wieckowska J. Adsorption of metal ions on humic acids extracted from brown coals. *Fuel Process. Technol.*, 2003; 84: 23-36.
9. Klučáková M, Pelikán P, Lapčík L, Lapčíková B, Kučerík J, Kaláb M. Study of structure and properties of humic and fulvic acids. I. Properties and reactivity of humic and fulvic acids. *J. Polym. Mater.,* 2000, 17: 337-356.
10. Klučáková M, Kaláb M, Pekař M, Lapčík L. Study of structure and properties of humic and fulvic acids. II. Study of adsorption of Cu^{2+} ions to humic acid extracted from lignite. *J. Polym. Mater.,* 2002; 19: 287-294.
11. Klučáková M, Pekař M. Study of structure and properties of humic and fulvic acids. IV. Study of interactions of Cu^{2+} ions with humic gels and final comparison, *J. Polym. Mater.,* 2003; 20: 155-162.
12. Klučáková M, Pekař M. Diffusion of metal cations in humic gels. In: Ghabbour EA, Davies G eds. *Humic substances: Nature's most versatile materials.* New York: Taylor & Francis, 2004: 263-273.
13. Crank J. *The mathematics of diffusion.* Oxford: Clarendon Press, 1956.
14. Klučáková M, Pekař M. Study of structure and properties of humic and fulvic acids. III. Study of complexation of Cu^{2+} ions with humic acid in Sols. *J. Polym. Mater.,* 2003; 20: 145-154.
15. Stevenson FJ. *Humus chemistry: Genesis, composition, reactions.* New York: Wiley, 1982.
16. Lide DR. *Handbook of chemistry and physics. 76th Ed.* New York: CRC Press, 1995.
17. Klučáková M. New approach to equilibrium sorption on solid humic acids. *XII International Meeting of IHSS*, Sao Pedro, Brazil, 2004, 377.

Chapter 12

MEASUREMENT OF FREE METAL ION IN EQUILIBRIUM WITH HUMIC - METAL COMPLEXES USING AN NMR SOLVENT SUPPRESSION TECHNIQUE

Raja Avanadula and David K Ryan

Department of Chemistry, University of Massachusetts Lowell, Lowell, MA 01854, USA

12.1. INTRODUCTION

Nuclear magnetic resonance (NMR) is a widely used technique for qualitative and quantitative analysis. NMR data for hydrogen (^{1}H NMR), carbon (^{13}C NMR), and certain other elements (Al, Cu, N, O, etc.) can be obtained with modern multinuclear instruments [1]. The present experimental procedure is designed to use ^{1}H NMR for quantitative analysis of a metal ion, Cu(II), by an indirect method using an NMR water suppression technique. This kind of technique is frequently used for analysis of biomolecules soluble in water, where the intensity of the water peak is of a magnitude higher than the analyte peaks. Experimental techniques to minimize the water signal include [2,3] (a) presaturation, (b) non-excitation using jump-return and binomial sequences, (c) spin-lock pulses in INEPT (Insensitive Nuclei Enhanced by Polarization Transfer sequences), and (d) suppression by pulsed field gradients combined with selective excitation and heteronuclear experiments using the PEP-HSQC (Heteronuclear Single Quantum Coherence) pulse sequence.

The presaturation technique using a composite pulse sequence is selected for this experiment. A composite pulse has the advantage of exciting the spin population over a large range of experimental parameters compared to a single radiofrequency pulse [4,5].

Copper(II) ion in an aqueous environment exists in a hydration shell surrounded by six water molecules bound via coordinate covalent bonds, Figure 12.1. Copper acts as a Lewis acid by accepting an electron pair from the oxygen of solvent water molecules [6,7]. As a result, hydrogens present in metal-bound water experience a deshielding effect compared to hydrogens of unbound solvent water molecules. When the solvent water peak is suppressed by NMR it results in a low intensity peak, representing the metal-bound water, which appears downfield from the solvent water peak. This criterion is considered to be the basis for differentiating solvent and metal-bound water peaks using NMR water suppression. NMR peaks resulting before and after suppression of solvent water are shown in Figure 12.2 (see Table 12.1).

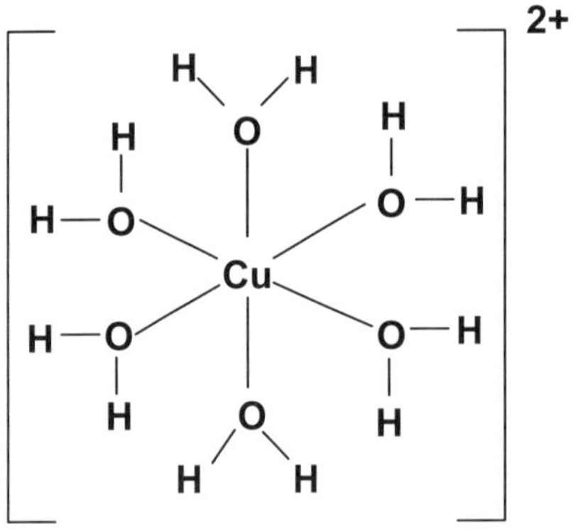

Figure 12.1 Schematic diagram of copper(II) ion in an aqueous solution forming a primary hydration shell with 6 water molecules

Ethylenediaminetetraacetate (EDTA) is chosen as a model ligand for this study and for the collection of preliminary data for quantitation of metal ion and validation of the method. EDTA at neutral pH acts as a hexadentate ligand as shown in Figure 12.3. The two amino nitrogens and four carboxyl oxygens of EDTA act as the binding site, contributing six coordinate covalent bonds to copper(II). When EDTA is added to an aqueous copper(II) solution, the hydration shell of the copper(II) ion is displaced by EDTA, releasing six metal-bound water molecules into solution [7], Eq. 12.1.

$$\mathrm{EDTA}^{4-} + \mathrm{Cu(H_2O)_6}^{2+} \longleftrightarrow \mathrm{Cu(EDTA)}^{2-} + 6\mathrm{H_2O} \qquad (12.1)$$

When this metal – EDTA complex is analyzed by NMR, a metal-bound water peak is not observed, which would otherwise be seen after solvent suppression of free metal ions in aqueous solution. Thus free metal ion is differentiated from complexed metal ion based on the metal-bound water concentration measured by NMR. Fulvic acid titration experiments similar to EDTA studies were also performed to provide a more challenging ligand system and demonstrate the utility of the technique. The binding reaction of $Cu(H_2O)_6^{2+}$ with fulvic acid (FA) is shown in Eq. 12.2 as adapted from Alvarez-Puebla et al. [8].

$$\mathrm{FA} + \mathrm{Cu(H_2O)_6}^{2+} \longleftrightarrow \mathrm{Cu}^{2+}(\mathrm{FA}) + 6\mathrm{H_2O} \qquad (12.2)$$

Metal-ligand titration curves were plotted as free copper ($C_{F\text{-}Cu}$) *vs* total copper ($C_{T\text{-}Cu}$) and a conditional stability constant (K) was obtained by using Eq. 12.3.

Charges are omitted for simplicity and square brackets denote molar concentration. [Cu] is the concentration of free metal ion, [L] is the concentration of all forms of ligand (EDTA or FA) not associated with copper and [CuL] is the concentration of the metal ligand complex.

$$K = \frac{[CuL]}{[Cu][L]} \tag{12.3}$$

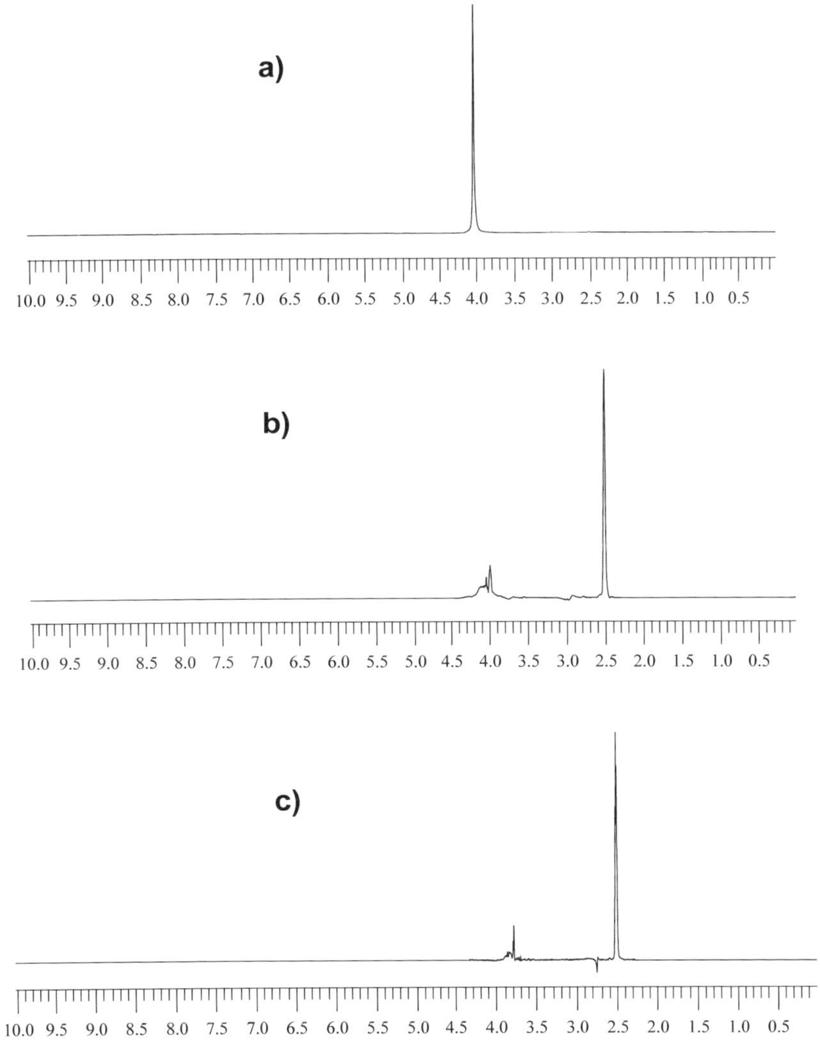

Figure 12.2 NMR spectra of 0.2 mM Cu(II) ions before and after water suppression using the parameters in Table 12.1. a) NMR spectrum before water suppression; b) NMR spectrum after water suppression; c) NMR spectrum of 0.0 mM Cu(II) after water suppression

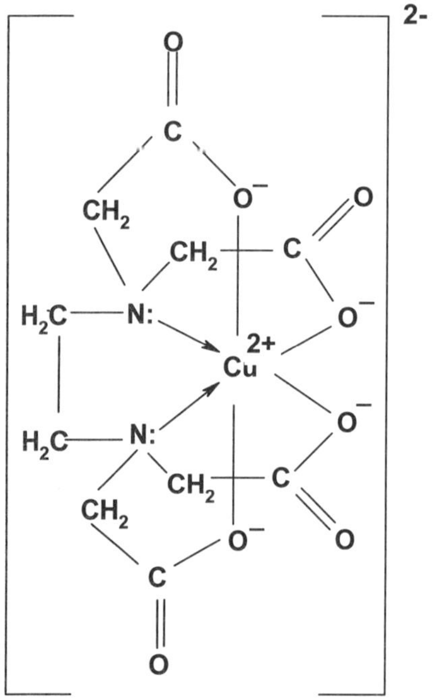

Figure 12.3 Schematic diagram of the EDTA-copper complex. The primary hydration shell of copper(II) is displaced by EDTA

A quadratic equation can be derived from the K expression and mass balances as described by Robinson et al. [9] with the positive root given by Eq. 12.4, where $C_{F\text{-}Cu}$ is the concentration of free metal ion, $Cu(H_2O)_6^{2+}$, with hydration shell intact, C_L is the total concentration of ligand and $C_{T\text{-}Cu}$ is the total copper concentration (free and bound). Equation 12.4 is fitted by non-linear regression [10] to the experimental data with suitable estimates of K and C_L.

$$C_{F-Cu} = \frac{-C_L + C_{T-Cu} - \frac{1}{K} + \sqrt{(C_L - C_{T-Cu} + \frac{1}{K})^2 + 4\frac{C_{T-Cu}}{K}}}{2} \qquad (12.4)$$

12.2. MATERIALS AND METHODS

Test solutions of copper(II) as copper perchlorate were prepared in deionized water. Different concentrations of Cu(II) ranging from 0.2 to 1 mM were analyzed by taking 500 μL of sample in 5 mm NMR tubes. The solution composition was 80:20 DMSO(D_6):water at neutral pH. Samples were analyzed on a Bruker ARX-250 MHz instrument using the zg30 pulse program. Water suppression was performed for each sample by the composite presaturation pulse sequence zgcppr using NMR parameters shown in Table 12.1.

NMR spectra of the water peak before and after suppression are shown in Figure 12.2. Water-suppressed NMR spectra of various concentrations of Cu(II) are shown in Figure 12.4. A calibration curve is plotted in Figure 12.5, with data given in Table 12.2 as the percentage area of metal-bound water peak *vs.* the concentration of Cu(II) ions. This plot is used to calculate the concentrations of free copper ion in metal-ligand titration experiments.

Table 12.1 NMR parameters for solvent water suppression on a Bruker ARX-250 MHz instrument

NMR Parameters	
Solvent lock	DMSO
Transmitter offset frequency	250.13847 MHz
Pre-saturation pulse length	4 s
90° ^{1}H transmitter pulse length (preset pulse length)	7.5 μs
Relaxation delay before pre-saturation	20 μs
Relaxation delay between pulse 1 and 2	4 μs
Number of scans	16
Preaquisition delay time	138.57 μs

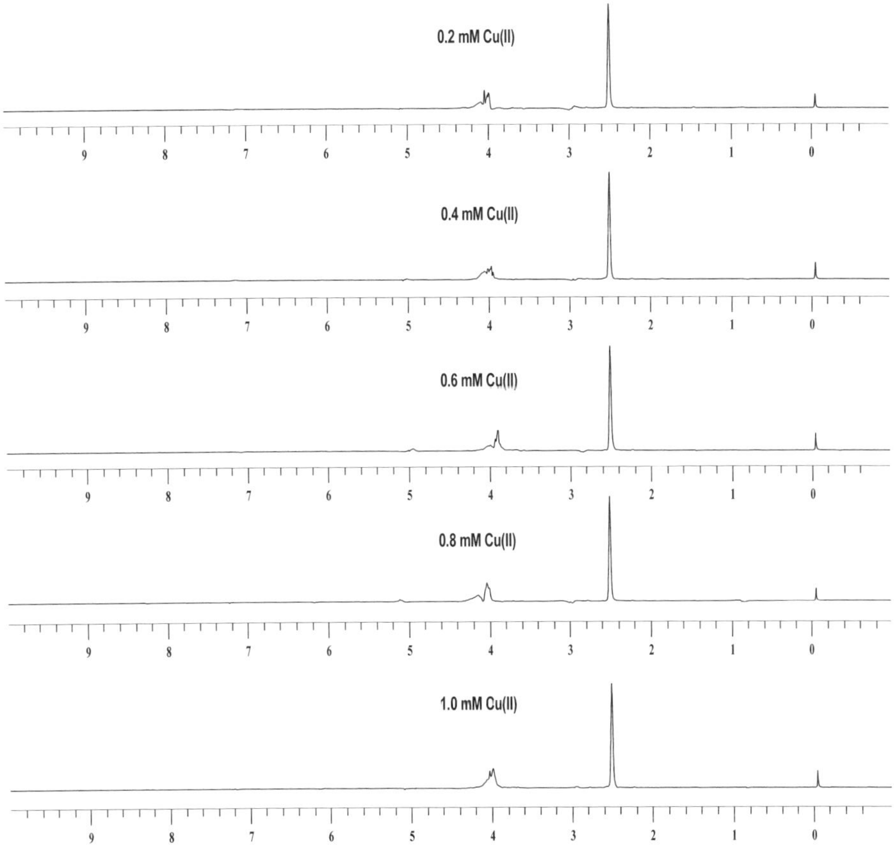

Figure 12.4 Water suppressed NMR spectra of Cu(II) ions ranging from 0.2 mM to 1 mM

For model ligand studies various concentrations of copper ion ranging from 0.2 to 2 mM were prepared in 1 mM EDTA solution at neutral pH. These solutions were analyzed by NMR using the same procedure and with the same parameters as given in Table 12.1. The percentage area of metal-bound water is plotted against total copper concentration in Figure 12.6, with data given in Table 12.3.

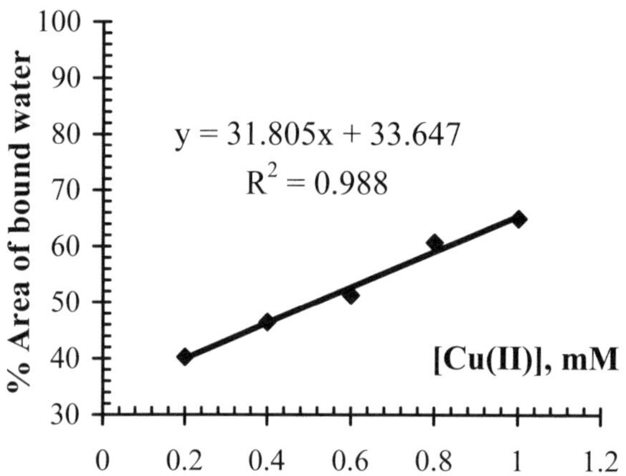

Figure 12.5 Calibration curve of Cu(II) as percent area of metal bound water *vs.* concentration of Cu(II)

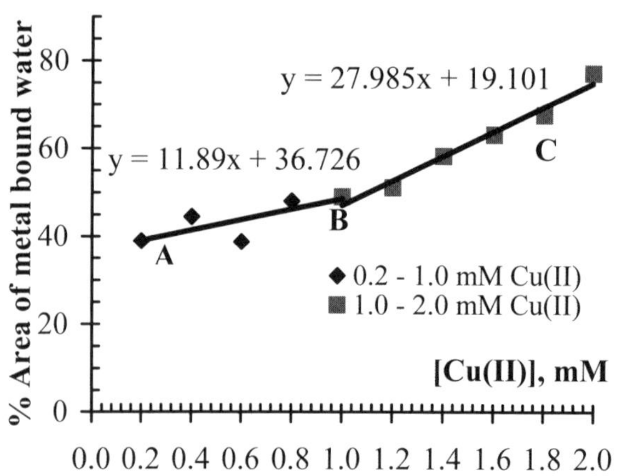

Figure 12.6 Titration curve with 1 mM EDTA at various concentrations of copper(II) from 0.2 to 2 mM

Table 12.2 Percent area of metal bound water at various concentrations of copper(II) ions ranging from 0.2 mM to 1 mM

$[Cu^{2+}]$ mM	Percent area of metal bound water
0.2	40.24
0.4	46.48
0.6	51.30
0.8	60.69
1.0	64.94

Table 12.3 Percent area of metal bound water at various concentrations of copper ions titrated with 1 mM EDTA

$[Cu^{2+}]$ mM in 1mM EDTA	Percent area of metal bound water
0.2	38.92
0.4	44.52
0.6	38.76
0.8	48.06
1.0	49.04
1.2	51.10
1.4	58.25
1.6	63.12
1.8	67.57
2.0	77.20

Finally, EDTA was replaced with 200 mg/L fulvic acid in different concentrations of copper ion ranging from 0.2 to 1 mM and the experiments were repeated in the same manner as for EDTA. The percentage area of metal-bound water *vs.* total copper concentration is plotted in Figure 12.7, with data given in Table 12.4.

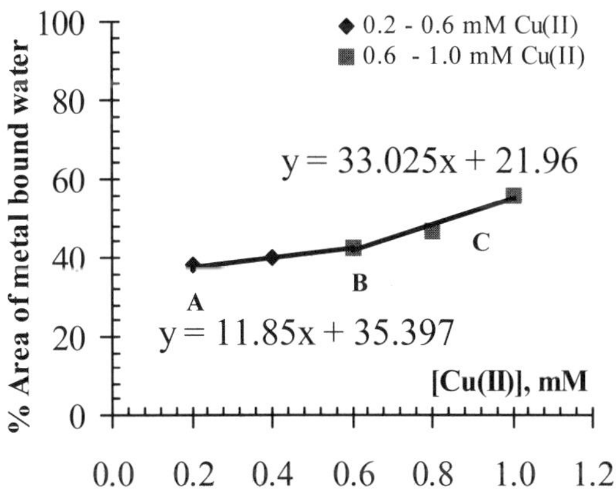

Figure 12.7 Titration of 200 mg/L fulvic acid with various concentrations of copper(II) ranging from 0.2 to 1 mM

Table 12.4 Percent area of metal bound water at various concentrations of copper ions titrated with 200 mg/L fulvic acid

$[Cu^{2+}]$ mM in 200ppm FA	Percent area of metal bound water
0.2	37.89
0.4	39.89
0.6	42.63
0.8	46.67
1.0	55.84

12.3. RESULTS AND DISCUSSION

Experiments were performed in a relatively low percent aqueous environment as high water content in the sample resulted in broader peaks and poor suppression of the solvent water signal. For these reasons the composition of aqueous solutions was optimized at 20% water. Shimming is an important consideration for this experiment. Poor shimming results in unsymmetrical peaks and improper signal suppression. Though all these considerations were taken into account, NMR water peaks could not be suppressed completely. The probable reason for this could be that the alignment of water molecules in the applied magnetic field is not uniform with high aqueous phase content. This limits the detection range of metal ions by this technique. The lowest possible concentration of metal ion measured is 0.2 mM.

Based on standard values from the literature, the peak at δ 2.5 ppm is identified as the methyl protons of DMSO. After suppression of the solvent water peak, theoretically there should not be any peak remaining, but the results show a peak of low intensity that is slightly downfield ($\Delta\delta$ - 0.1 to 0.3 ppm) from the solvent water peak as shown in Figure 12.2. Since protons of the metal-bound water experience a deshielding effect, it is likely that the resulting peak after solvent suppression would be located slightly downfield. Further evidence for the relationship between the downfield shifted water peak and the Cu(II) ion concentration is the observation that, with increasing metal ion concentration, the percentage area of metal-bound water increased according to Eq. 12.5. The resultant peaks are shown in Figure 12.4.

$$[\text{Cu(II)}] \ \alpha \ [\text{Metal bound water}] = \text{Percent area of suppressed water peak} \qquad (12.5)$$

This procedure was applied to quantitate free metal ion in the presence of ligands like EDTA and fulvic acid. Varying concentrations of Cu(II) ion ranging from 0.2 to 2 mM were prepared in 1 mM EDTA. The reaction between copper(II) and EDTA is a simple 1:1 reaction as shown in Eq. 12.1. One mole of EDTA reacts with one mole of copper(II) resulting in a metal-ligand complex and free water molecules.

A percentage area *vs.* Cu(II) ion concentration plot is shown in Figure 12.6. The first part of the curve from A to B is before the endpoint of the reaction between metal and ligand (i.e., almost all of the copper ion is bound by excess ligand). Theoretically, the AB portion of the curve should have a slope near zero, but due to a slight precipitation of EDTA in the DMSO-water solvent system, the result is a small positive slope. The stoichiometric endpoint between EDTA and copper(II) ion is achieved at point B, where all of the EDTA ligand is associated with copper(II) ion. Excess copper(II) ion added after the endpoint (B) of the reaction exists as free metal ion not bound to EDTA and with its hydration shell undisturbed. The slope of the BC portion of the curve Figure 12.6 is nearly the same as the slope of the calibration curve in Figure 12.5 because each increment of copper(II) ion added after the endpoint remains as free copper(II) ion in solution and gives rise to an increase in the NMR signal. Non-linear regression analysis of the data using Eq. 12.4 with estimates of K and concentration of ligand resulted in an endpoint value of 1.08 mM compared to the actual experimental concentration of 1mM EDTA. This is only an 8% excess of ligand concentration, which is within experimental error for this preliminary experiment. The value of K was much lower than theoretically predicted for EDTA with copper(II) at pH 7, probably due to the high organic (DMSO) content.

The same procedure was followed for various concentrations of Cu(II) ion ranging from 0.2 to 1 mM in 200 mg/L fulvic acid. The reaction is given by Eq. 12.2.

A titration curve is plotted as percentage area of metal-bound water *vs.* Cu(II) concentrations in Figure 12.7 and the data are given in Table 12.4. The end point observed in Figure 12.7 is approximately 0.6 mM Cu(II) for 200 mg/L FA, which is quite similar to previous data for this sample when corrected for concentration [11].

In Figure 12.7, the BC portion of the titration curve shows nearly the same slope as the calibration curve, indicating that further increase in the concentration of Cu(II) ion after point B resulted in free copper with an intact hydration shell. Additional work remains to be done with this sample to fully realize the benefits of the NMR method with non-linear regression analysis to determine both K and C_L.

12.4. CONCLUSIONS

Quantitation of free metal ion by monitoring metal-bound water with solvent-suppressed NMR was successful in metal-ligand equilibrium systems. Using this method it may be possible to analyze free metal ion in environmental samples from lakes, sediments, river water and contaminated areas like harbors and industrial sites. Unlike electrochemical methods, a disadvantage of this technique is that it is not species-specific and potentially detects all free metal ions that are capable of forming a primary hydration shell in an aqueous environment. Another disadvantage is that detection limits are not comparable with other speciation techniques. The lowest concentration detected by this method is 0.2 mM, whereas electroanalytical techniques can detect at the μM concentration range. However, this method does present certain advantages, such as an inherent ability to measure free metal without disturbing or shifting the equilibrium and freedom from electrode adsorption artifacts that plague other methods. Refinements to the technique using higher field strength NMR instruments will lower detection limits and the replacement of DMSO with D_2O will broaden the applicability and clarify the interpretation of results.

Copper(II) is a paramagnetic metal ion and NMR analysis of paramagnetic materials affects peak width and chemical shift values. It is observed that with increasing concentration of copper, the water peak shifts downfield. However, it is assumed that this effect is uniform throughout the experiment and does not affect relative areas of suppressed water peaks. Copper(II) exhibits octahedral structural distortion due to the Jahn-Teller effect. Also, there is a constriction of axial bonds in Cu^{2+}-FA complexes as explained by Korshin et al. [12]. It may be possible that the primary hydration shell of copper ion is not displaced completely and is detected in measuring free metal ion. Experimental work is in progress to analyze this effect. This technique is still in its infancy and more advances are expected by modifying the pulse sequence program, solvent system and analyzing different metal ligand systems to overcome the effects explained above.

REFERENCES

1. Braun S, Kalinowski H-O, Berger S. *150 and more basic NMR experiments*. NewYork: Wiley, 1998.
2. Gunther UL, Ludwig C, Ruterjans H. WAVEWAT-improved solvent suppression in NMR spectra employing wavewat transforms. *J. Magn. Reson.*, 2002; 156: 19-25.
3. Lee GSH, Michael WA, Brent RY. The application of watergate suppression technique for analyzing humic substances by NMR. *Org. Geochem.*, 1998; 28: 549-559.
4. Mao X, Chao-hui Y. Phase shift presaturation in one and two dimensional NMR experiments. *Chem. Phys. Lett.*, 1994; 227: 645-650.
5. Millot Y, Man PP. Determination of quadrupole parameters with a composite pulse for spurious signal suppression. *J. Magn. Reson.*, 2001; 150: 10-16.
6. Wu Y, West LJ, Stewart DI. Effect of humic substances on Cu (II) solubility in kaolin-sand soil. *J. Hazard. Mater.*, 2002; B94: 223-238.
7. Morel MMF, Hering GJ. *Principles and applications of aquatic chemistry*. NewYork: Wiley Interscience, 1993.
8. Alverez-Puebla RA, Valenzuela-Calahorro C, Garrido JJ. Cu(II) retention on a humic substance. *J. Colloid. Interf. Sci.*, 2004; 270: 47-55.
9. Robinson WE, Ryan DK, Sullivan PA, Boggs CC. Cadmium binding in the blood plasma of two marine bivalves. *Environ. Toxicol. Chem.*, 1997; 16: 1195-1202.
10. Levie DR. *How to use Excel in analytical chemistry and general scientific data analysis*. Cambridge: Cambridge University Press, 1999.
11. Ryan DK, Weber JH. A fluorescence quenching titration technique for the determination of complexing capacities and stability constants of fulvic acid. *Anal. Chem.*, 1982; 53: 969-973.
12. Korshin GV, Frenkel AI, and Stern EA, EXAFS study of the inner shell structure in copper (II) complexes with humic substances. *Environ. Sci. Technol.*, 1998; 32: 2699-2705.

Chapter 13

QUANTITATIVE ANALYSIS OF ALUMINUM AND SOIL FULVIC ACID COMPLEXES BY SOLUTION STATE ALUMINUM-27 NUCLEAR MAGNETIC RESONANCE SPECTROSCOPY

Norman C.Y. Lee, David K. Ryan and Geetha Rajesh

Department of Chemistry, University of Massachusetts Lowell,
Lowell, Massachusetts 01854, USA

13.1. INTRODUCTION

The study of humic material has been ongoing for decades, yet scientists around the world still do not understand it fully [1]. It can be found in both terrestrial and aquatic environments. Humic material is an important class of molecules because it is a major reservoir of carbon in the global carbon cycle, can influence bioavailability of metals, controls solubility of both organic and inorganic compounds, binds metals and organics, and plays a major role in coagulation and aggregation in aquatic systems [2]. Therefore, having a good understanding of humic material and its properties is important in many environmental applications.

Aluminum is one of the most abundant metallic elements in the Earth's crust. It is also believed to be a leading cause of Alzheimer's Disease [3-8]. Under acidic conditions, aluminum takes its most mobile and toxic form, hexaquoaluminum(III), $Al(H_2O)_6^{3+}$, and can be transported easily in the environment. There is a growing concern with more frequent episodes of acid rain around the world and acidification of lakes, especially in countries where less attention has been paid to environmental issues. Aluminum undergoes strong hydrolysis when the pH approaches 5. Previous study [9] has shown that below pH 5, aluminum exists primarily as labile $Al(H_2O)_6^{3+}$. Between pH 5 and 7, aluminum hydrolysis products become more abundant, namely $Al(OH)_{aq}^{2+}$ and $Al(OH)_{2aq}^{+}$. The white precipitate aluminum hydroxide, $Al(OH)_3$, starts to form as pH is increased above 7. Beyond pH 8, aluminate anions $Al(OH)_4^-$ dominate, and aluminum hydroxide is re-dissolved and becomes soluble in basic media. Hence, our study investigates aluminum speciation at pH $\leq$ 5 before aluminum hydrolysis products dominate the system.

Nuclear magnetic resonance spectroscopy (NMR) has emerged as a powerful analytical technique in humic material research. Traditionally, most humic NMR studies have been concerned with ^{1}H and ^{13}C, which reveal extensive functionality on humic molecules. With new NMR instrument development and more sophisticated pulse programs, multinuclear NMR techniques are becoming popular [10 11]. In this study, ^{27}Al NMR is applied to investigate binding between aluminum ions and soil fulvic acid in the pH range 3 to 5 and stability constants are obtained using both experimental NMR results and graphical Scatchard and Langmuir regressions. A model compound, oxalic acid, was also analyzed by the same NMR technique to evaluate the ability of this new method to analyze a known system.

13.2. MATERIALS AND METHODS

Humic material used in this study is a soil fulvic acid (SFA) from the B2 horizon of an organic-rich Podzol from Conway, New Hampshire, USA, which was isolated and generously given to us by Dr. James Weber. The elemental composition of this SFA is 53.1 % carbon, 3.24 % hydrogen, 0.90 % nitrogen and 0.8 % ash [15].

Stock solutions of aluminum (270 mgL^{-1}) and SFA (10,000 mgL^{-1}) were prepared by dissolving $AlCl_3 \cdot 6H_2O$ and SFA in de-ionized water. Complexes of aluminum and SFA were prepared by adding 0.100 mL of aluminum stock solution and 1.00 mL of SFA stock solution in a 10.0 mL volumetric flask. Exactly 0.200 mL of 5.0 M $NaNO_3$ (Fisher) was also added to ensure an ionic strength equal to 0.1 M and 0.01 M NaOH (VWR) was added drop-wise to adjust the pH to 3.0, 4.0 or 5.0 before diluting to the mark with de-ionized water. The mixture was kept stirring for 1-12 h before NMR analysis.

In this study, oxalic acid (Fisher) was used as a model compound. The oxalic acid (Ox) solution was prepared by dissolving oxalic acid in de-ionized water to make a 100 mM stock solution. Aluminum oxalate solutions (molar ratio 1 to 1) were prepared at the 2 mM concentration level by adding 1.00 mL of each stock solution into a 50 mL volumetric flask. Also, $NaNO_3$ was added to adjust the ionic strength to 0.1 M and NaOH was added to adjust the pH to 3.0, 4.0 or 5.0. Similarly, aluminum oxalate solutions with a molar ratio of 1 to 2 were prepared.

^{27}Al NMR measurements were carried out on a Bruker DRX 500 MHz instrument operated at 130 MHz and equipped with a 5 mm broadband probe. All ^{27}Al NMR spectra were referenced to a 1 M $AlCl_3$ solution at pH 2 in 20 % D_2O. To 0.500 mL of sample solution, 3 drops of D_2O were added to provide a signal lock and pH was again measured by an NMR pH electrode purchased from Cole-Parmer. Due to the broad background interference signal from metal alloy in the construction materials of the NMR probe, all spectra show a broad peak centered at 85 ppm with a width of ~100 ppm, which interferes with peaks from 40 to 135 ppm. The peaks of interests for aluminum-soil fulvic acid generally appear between 0 ppm and 30 ppm [16,17]. Therefore, a quantitative reference must appear outside the 0-135 ppm range to be useful.

In this study, a quantitative reference was achieved by using AlI_3 in acetonitrile. As reported previously [18], AlI_3 in acetonitrile forms a set of distinct peaks in the negative or upfield region on a ^{27}Al spectrum, depending on the degree of saturation, and they do not interfere with peaks appearing in the positive or downfield region. AlI_3 is highly reactive with H_2O and O_2. Therefore, the AlI_3 solution was prepared in a glove bag (NPS) under argon for each experimental day.

Before reaching the glove bag, argon is passed through drierite ($CaSO_4$) to remove moisture and the glove bag is flushed with argon three times. After the AlI_3 solution had been stirred for 20 min, 0.200 mL of AlI_3 solution was transferred into an NMR insert (Wilmad). This insert was used as the quantitative reference between sample solutions. All manipulations were carried out in the atmospheric glove bag under argon. A sample of 1.0 mM aluminum solution was prepared from the stock solution and this solution was used to determine the concentration of AlI_3 complexes in the NMR insert. All subsequently generated NMR peak integrals were referenced to the same NMR insert containing the AlI_3 complexes. This was done every time before Al-SFA and Al-Ox solutions were analyzed. A solution of $Al(H_2O)_6^{3+}$ was also run with the insert as the first and the last experiment of each day to ensure the AlI_3 complex integrals remained constant. A typical spectrum for the AlI_3 reference is shown in Figure 13.1.

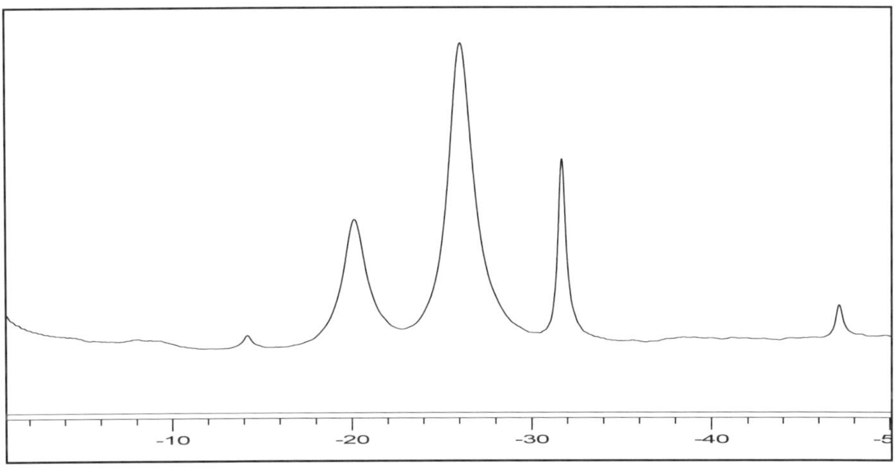

Figure 13.1 ^{27}Al NMR of anhydrous AlI_3 (0.010 mM) in dry acetonitrile

All NMR spectra were manipulated with MestRe-C v3.0 software. A polynomial function was applied to correct the baseline. AlI_3 reference peaks in the negative region were integrated as one single integral and calibrated to be 1.000. The remainder of the peaks in the positive region were also integrated together. Since these peaks severely overlapped each other, they were subjected to deconvolution and the intensity of each peak was then determined.

13.3. RESULTS AND DISCUSSION

13.3.1. Aluminum-Soil Fulvic Acid NMR

As we have reported previously [19], aluminum-soil fulvic acid (Al-SFA) spectra give rise to four distinct peaks that are observable between pH 3 and 5 and can be seen in Figure 13.2. The peak at approximately 0 ppm is free $Al(H_2O)_6^{3+}$ ion. The peaks at 7, 12 and 17 ppm are believed to be aluminum ions bound with one oxalic acid type site on SFA (Al-SFAox), aluminum ions bound with two oxalic acid type sites on SFA (Al-SFAox$_2$), and aluminum ions bound with three oxalic acid type sites on SFA (Al-SFAox$_3$), respectively. In Figure 13.2A at pH 3, peaks for Al^{3+} (0 ppm) and Al-SFAox (7 ppm) dominate the system. As pH is increased to 4 and 5, peaks for Al-SFAox (7 ppm) and Al-SFAox$_2$ (12 ppm) become more prominent. The same behavior is observed in Figure 13.2B. Since there is more SFA in the system, peaks for Al-SFAox (7 ppm) and Al-SFAox$_2$ (12 ppm) dominate at pH 3. When pH is increased to 4 and 5, most of the peak intensity is shifted to peaks for Al-SFAox$_2$ (12 ppm) and Al-SFAox$_3$ (17 ppm).

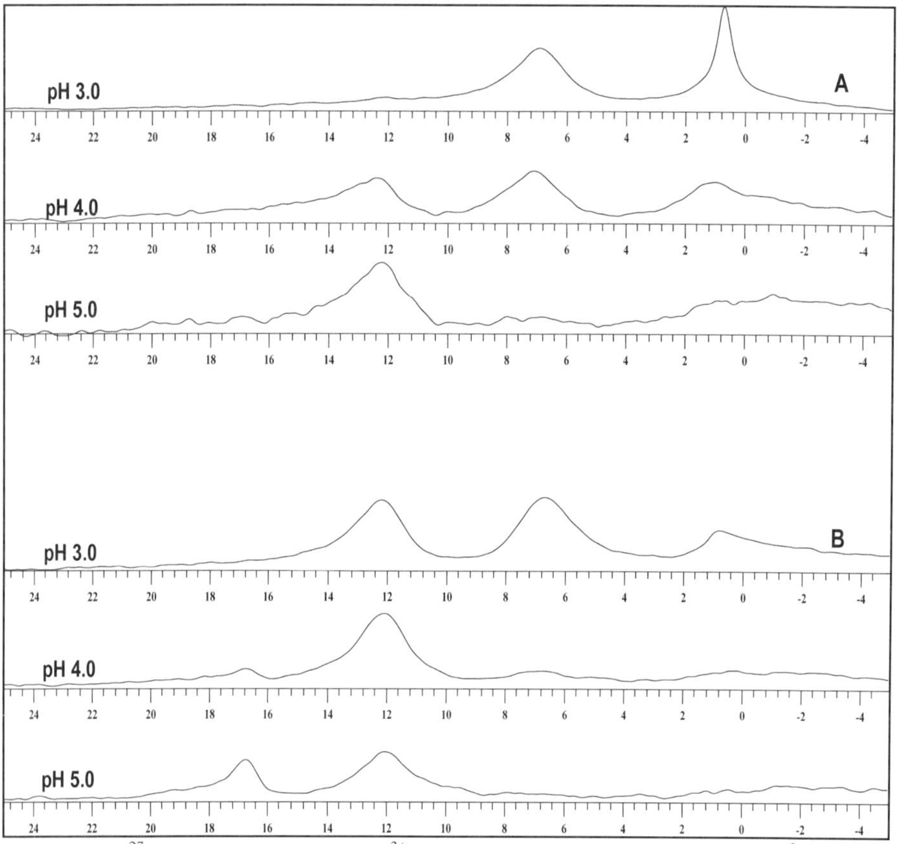

Figure 13.2 ^{27}Al NMR spectra of (A) Al^{3+}(135 mg/L)-SFA(5000 mg/L) and (B) Al^{3+}(135 mg/L)-SFA(12500 mg/L) at pH 3, 4 and 5

Peak assignments were confirmed by studying oxalic acid as a model compound, where similar chemical shifts are observed [20]. NMR measurements were carried out on the model compound oxalic acid with aluminum. The results are shown in Figures 13.3 and 13.4 and the data are summarized in Table 13.1. Similar to the Al-SFA system, four distinct peaks are observed, namely peaks at 0, 7, 12 and 17 ppm, and they are assigned to Al^{3+}, Al-Ox, Al-Ox$_2$ and Al-Ox$_3$, respectively. In Figure 13.3 at pH 3, peaks for Al^{3+} (0 ppm) and Al-Ox (7 ppm) dominate and as pH goes to 4 and 5, peaks for Al-Ox (7 ppm) and Al-Ox$_2$ (12 ppm) become more intense. In Figure 13.4, since more oxalic acid is in the system, peaks for Al-Ox$_2$ (12 ppm) and Al-Ox$_3$ (17 ppm) dominate. Aluminum oxalate speciation was modeled by MINTEQ [21] with conditional stability constants reported by Sillen and Martell [22]. MINTEQ was able to predict similar aluminum speciation within approximately 10 %, as shown in Table 13.1.

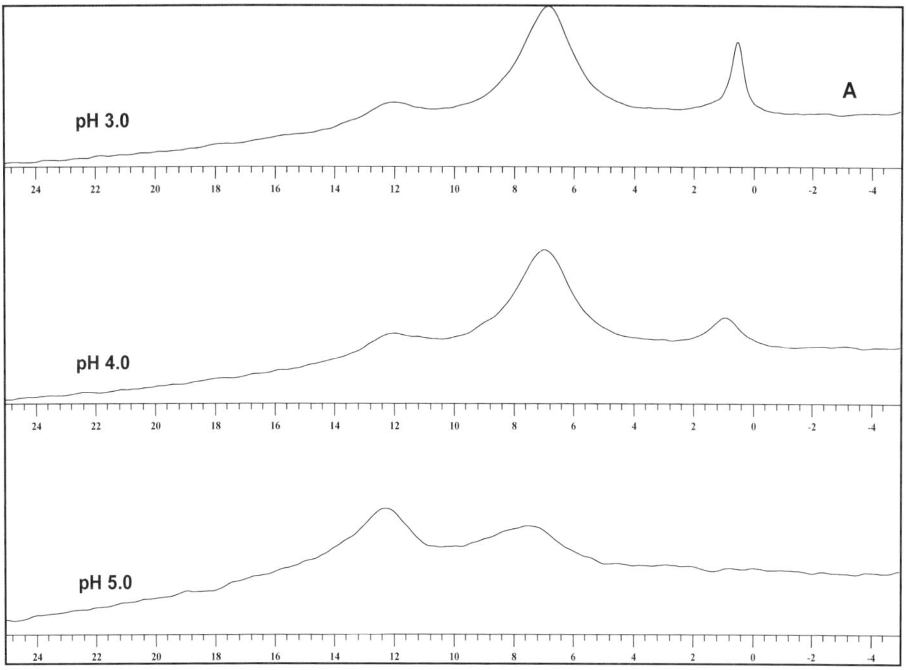

Figure 13.3 ^{27}Al NMR spectra of Al^{3+}(2 mM)-Ox(2 mM) complexes

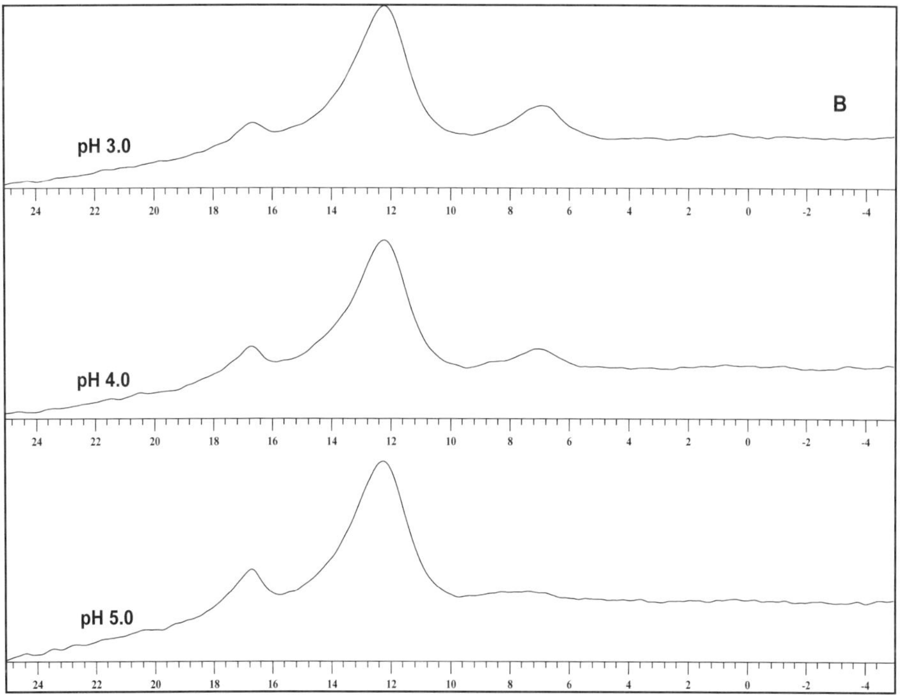

Figure 13.4 ^{27}Al NMR spectra of Al^{3+}(2 mM)-Ox(4 mM) complexes

13.3.2. Stability Constants

In order to determine stability constants for each of the Al-SFA binding sites, namely Al-SFAox (7 ppm), Al-SFAox_2 (12 ppm) and Al-SFAox_3 (17 ppm), quantitation of these peaks is required. Equation 13.1 allows calculation of the concentration of aluminum for peaks at 7, 12 and 17 ppm from the reference AlI_3 solution of known concentration in the NMR insert.

$$\frac{[Al_{sample}] \cdot V_{sample}}{[Al_{reference}] \cdot V_{reference}} = \frac{I_{sample}}{I_{reference}} \qquad (13.1)$$

Here, V_{sample} and $V_{reference}$ are the volume in the NMR tube and NMR insert, respectively. I_{sample} and $I_{reference}$ are the integrated peak areas for the peaks of interest and AlI_3 reference peaks in a spectrum. $[Al_{reference}]$ is the concentration of AlI_3, which is known. Therefore, the concentration corresponding to each peak [Al sample] can be subsequently determined. After each Al-SFA complex concentration is known, a stability constant, K, for the Al-SFA complexes can be calculated [23].

Table 13.1 Percentage of each aluminum species in equilibrium with the oxalate ligand as measured experimentally by NMR and calculated with MINTEQ

Al:Ox = 1:1 at 2 mM Al and 0.1 M Ionic Strength					
		Al^{3+}	$Al\text{-}Ox^{+}$	$Al\text{-}Ox_2^{-}$	$Al\text{-}Ox_3^{-3}$
pH 3.0	NMR	14.2	71.4	14.3	0.0
	MINTEQ	13.6	74.0	12.4	0.0
pH 4.0	NMR	6.8	69.4	23.9	0.0
	MINTEQ	13.0	73.8	12.9	0.0
pH 5.0	NMR	0.0	75.3	24.7	0.0
	MINTEQ	10.0	70.8	14.4	0.0
Al:Ox = 1:2 at 2 mM Al and 0.1 M Ionic Strength					
		Al^{3+}	$Al\text{-}Ox^{+}$	$Al\text{-}Ox_2^{-}$	$Al\text{-}Ox_3^{-3}$
pH 3.0	NMR	3.9	13.6	76.4	6.2
	MINTEQ	0.0	24.2	75.0	0.0
pH 4.0	NMR	0.0	11.4	73.8	14.8
	MINTEQ	0.0	15.1	83.8	1.1
pH 5.0	NMR	0.0	8.0	80.2	11.8
	MINTEQ	0.0	13.7	85.0	1.2

Two selected data sets are shown for illustration. Table 13.2 shows the experimental NMR results in molar concentration for the speciation of Al^{3+}(1 mM) with SFA(1000 mg L^{-1}). At pH 3 and 4, more than 92% of the aluminum species are accounted for in the system observed by NMR. However, at pH 5, only 57 % is observed, meaning that the remaining 43 % has been removed from the system, probably by precipitates forming at pH 5. Looking at the individual peaks in Figure

13.2, as pH is increased, peak intensity shifts from Al^{3+} (0 ppm) and Al-SFAox (7 ppm) to Al-SFAox$_2$ (12 ppm) and Al-SFAox$_3$ (17 ppm). Table 13.3 shows the results in molar concentration of the speciation of Al^{3+}(5 mM) with SFA(7500 mg L^{-1}). At pH 3, 4.49 out of 5 mM or about 90 % of the aluminum species are observed. However, at pH 4 and 5, some aluminum species are not observed due to precipitates forming in solution. As for individual peak intensity, since more SFA is present in the solution, peaks for Al-SFAox (7 ppm) and Al-SFAox$_2$ (12 ppm) are dominant (Figure 13.2). Also, increasing the pH from 3 to 5 causes peak intensities to shift because increasing pH promotes aluminum binding with SFA to form complexes. At higher concentrations and pH, precipitates create a problematic situation. At this time, investigation is still ongoing to determine methods to account for loss of aluminum and SFA due to precipitates in solution. A stability constant K can be defined as in Eq. 13.2.

$$K = \frac{[ML]}{[M_f][L_f]} \tag{13.2}$$

Here, [ML] is the molar concentration of a metal-ligand complex, $[M_f]$ is the molar concentration of free metal and $[L_f]$ is the free ligand molar concentration in the equilibrium system. ^{27}Al NMR allows a direct measure of [ML] and $[M_f]$. However, some calculations are necessary to obtain $[L_f]$.

Table 13.2 Speciation (in mM) of Al^{3+}(1 mM)-SFA(1000 mg/L)

	Al^{3+} (0 ppm)	Al-SFAox (7 ppm)	Al-SFAox$_2$ (12 ppm)	Al-SFAox$_3$ (17 ppm)	Total Observed
pH 3.0	0.67	0.31	0.00	0.00	0.98
pH 4.0	0.43	0.39	0.10	0.00	0.92
pH 5.0	0.17	0.22	0.18	0.00	0.57

Table 13.3 Speciation (in mM) of Al^{3+}(5 mM)-SFA(7500 mg/L)

	Al^{3+} (0 ppm)	Al-SFAox (7 ppm)	Al-SFAox$_2$ (12 ppm)	Al-SFAox$_3$ (17 ppm)	Total Observed
pH 3.0	2.01	1.80	0.63	0.00	4.49
pH 4.0	1.30	1.30	0.87	0.00	3.42
pH 5.0	0.23	0.52	0.83	0.41	2.00

Based on our observations from the NMR data presented above and previously [19], Al-SFA binding can be described by oxalic acid type binding sites. We define L as a bidentate oxalic acid binding site on the soil fulvic acid. For our sample of soil fulvic acid, it has been reported that there are 8.2 carboxyl groups (meq)/soil fulvic acid (g) [15]. Since one oxalic binding site requires two carboxyl groups (bidentate), we can then calculate the total potential oxalic binding sites $[L_T]$ on the fulvic acid from the mass in grams of fulvic acid in our solutions. To obtain free ligand concentration $[L_f]$, ligands already complexed with metals are subtracted from the total oxalic binding site concentration as shown in Eq. 13.3.

$$[L_f] = [L_T] - [ML] - [ML_2] - [ML_3] \tag{13.3}$$

Also, because fulvic acid is a weak acid, the degree of acid dissociation α is important. Our model is a bidentate oxalate ligand. Hence, acid dissociation can be described by a pH-dependent α_2, Eq. 13.4 [24].

$$\alpha_2 = \frac{K_{a1}K_{a2}}{[H^+]^2 + [H^+]K_{a1} + K_{a1}K_{a2}} \tag{13.4}$$

The acid dissociation constants assuming a diprotic SFA, K_{a1} (4×10^{-3}) and K_{a2} (2×10^{-5}) have been reported previously [15] and can be used to calculate acid dissociation for pH 3, 4 and 5. In addition, since α_2 $[L_f]$ represents the diprotic dissociated portion of the total free ligand concentration, the pH-dependent diprotic stability constant K_1 can be calculated from Eq. 13.5.

$$K_1 = \frac{[ML]}{[M_f]\alpha_2[L_f]} \tag{13.5}$$

Similarly, the step-wise second stability constant K_2 can be calculated as Eq. 13.6. The stability constants obtained are summarized in Table 13.4.

$$K_2 = \frac{[ML_2]}{[ML]\alpha_2[L_f]} \tag{13.6}$$

Table 13.4 Comparison of stability constants calculated from experimental ^{27}Al NMR and graphically approximated from Scatchard plots and Langmuir isotherms

	pH 3.0		pH 4.0		pH 5.0	
	Log K_1	LogK_2	Log K_1	Log K_2	Log K_1	Log K_2
NMR	3.6	2.8	2.8	2.4	2.5	2.4
Scatchard	2.4	n/a	3.0	2.7	3.5	3.6
Langmuir	2.8	n/a	2.6	3.5	3.7	4.0

K_1 is larger than K_2 since it is easier for aluminum to be bound with one SFAox than with two $SFAox_2$ as the ligands may hinder each other. Both K_1 and K_2 are lower than expected, which may occur because of the removal of aluminum-SFAox complexes (ML, ML_2 and ML_3) by precipitates in solution and as pH approaches 5, forming hydrolysis products that can affect the binding affinity of aluminum, which in turn causes the denominator in the stability equation to be higher and K to be smaller. Also, the acid dissociation constants (K_{a1} and K_{a2}) used in this calculation were published [15] almost thirty years ago, so the accuracy of these numbers may be questionable. Nevertheless, preferential binding with aliphatic sites on SFA by aluminum is demonstrated.

Scatchard plot and Langmuir regression analyses were also applied to graphically deduce conditional stability constants for comparison. They are also summarized in Table 13.4. The results are in general slightly larger than the theoretical calculations and agree with each other within a reasonable discrepancy. The small difference may result from the preliminary and limited nature of the data sets, the fact that precipitates are forming at pH 5 and the relatively small concentration range used. Both Scatchard and Langmuir regressions have been traditionally applied to determine the total bound versus free or unbound metal to organic species [25,26]. Here, since we have a measure of each bound species, we can plot each bound species versus free metal to determine a conditional stability constant for that species. Having done this, the slope of the Scatchard plot is actually positive instead of negative. This suggests that in this concentration range the cooperative nature of the binding between sites is positive. In other words, the affinity of one site is affected by the affinity of other sites, which is expected since the aliphatic oxalic type of site on SFA needs to be spatially close enough to bind with aluminum. Interestingly, the slope of the Langmuir isotherm is also opposite to traditional regressions, showing a negative slope. The details of these observations are still under investigation.

13.4. CONCLUSIONS

Solution state ^{27}Al nuclear magnetic resonance spectroscopy is demonstrated to be a powerful tool in investigating the binding site environment of aluminum ions with soil fulvic acid. Four distinct peaks were observed in an Al-SFA system, namely free Al^{3+} (0 ppm), aluminum ion bound with one oxalic acid type binding site Al-SFAox (7 ppm), aluminum ion bound with two oxalic acid sites Al-$SFAox_2$ (12 ppm) and aluminum ion bound with three oxalic acid sites Al-$SFAox_3$ (17 ppm). Stability constants were calculated from the experimental results and were compared with results obtained by Scatchard and Langmuir regressions. They agreed with each other reasonably well despite the limited number of data points. Currently, we are conducting experiments to minimize effects from precipitates formed in solution. In the future, we plan to investigate other fulvic acids using ^{27}Al NMR to better understand the binding behavior of aluminum in the natural environment.

ACKNOWLEDGEMENTS

The authors wish to thank Mr. Xiao-Dong Wu and Dr. Jeffery Njus for NMR instrumental setup, Dr. Melisenda McDonald and Dr. Gayathri Vasudevan for regression analysis, and Erick Estime, Qi Zhang, Raja Manjeeth Avanadula, Haengah Lee and Julie Nobrega for helpful discussions.

REFERENCES

1. Schnitzer M, Skinner S. Gel filtration of fulvic acid, a soil humic compound. In: UNIPUB eds. *Isotopes and radiation in soil organic matter studies*, Vienna: Int. Atomic Energy Agency, 1968: 41-55.

2. Dempsey BA. Reaction between fulvic acid and aluminum: effects on the coagulation process. In: Seilacher AA, Bhattacharji S eds. *Aquatic humic substances: Influence on fate and treatment of pollutants*. Washington DC: Library of Congress Cataloging-in-Publication Data, 1989: 409-424.
3. Harris WR, Berthon G, Day JP, Exley C, Flaten TP, Forbes WF, Kiss T, Orvig C, Zatta PF. Speciation of aluminum in biological systems. *J. Toxicol. Environ. Health*, 1996; 48: 543-568.
4. Makjanic J, McDonald B, Chen CP, Watt F. Absence of aluminium in neurofibrillary tangles in Alzheimer's disease. *Neurosci.*, 1998; 240: 123-126.
5. Klein GL. Metabolic bone disease of total parenteral nutrition. *Nutrition*, 1998; 14: 149-152.
6. Exley C. A molecular mechanism of aluminum-induced Alzheimer's disease? *J. Inorg. Biochem.*, 1999; 133-140.
7. Solomon B, Koppel R, Jossiphov J. Immunostaining of calmodulin and aluminium in Alzheimer's disease-affected brains. *Brain Res. Bull.*, 2000; 55: 253-256.
8. Flaten TP. Aluminum as a risk in Alzheimer's disease, with emphasis on drinking water. *Brain Res. Bull.*, 2001; 55: 187-198.
9. Ryan DK, Shia CP, O'Conner DV. Fluorescence spectroscopic studies of Al-fulvic acid complexation in acidic solutions. In: Gaffney JS, Marley NA, Clark SB eds. *Humic and fulvic acids: Isolation, structure, and environmental role.* American Chemical Society, 1996: 125-139.
10. Otto W, Carper W, Larive C. Measurement of cadmium(II) and calcium(II) complexation by fulvic acids using ^{113}Cd NMR. *Environ. Sci. Technol.*, 2001; 35:1463-1468.
11. Otto W, Burton S, Carper W, Larive C. Examination of cadmium(II) complexation by the Suwannee river fulvic acid using ^{113}Cd NMR relaxation measurements. *Environ. Sci. Technol.*, 2001; 35: 4900-4904.
12. Yokoyama T, Abe H, Kurisaki T, Wakita H. ^{13}C and ^{27}Al NMR study of the interaction between aluminum ion and iminodiacetic acid in acidic aqueous solutions. *Anal. Sci.*, 1999; 15: 393-395.
13. Bertsch PM, Barnhisel RI, Thomas GW, Layton WJ, Smith SL. Quantitative determination of aluminum-27 by high-resolution nuclear magnetic resonance spectrometry. *Anal. Chem.*, 1986; 58: 2583-2585.
14. Karweer SB, Mhatre SN, Pillai BP, Lyer RK, Moorthy PN. Studies on speciation of aluminum complexes with malic acid by high field ^{27}Al NMR spectroscopy. *Ind. J. Chem.*, 1993; 32A: 502-505.
15. Weber JH, Wilson SA. The isolation and characterization of fulvic acid and humic acid from river water. *Water Res.*, 1975; 9: 1079-1084.
16. Hook J, Lu XQ, Howe RF. A ^{27}Al NMR study of aluminum: Humic substances interactions. *Bull. Mag. Reson.,* 1995; 18: 186-187.
17. Howe RF, Lu XQ, Hook J, Johnson WD. Reaction of aquatic humic substances with aluminum: a ^{27}Al NMR study. *Mar. Freshwater Res.*, 1997; 48: 377-383.
18. Haraguchi H, Fujiwara S. Aluminum complexes in solution as studied by alumminum-27 nuclear magnetic resonance. *J. Phys. Chem.*, 1969; 73: 3467-3473.
19. Lee NCY, Ryan, DK. Study of fulvic-aluminum(III) ion complexes by ^{27}Al solution NMR. In: Ghabbour EA, Davies G eds. *Humic substances: Nature's most versatile materials*, New York: Taylor & Francis, 2003: 219-228.

20. Kerven GL, Larsen PL, Bell LC, Edwards DG. Quantitative ^{27}Al NMR spectroscopic studies of Al(III) complexes with organic acid ligands and their comparison with GEOCHEM predicted values. *Plant Soil,* 1995; 171: 35-39.
21. Allison JD, Brown DS, Novo-Gradac KJ. *MINTEQA2/PRODEFA2, A geochemical assessment model for environmental systems: version 3.0 user's manual.* Athens, GA: U.S. Environmental Protection Agency, EPA/600/3-91/021.
22. Sillen LG, Martell AE. *Stability constants of metal-ion complexes.* London: Metcalfe & Cooper, 1964.
23. Lambert J, Buddrus J, Burba P. Evaluation of conditional stability constants of dissolved aluminum/humic substance complexes by means of ^{27}Al nuclear magnetic resonance. *Fres. J. Anal. Chem.,* 1995; 351: 83-87.
24. Skoog DA, West DM, Holler FJ. *Fundamentals of analytical chemistry,* 7th Ed., Philadelphia: Saunders College Publishing, 1996.
25. Pott DB, Alberts JJ, Elzerman AW. The influence of pH on the binding capacity and conditional stability constants of aluminum and naturally-occurring organic matter. *Chem. Geol.*, 1985; 48: 293-304.
26. Scatchard G. *The attractions of proteins for small molecules and ions.* New York: Acad. Sci., 1949; 51: 660-672.

Chapter 14

AFFINITY OF SOIL AND LIGNITIC HUMIC ACIDS FOR Cu(II) AND Cd(II) IONS

Miloslav Pekař,[1] Martina Klučáková,[1] Gabriela Barančíková,[2] Mikuláš Madaras[2] and Jarmila Makovníková[3]

[1] Institute of Physical and Applied Chemistry, Faculty of Chemistry, Brno University of Technology, Purkyňova 118, 612 00 Brno, Czech Republic
[2] Soil Science and Conservation Research Institute, Research Station, Reimannova 1, 080 01 Prešov, Slovakia
[3] Soil Science and Conservation Research Institute, Research Station, Mládežnícka 36, 974 05 Banská Bystrica, Slovakia

14.1. INTRODUCTION

Nature has created a number of fascinating materials regarding both their structures and functions. Soil and coal may serve as nice examples. Soil belongs among indispensable conditions for life on Earth. It is closely related to the plant population but also the lives of animals depend on it, ranging from microorganisms to great mammals. Soil is really a dynamic system, continuously changing its state, composition, properties, reflecting the effects of the surrounding environment and sensitive to it. On the other hand, coal may be from the soil viewpoint considered as an old and essentially static system. Coal has been forming for millions of years and very, very slowly maturing from peat or lignite to anthracite. Remember that, according to our ideas on its origin, coal is intimately connected to plants. No wonder then that soil and coal, especially its young form – lignite – meet in another beautiful natural material – humic substances (HSs).

Humic substances are well known to be one of the most important soil constituents. They are the principal components of soil organic matter and have indispensable roles for soil and the environment in general. They participate in both nutrient and toxin transport, and they support soil structure and pH. From this viewpoint HSs present in coal deposits possess "no" function and are in fact being destroyed during continuing coalification. Huge coal basins are perhaps too deep and compact to participate, like soil, in environmental dynamics and to attract scientific attention aimed at investigating their role in subtle environmental events.

Of course, low rank (this strange word in the humic world stems only from the fuel point of view) coal is a good source to produce HSs for practical technical applications. It seems like a small natural miracle that materials of such different age and of various yet similar origins have such close properties and may in some sense substitute for each other. It has been proved that soil organic matter is capable of binding both inorganic and organic pollutants and thus suppress their mobility and harmful effects. This natural capability is often not able to cope with human originated pollutant amounts. The starting idea of our work was to formulate some organic-mineral preparations as an aid to the natural soil remediation function. These preparations are to be based on natural materials and should synergistically combine the properties of organic and mineral constituents.

The organic component would, of course, be humic acids (HAs). Humic acids can be isolated from some sources in sufficient supply and humics content: lignite. The first logical step of our effort was to compare the affinity of soil and lignitic HAs. We have started with simple tests of affinity for two metal cations – copper and cadmium. The former is known for general high affinity for HSs. Cadmium represents one of the potential contaminants.

Metal binding by HSs was and is the subject of many studies. Our aim was perhaps rather different. The affinity of Slovak soil HAs, constituents of potential contaminated areas to be remediated, should be compared with the affinity of Moravian lignite-derived HAs having potential for soil amendment. The metal-binding capacity of HAs isolated from North-Bohemian coals, another deposit in the Czech Republic, was studied by Čežíková et al. [1]. It was found that soil HAs isolated from chernozem had markedly higher metal-binding capacity in comparison with coal-derived preparations.

Davies et al. [2] published a thorough study on metal binding by solid peat and soil HAs and its tightness as revealed by water and acid washing. It was demonstrated with 10 metals that the tight metal binding capacities of HAs from different origins are similar under the same loading conditions.

Another comprehensive investigation of adsorption of metal ions by HAs extracted from brown coals was published by Martyniuk and Więckowska [3]. They also included a brief review of the literature data on this problem. The main conclusion was that although the affinity of HAs of various origins for various heavy metals is indisputable, opinions on the structures of metal-humic complexes and the involvement of various functional groups in them are not unified. Also, there are differences in published comparative series of adsorption capacities for various metals. From their own results the authors concluded that brown coal HAs have the highest affinity for Pb, Ag, Hg, Cu, Ba and Cd ions from 17 metals tested and that metal binding is realized mainly through carboxylic groups. Pb, Ag, Hg and Cu ions are able to interact with phenolic groups, even at low pH.

Close to our ultimate goal is, for example, the work of Arias et al. [4], who investigated the enhancement of copper and cadmium adsorption on kaolin by the presence of HAs. Enhancement was clearly demonstrated for both single and binary metal ion solutions.

14.2. MATERIALS AND METHODS

14.2.1. Materials

Soils. Five soils from Slovakia were selected as sources of HAs. They come from the long-term soil quality monitoring project made by the Soil Science and Conservation Research Institute and have been selected to represent soils of various quality:

1. Calcaric Fluvisols on carbonated meadow sediments
2. Gleyic Fluvisols on non-carbonated meadow sediments
3. Calcaric Chernozems on loess
4. Haplic Luvisols on loess loams
5. Dystric Cambisols on flysch

Lignite. Lignite mined in the South Moravia, Czech Republic, locality Mikulčice was used. Prior to extraction of HAs, the lignite was dried to an equilibrium moisture level of about 7% (by weight) and crushed to particle sizes smaller than about 3 millimeters. Typical compositions [5] of this material and its combustible part are given in Tables 14.1 and 14.2, respectively.

Table 14.1 Average composition of lignite (weight %)

Water	Ash	Combustible matter	C	H	O	S
48	12	40	26	2	11	0.8

Table 14.2 Average composition of lignite combustible matter (weight %)

C	H	O	S	N
65	5.2	28	1.2	0.9

Other Materials. Cadmium nitrate and copper sulfate were used as sources of metal ions. Ammonium nitrate and Titriplex II (Merck's trademark of ethylenedinitrilo-tetraacetic acid, EDTA) were used in desorption studies.

14.2.2. Methods

Extraction of Humic Acids. Humic acids were isolated by the standard alkaline extraction method used in Slovak pedological procedures [6]. Briefly, HAs were extracted with 0.1 M NaOH, precipitated with hydrochloric acid, purified by two-fold re-dissolving and precipitating, and finally by treating with an HCl-HF mixture. The washed product was dried at 50°C. HAs were characterized by standard elemental analysis, determination of moisture and ash content, and solution NMR.

Sorption Experiments. Sorptions were made at laboratory temperature using a simple batch method. The solid sample was weighed (0.1 g) into a test tube and 5 mL of metal ion solution was added. Cadmium nitrate and copper sulfate were used in concentration ranges of 0.3-10 mg/dm^3 and 0.06-12.7 g/dm^3, respectively. Test tubes were shaken for 24 hours, the solid was separated by centrifugation and the supernatant was analyzed for metal ion content.

Desorption Experiments. Desorptions were made at laboratory temperature in a batch system with a two-step sequential procedure. In the first step, 1 mol/dm^3 ammonium nitrate solution was used while the second step was made with 0.025 mol/dm^3 NH_4^+-EDTA. The latter was prepared by dissolving 7.31 g of Titriplex II in 1 liter of deionized water and adjusting the pH to 4.6 with a diluted (1:1, v/v) solution of ammonia. These are the first two steps of the seven-step leaching procedure [7] for speciation of the binding form of heavy metals in soils. In the first step, the freely exchangeable or mobile metal fraction should be released, whereas in the second step the fraction of potentially or easily available ions should be leached. The solid sample of the preceding sorption experiment was washed with water and added to 5 mL of ammonium nitrate solution. The sample was shaken for 24 hours, the solid was separated by centrifugation and the supernatant analyzed for metal ion content. The procedure was repeated with the NH_4^+-EDTA complex in the same manner.

Concentrations Determination. Copper concentrations were determined spectrophotometerically, whereas those of cadmium were determined by polarography using calibration curves.

14.3. RESULTS

14.3.1. Parameters of Humic Acids

Basic parameters of isolated HAs are collected in Table 14.3. Percentages of aromatic and aliphatic carbon were determined from NMR spectra [8].

Table 14.3 Parameters of isolated humic acids

Source	C	H	O	N	C_{aliph}	C_{arom}
			weight %			%	
Fluvisol 1	58.1	5.46	31.0	5.3	43.1	34.4
Fluvisol 2	58.3	4.99	31.2	5.5	46.9	32.8
Chernozem	59.9	3.07	32.1	5.0	33.7	42.2
Luvisol	58.0	4.13	32.9	4.9	41.8	37.2
Cambisol	56.2	5.24	32.4	6.1	48.6	29.0
Lignite	63.7	3.02	31.7	1.6	34.7	44.0

The HA isolated from lignite is indeed rather different from the whole group of soil materials, particularly in respect to nitrogen contents. Lignite HAs also possessed the highest percentage of carbon.

14.3.2. Sorption Isotherms and Adsorbed Amounts

Adsorption isotherms for cadmium and copper on HAs of various origins are shown in Figures 14.1 and 14.2, respectively. For the purposes of our work, relative adsorbable amounts are more important and are illustrated in Figures 14.3 and 14.4.

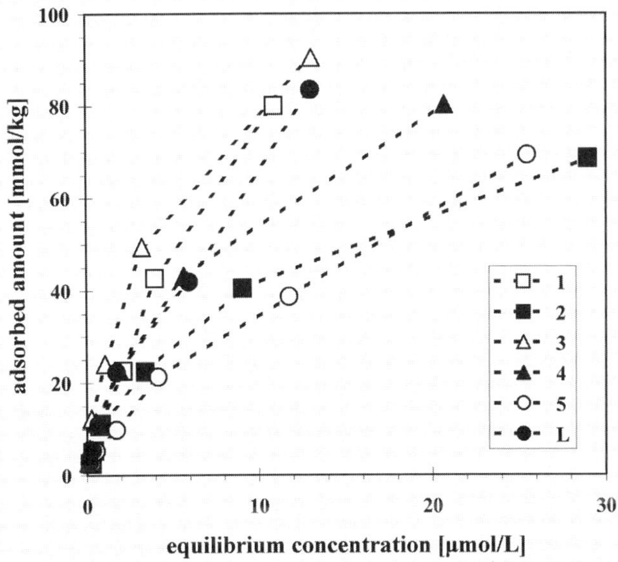

Figure 14.1 Cadmium adsorption isotherms on HAs isolated from soils 1-5 and lignite (L)

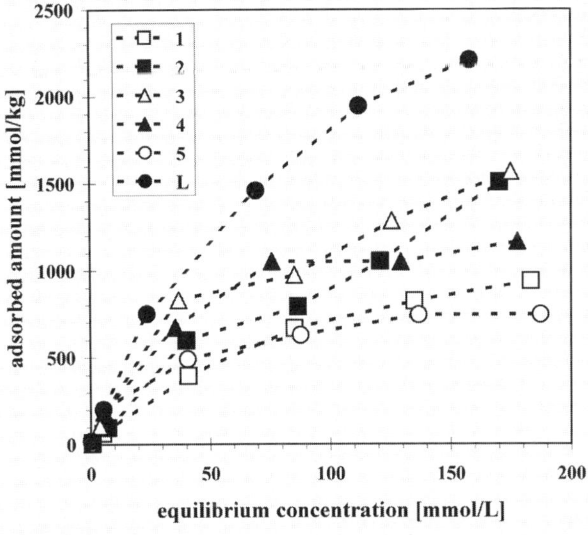

Figure 14.2 Copper adsorption isotherms on HAs isolated from soils 1-5 and lignite (L)

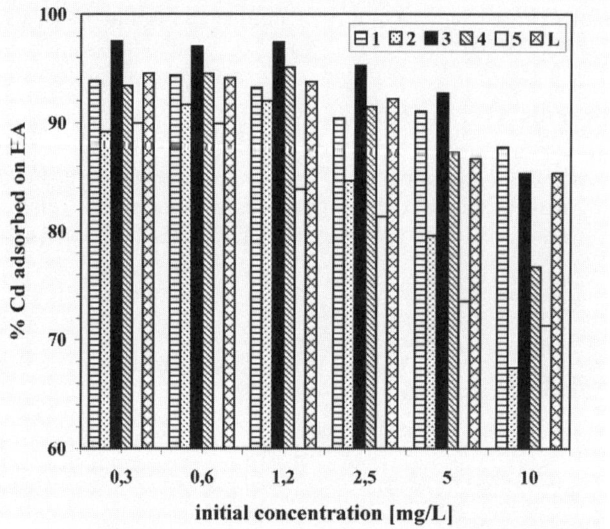

Figure 14.3 Cadmium uptake on HAs isolated from soils 1-5 and lignite (L)

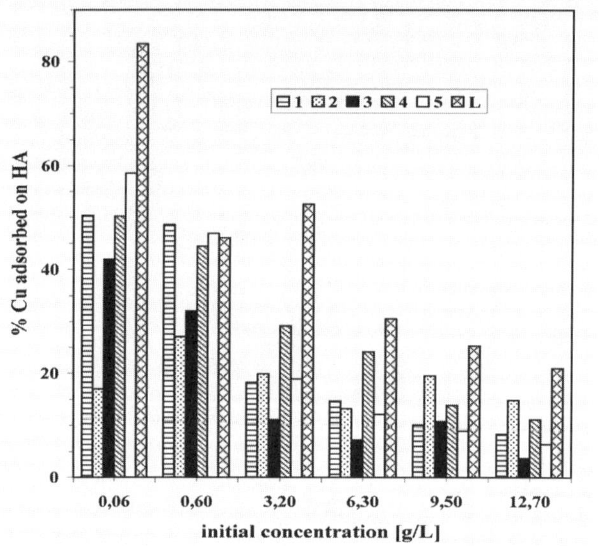

Figure 14.4 Copper uptake on HAs isolated from soils 1-5 and lignite (L)

Adsorption isotherms were fitted with both Langmuir (Eq. 14.1) and Freundlich (Eq. 14.2) models.

$$a = a_{\max} bc_{eq} /(1 + bc_{eq}) \tag{14.1}$$

$$a = K(c_{eq})^n \tag{14.2}$$

Here, a is the adsorbed amount at equilibrium and c_{eq} is the equilibrium concentration in solution. Corresponding parameters are given in Tables 14.4 and

14.5 together with squared correlation coefficients of non-linear data and Eqs. 14.1 and 14.2.

From the latter it is seen that both equations fit the data quite well. As expected, isotherms showing at least the beginning of saturation are slightly better described by the Langmuir equation, for copper data in particular.

14.3.3. Desorption Experiments

Results of both desorption steps for cadmium and copper are given in Figures 14.5 and 14.6, respectively. Cumulative plots are shown in Figures 14.7 and 14.8.

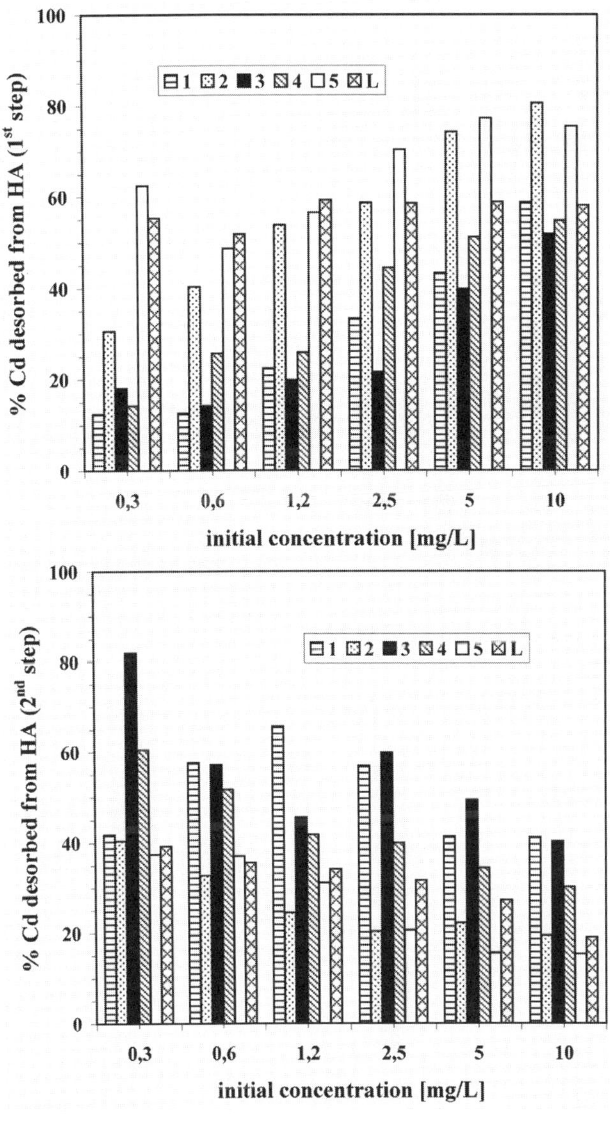

Figure 14.5 Cadmium desorption from HAs isolated from soils 1-5 and lignite (L) during the first (a) and second (b) steps

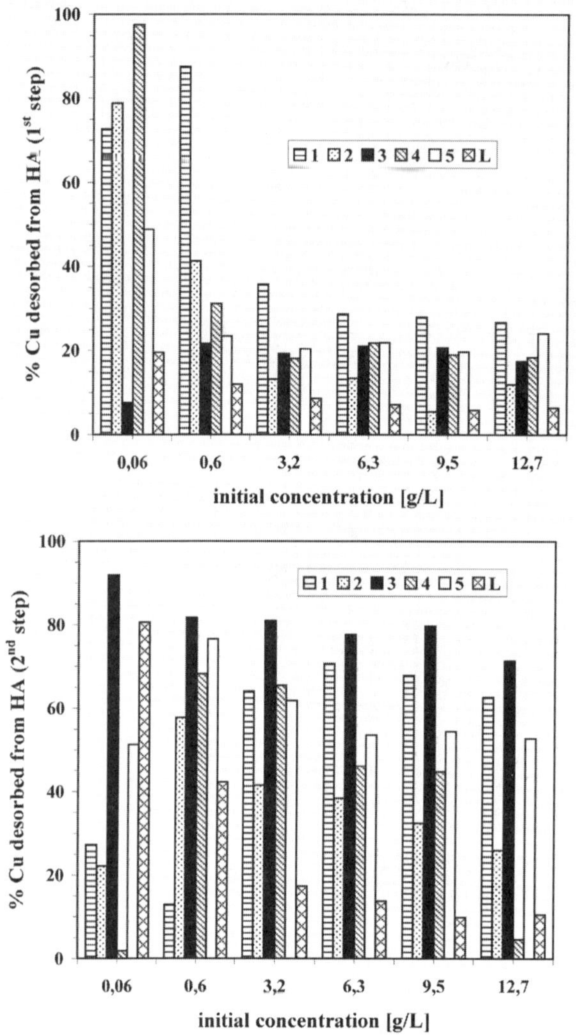

Figure 14.6 Copper desorption from HAs isolated from soils 1-5 and lignite (L) during the first (a) and second (b) steps

Table 14.4 Parameters of adsorption isotherms for cadmium binding and their squared correlation coefficients

Source of HAs	Langmuir			Freundlich		
	a_{max}, mmol/g	b, dm^3/mmol	R^2	K, mol^{1-n} dm^{3n}/kg	n	R^2
Fluvisol 1	0.146	106	0.998	874	0.806	0.996
Fluvisol 2	0.0862	122	0.994	90.4	0.669	0.989
Chernozem	0.109	339	0.993	157	0.646	0.989
Luvisol	0.100	176	0.993	127	0.666	0.991
Cambisol	0.0973	77.7	0.986	116	0.701	0.999
Lignite	0.119	140	0.980	323	0.733	0.996

Table 14.5 Parameters of adsorption isotherms for copper binding and their squared correlation coefficients

Source of HAs	Langmuir a_{max}, mmol/g	b, dm^3/mol	R^2	Freundlich K, $mmol^{1-n}$ dm^{3n}/g	n	R^2
Fluvisol 1	1.50	9.10	1.00	4.38	0.806	0.985
Fluvisol 2	3.63	3.90	0.988	8.56	0.932	0.989
Chernozem	2.26	11.6	0.986	5.49	0.670	0.983
Luvisol	1.35	34.3	0.995	9.98	0.938	0.898
Cambisol	0.884	32.6	0.994	3.62	0.712	0.944
Lignite	3.00	15.7	0.998	8.86	0.698	0.993

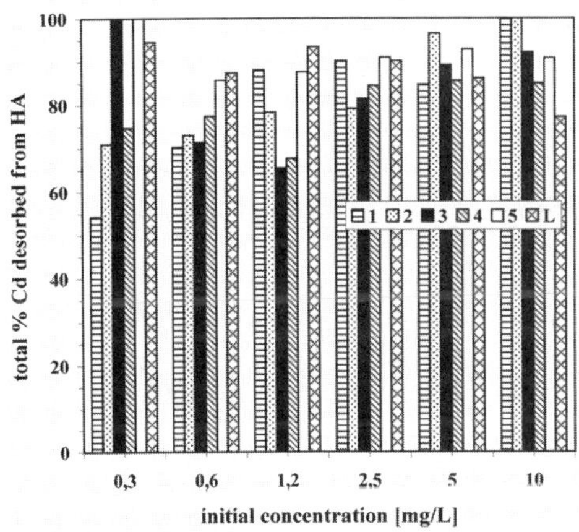

Figure 14.7 Total desorbed amount of cadmium from HAs isolated from soils 1-5 and lignite (L)

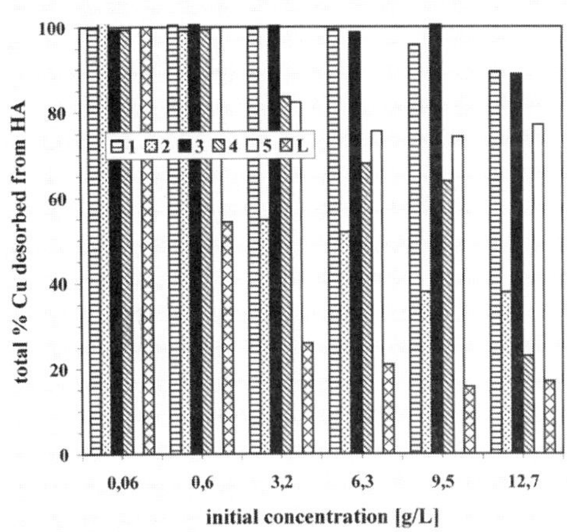

Figure 14.8 Total desorbed amount of copper from HAs isolated from soils 1-5 and lignite (L)

14.4. DISCUSSION

14.4.1. Sorption Behavior

Adsorption isotherms measured for both metals show clear differences among HAs of different origin. The cadmium sorption isotherm on lignitic HAs lies somewhere in the middle of all measured isotherms. The sorption capacity and behavior of humics isolated from such an old deposit is thus fully comparable with humics originating from a much younger and dynamic environment, soil.

No cadmium isotherm shows saturation of the humic sorbent, undoubtedly due to the low concentrations used, but a curve can be seen in some of them, indicating the start of achieving a saturation limit. From the position of the adsorption isotherm it can be deduced that chernozem HAs have the highest whereas cambisols have the lowest affinity for cadmium ions in the respective concentration regions.

Lignite HAs possess superior affinity for copper ions as compared to the other humics, (Figure 14.2). The chernozem product follows while cambisol and fluvisol 1 HAs have the lowest capacity. Most copper isotherms show clear approaching of the saturation limit, which is expected in the high concentration range employed.

Whereas the Langmuir isotherm has some theoretical basis, namely a kinetic background, the Freundlich equation is strictly empirical and its parameters can hardly be given some physical interpretation. Values of parameters K cannot be compared unless the value of the exponent n is the same because the K parameters have different dimensions and units. Exponent n is sometimes used to discuss the degree of non-linearity of adsorption isotherms, which in our opinion is of little value. Nevertheless, all exponents showed clear non-linear isotherm behavior, except perhaps for sorption of copper on fluvisol 2 and luvisol HAs.

Parameter a_{max} of the Langmuir isotherm estimates the maximum sorption capacity, whereas b is the equilibrium constant for adsorption and can be considered to be related to the adsorption strength. Accepting this interpretation of parameters, lignite humic material is evaluated as having the second highest cadmium sorption capacity and moderate strength. Fluvisol 1 humics have the highest capacity and chernozems the highest adsorption strength. Ranking of the lignite preparation remains unchanged when data for copper are considered (Table 14.5). Fluvisol 2 HAs have the highest sorption capacity for copper whereas luvisols have the highest adsorption strength.

For our ultimate aim the relative adsorbed amounts perhaps are more significant. From Figure 14.3 it can be concluded that all humic preparations adsorbed most of the cadmium initially present in the solution. However, percentage uptakes slowly decrease with increasing initial cadmium concentration. This is another indication of slowly attaining saturation. Based on Figure 14.3, chernozem, fluvisol 1 and lignite can be ranked as containing HAs with the highest affinity for cadmium and cambisol and fluvisol 2 the lowest.

Figure 14.4 apparently shows much lower capacity for copper sorption compared to cadmium. Note, however, the significantly higher concentration levels of copper solutions in the sorption experiments. For the same reason, uptake percentages decrease with increasing initial copper concentration because the sorption system gets closer to saturation. Whereas lignite HAs can be viewed as having the highest affinity for copper, there is no clear indication for material with the lowest affinity. Ranking depends on the initial metal concentration.

Affinity ordering made according to parameters of the Langmuir isotherm thus does not fully accord with conclusions made from percentage uptakes. Interpretation of fitted parameters should be done carefully and preferably after the whole isotherm, i.e. up to the saturation limit has been measured.

Maximum adsorption capacities estimated from our measurements using Langmuir isotherm analysis are comparable with published data. From Tables 14.4 and 14.5, capacities of 0.086-0.15 mmol/g for cadmium and 0.88-3.6 mmol/g for copper follow, respectively. Čežíková et al. [1] reported for cadmium and copper in the range 0.5-0.8 mmol/g for different Czech brown coals derived HAs capacities. Davies et al. [2] found capacities of HAs from various sources to be about 1 mmol/g for cadmium and 1.5-2.3 mmol/g for copper. The review by Martyniuk and Więckowska [3] reports capacities of various metals to be within the range 0.03-0.5 mmol/g, cadmium being at the lower limit and copper at the higher limit. Their own experiments gave capacities of 0.9-1.5 mmol/g for cadmium and 1.1-2.4 mmol/g for copper.

Capacities for copper determined in this work are thus in very good agreement with published information and even higher. On the other hand, capacities determined for cadmium seem to be lower than in most cited works. However, our experiments demonstrated that most of the cadmium was removed from tested solutions under our experimental conditions, that is, the actual metal capacity was quite sufficient. Maximum capacity was estimated from the Langmuir isotherm and its value is thus only an extrapolation from measured data because saturation was not achieved under our experimental conditions. Studies reporting higher capacities for cadmium were performed with much more concentrated solutions, which are more appropriate to assess maximum adsorption capacity with no need for extrapolation. In any case, capacities of HAs for cadmium and copper are fully comparable and usually much better than capacities of all alternative cost-effective sorbents tested by Ulmanu et al. [9]; their best sorbents had estimated capacity (from Langmuir isotherms) of only 0.1 mmol/g for cadmium and only 0.2 mmol/g for copper.

Capacities determined in this work are also in line with results obtained in our laboratory for capacities of the original South Moravian lignite [10].

14.4.2. Desorption Behavior

Let us compare overall behavior first. The amount of cadmium desorbed in the first step increases or stays approximately constant with increasing initial cadmium concentration (Figure 14.5a). By contrast, the amount desorbed during the second step usually decreases with increasing initial concentration (Figure 14.5b). The first step desorbing agent should attack the weakest bonds whereas during the second step tighter bonds should be broken. Therefore, cadmium at its low concentrations preferably occupies tight binding sites and upon increasing its concentration more and more weak sites are occupied.

Regarding individual HAs and the total desorption results (Figure 14.7), there is no clear indication of the best HA. At low initial cadmium concentrations, humics prepared from fluvisol 1 and chernozem show the highest affinity, while at high initial concentrations lignite and luvisol materials are the best.

Copper showed different behavior, particularly in the first step, where for most HAs a decrease of desorbed amount with increasing initial concentration was observed (Figure 14.6a). Trends in the second step are not so clear (Figure 14.6b), but most HAs showed a similar decrease, which is somewhat in contrast with what could be expected from the results of the first step. Tight binding sites should be saturated at the lowest initial concentration used and weak binding sites are occupied upon increasing the initial metal concentration.

Considering the total desorbed amount of copper, superior affinity of lignitic HAs is observed. On the other hand, the chermozem HA with the closest parameters (see Table 14.3) exhibits low affinity while luvisol and fluvisol 2 HAs possess good affinity.

Again, desorbed amounts do not agree very well with adsorption strength evaluated from the Langmuir isotherm parameter *b*. However, it should be remembered that desorptions were made using specific agents whereas the Langmuir isotherm was determined with aqueous solutions.

14.4.3. Comparison of Affinity

Interpretation and explanation of the whole data set is not an easy task. Sometimes the trends are not unambiguously clear and there is no clear affinity favorite among all the HAs. Nevertheless, it can be roughly stated that lignite HAs have in general very good overall affinity for Cd(II) and Cu(II) ions when compared with soil preparations. The most distinct parameter of lignite materials is low nitrogen content followed by the highest carbon content. Also, the ratio of aromatic to aliphatic carbon is the highest. It can be therefore speculated that lignite HAs contain high amounts of aromatic structures with oxygen rich (relatively low nitrogen) groups giving them high affinity. This confirms general ideas on the structure and function of HSs. The ratio of oxygen to carbon contents is high for lignite HAs, so phenolic and carboxylic groups should be mainly responsible for the good affinity for metals.

14.5. CONCLUSIONS

Our experiments confirm the old truth – although HAs denote some specific class of substances, there are differences in details of their properties. They are all capable of binding metals in general, but the quality and quantity of such interactions depend on the HA source and origin. This may have important consequences – the functions of particular HAs in a particular environment may depend on its particular state and be quite different from those of similar matter elsewhere, like the different behavior of individual humans who still belong to the same class of beings.

Regarding our main aim – supporting the soil immobilization function – the experiments gave very promising results. In our model experiments, lignite HAs possessed comparable or even higher affinity for the tested metal ions. At the same time, the experimental results pose new questions that should be answered: how can we explain the differences in adsorption behavior, what are the causes of different relative amounts desorbed in both steps for different HAs, and so on. The causes may not only be chemical – differences in chemical composition, functional groups content, functional groups accessibility, and adsorption/desorption rates.

REFERENCES

1. Čežíková J, Kozler J, Madronová L, Novák J, Janoš P. Humic acids from coals of the North-Bohemian coal field II. Metal-binding capacity under static conditions. *React. Funct. Polym.*, 2001; 47: 111-118.
2. Davies G, Ghabbour EA, Cherkasskiy A, Fataftah A. Tight metal binding by solid phase peat and soil humic acids. In: Clapp CE, Hayes MHB, Senesi N, Bloom PR, Jardine PM eds. *Humic substances and chemical contaminants.* Anaheim: Soil Science Society of America, 2001: 371-395.
3. Martyniuk H, Więckowska J. Adsorption of metal ions on humic acids extracted from brown coals. *Fuel Process. Technol.*, 2003; 84: 23-36.
4. Arias M, Barral MT, Mejuto JC. Enhancement of copper and cadmium adsorption on kaolin by the presence of humic acids. *Chemosphere*, 2002; 48: 1081-1088.
5. Kučerík J, Pekař M, Klučáková M. South-Moravian lignite – potential source of humic substances. *Petroleum and Coal*, 2003; 45: 58-62.
6. Fiala K. *Obligatory methods of soil analyses* (in Slovak). Bratislava: VÚPOP, 1999.
7. Zeien H, Brűmmer GW. Chemische Extraktionen zur Bestimmung von Schwermetallbindungsformen in Bőden. *Mitteilgn. Dtsch. Bodenkundl. Gesellsch.*, 1989; 59: 505-510.
8. Hayes MHB, MacCarthy P, Malcolm RL, Swift RS eds. *Humic substances: II. In search of structure.* Chichester: Wiley, 1985: 642-686.
9. Ulmanu M, Marañón E, Fernández Y, Castrillón L, Anger I, Dumitriu D. Removal of copper and cadmium ions from diluted aqueous solutions by low cost and waste material adsorbents. *Water, Air Soil Pollution*, 2003; 142: 357-373.
10. Mikulášková B, Lapčík Ľ, Mašek I. Lignite – structure, properties, use (in Czech). *Chem. Listy*, 1997; 91:160-168.

Chapter 15

TRACE METAL SPECIATION INVESTIGATION OF SOIL HUMIC ACIDS ISOLATED FROM NEW ENGLAND SOILS: A CAPILLARY ELECTROPHORESIS-INDUCTIVELY COUPLED PLASMA-MASS SPECTROMETRY (CE-ICP-MS) STUDY

Anahita Dua,[1] John Danku,[2,4] Dula Amarasiriwardena,[1] Baoshan Xing[3] and Ramon Barnes[2,4]

[1] School of Natural Science, Hampshire College, Amherst, MA 01002, USA
[2] Department of Chemistry, 701 Lederle Graduate Research Center, University of Massachusetts, 710 N. Pleasant St., Amherst, MA 01003-9336, USA
[3] Department of Plant and Soil Sciences, University of Massachusetts, Amherst, MA 01003, USA
[4] University Research Institute for Analytical Chemistry, 85 North Whitney St., Amherst, MA 01002-1869, USA

15.1. INTRODUCTION

Soil-derived humic acids (HAs) are dark colored macromolecular polyelectrolytes [1-3]. They tend to bind with trace metals by ion exchange reactions (*i.e.,* with carboxylic groups -COO^-H^+) and ligand binding with oxygen, nitrogen, and sulfur containing functional groups [1,3,4]. Humic acids are polydispersed and contain ranges of molar mass fractions with various hydrodynamic diameters [5,6], making macromolecular or supramolecular aggregates [7] formed by low mass components [8]. The chemical functionality and structure, molar mass fraction sizes, and the physicochemical features influence the metal binding characteristics [9], and they play a vital role in complexing and transportation of nutritionally and toxicologically important trace metals in soil and aqueous environments.

Analytical approaches that can separate humic and fulvic acid molar mass fractions and simultaneously determine trace metals bound to those fractions are invaluable. To perform speciation analyses of humic substances based on molar mass fractions, separation techniques such as high performance size exclusion chromatography (HPSEC) [10,11] and flow field-flow fraction (FlFFF) [5,12] have been used. The amounts of trace metals bound to these individual mass fractions are minute, and a sensitive elemental detector like an inductively coupled plasma-mass spectrometer (ICP-MS) is required [13,14]. The SEC separation mechanism is related to pore sizes of the solid support material (that is, a gel), and the resulting elution order is directly linked to the molecular dimensions of various humic molecular size fractions [15]. In FlFFF, the separation mechanism is based on the diffusion characteristics of humic substances in an open channel under a field force [16-18]. Since mild forces are applied, FlFFF separation is less prone to artifacts than HPSEC [5, 16-18]. A third separation technique used for humic acid characterization is capillary electrophoresis (CE) [19, 20]. The CE separation principle is based on the ability of humic substances to separate through a conductive buffer medium under the influence of an applied electric field [15]. The capillary column has more theoretical plates per unit length than SEC; therefore, CE is capable of making efficient and rapid separations [15]. Moreover, CE uses small sample and buffer solution volumes (nano- and micro-liter). Well-resolved peaks result from its high separation efficiency. The CE fractions often suffer from several orders of magnitude poorer detection limit than competing methods [15]. However, the successful on-line coupling of a powerful size separation with an element-selective detection technique has provided analytical information about trace metals complexed to humic mass (or size) fractions. Among the hyphenated methods, HPSEC-ICP-MS has been applied to identify aquatic humic substances [10,13] and soil humic acids [14, 21, 22]. Flow FFF-ICP-MS has been used to identify soil-derived humic acids [6, 23] in sediments, and particles [23-26] and fulvic acid [27]. Since the development of an initial interface between CE and ICP-MS [28, 29], CE-ICP-MS has been applied for speciation studies including aquatic fulvic acids [20].

In this study we have investigated CE-ICP-MS to determine trace metals bound to soil humic acids that were successively extracted from two soils from a local mountain.

15.2. MATERIAL AND METHODS

15.2.1. Materials

Humic Acid Samples. The soils used for humic acid extractions were obtained from the south- and north-facing slopes of Ox Hill (~ 280 m above seal level, 42° 30' N, 72° 31') of Mount Toby situated in the Connecticut River valley, Massachusetts, USA [30]. These slopes are ecologically different owing to contrasting microclimatic conditions and ecology. Therefore, the resulting soil organic matter may have different characteristics. Soil-derived humic acids used in this study were extracted for a previous study by Kang et al. [30]. The two soils were extracted by eight repetitive treatments with 0.1M $Na_4P_2O_7$ followed by two extractions with 0.1 M NaOH [30]. Humic acid extraction protocols described in the literature [31] were applied with minor changes [30]. Thus, 10 humic acid fractions were obtained for each soil. For convenience, the fractions were numbered N1 to N10 and S1 to S10. In this study we used N1, N3, N4 and N9 humic acids extracted from the north slope soil sample (N) and S1, S3, S8 and S9 humic acids from the south slope soil sample (S).

15.2.2. Methods

Humic Acid Characterization. Humic acid fractions were analyzed to identify their functional groups using diffuse reflectance infrared Fourier transform spectroscopy (DRIFTS, Midac M2010 series, Irvine, CA, USA). Two-mg aliquots of freeze-dried humic acid were mixed with 98 mg of ground, spectroscopic grade potassium bromide (Thermo Spectra-Tech Inc., Shelton CT, USA) using an agate mortar and pestle. Pure spectroscopic grade KBr was used as a blank. The sample mixture and blank were gently transferred to the sample holder cups. The top surface of the sample or blank mound was smoothened using a glass microscope cover slip without tapping the holder. Each sample and blank was purged with a stream of dry nitrogen during the spectral recording to minimize interference from moisture in the sample compartment and to eliminate carbon dioxide peaks. The DRIFT spectra were obtained at 16 cm^{-1} resolution, and 100 scans were collected between 400-4000 cm^{-1}. The collected absorbance spectra were converted to the Kubelka-Munk (KM) function using the Grams/32 spectroscopic software package (Galactic Corporation, Salem, NH, USA), and the wave numbers were identified using the same software package. The wave number and KM intensity data were subsequently transferred into a Microsoft Excel® spreadsheet software program (Redmond, WA, USA) to plot stacked DRIFT spectra.

Flow Field-flow Fractionation (Flow FFF) On-channel Preconcentration. The FlFFF on-channel preconcentration technique [32,33] was used to concentrate the HA samples before CE separation. The operating conditions are listed in Table 15.1.

Briefly, the procedure comprises three steps: sample loading and focusing; sample relaxation; and elution (fraction collection). Sample (49.5 mL) was loaded through the front end of the FlFFF channel while the carrier liquid entered simultaneously through both the front and back ends with a constant ratio between the forward and backward flow rates. During relaxation only the cross flow was introduced through the channel. Finally, the channel and cross flows revert to the standard FlFFF configuration during the elution step when the focused fraction (usually about 0.5 mL) was collected.

Capillary Electrophoresis-Inductively Coupled Plasma-Mass Spectrometry (CE-ICP-MS). Humic acid (HAs) samples (100 mg/L) were dissolved in 30 mM TRIS-HNO_3 (pH 7.5) buffer, which was prepared by dissolving about 3.6 g of tris(hydroxymethyl)aminomethane (Aldrich Chemical Co., Milwaukee, WI, USA) in 1 L deionized water (Millipore, Bedford, MA, USA). The pH was adjusted to 7.5 with concentrated sub-boiled HNO_3. Basic media like 0.1M sodium bicarbonate or sodium hydroxide solutions were also used in attempts to dissolve sequentially extracted humic acids.

The CE-ICP-MS operating conditions are listed in Table 15.1. The CE interface (not shown) consisted of a PEEK T-piece with PTFE tubing through which the CE capillary (J&W Scientific, 50 μm i.d.) passed from the CE system to a low-flow nebulizer (MicroMist, Glass Expansion, Pocasset, MA, USA) with a custom-made, 40 mL spray chamber. The makeup liquid (3 mM TRIS-HNO_3, containing 10 mM ^{115}In) was pumped (using the ICP-MS peristaltic pump) through stainless steel tubing adjoining the remaining arm of the T-piece. This piece also served as the CE system electrical ground. The CE run buffer (30 mM TRIS-40 mM Tricine (N-[Tris(hydoxymethyl)methyl]glycine, pH 8.0) and humic acid samples were filtered through 0.2 and 0.45 μm membrane disks, respectively.

Trace metals associated with the separated humic acid species were analyzed with an Agilent 7500a (Agilent Technologies, Palo Alto, CA, USA) ICP-MS spectrometer. Trace elemental analyte masses (m/z) ^{27}Al, ^{55}Mn, ^{59}Co, ^{60}Ni, ^{63}Cu, ^{64}Zn, ^{65}Cu, ^{66}Zn, ^{206}Pb, and ^{208}Pb, were monitored. The ^{115}In signal was use as a sheath flow liquid monitor. A solution of analyte metal ions was prepared in 30 mM

TRIS buffer and analyzed under the same conditions to determine the migration times of the uncomplexed ions.

Table 15.1 Flow FFF preconcentration, capillary electrophoresis and ICP-MS instrument operating parameters

Flow Field-flow fractionation preconcentration conditions (FlFFF Model F-1000-FO)	
FFF channel dimensions (cm)	27.7 long x 0.0254 thick x 2.0 wide
Carrier liquid	30 mM TRIS-HNO_3, pH 7.5
Channel flow rate (mL/min)	1.0
Cross flow rate (mL/min)	2.0
Membrane	1 kDa MWCO poly (regenerated cellulose acetate)
Preconcentration conditions:	
Forward flow rate (sample intro, mL/min)	5.00
Backward flow rate (carrier, mL/min)	0.50
Focusing time (min)	8
UV detector wavelength (nm)	254
Capillary Electrophoresis (ATI UNICAM Crystal CE system model 300)	
Capillary dimensions (uncoated)	50-µm i.d., 65 cm long
High voltage (kV)	30
Buffer	30 mM TRIS-40 mM Tricine, pH 8
Makeup liquid (sheath)	3 mM TRIS-HNO_3, pH 7.5
Injection (hydrodynamic)	80 mbar 6 s (11.4 nL)
ICP-MS: Agilent 7500a	
Forward RF power (W)	1240
Argon gas flows (L/min)	14.95 (coolant), 0.92 (auxiliary), 1.40 (carrier)
Nebulizer	MicroMist (Glass Expansion)
Spray chamber	Custom machined 40 mL (glass)
Sheath flow rate (µL/min)	33 (from peripump, 0.01 rps)
Dwell time (ms)	200
Isotopes monitored (m/z)	^{27}Al, ^{55}Mn, ^{59}Co, ^{60}Ni, ^{63}Cu, ^{64}Zn, ^{65}Cu, ^{66}Zn, ^{206}Pb, ^{208}Pb, ^{115}In (sheath flow monitor)

15.3. RESULTS

15.3.1. Spectroscopic Properties of Humic Acids

The DRIFT spectra obtained for some north slope and south slope humic acid fractions are shown in Figures 15.1 and 15.2. The spectra contain three common major spectral regions. The major hump between 3400 and 3300 cm^{-1} is due to hydrogen-bonded O-H stretching or phenolic and amide N-H groups. Asymmetric and symmetric stretching vibrations of aliphatic C-H bands in $-CH_3$ and $-CH_2$ groups at 2923 cm^{-1} and 2862 cm^{-1}, respectively, are prominent features in these spectra. The second fingerprint region is rich with many functional groups characteristic of humic acids. The strong band at 1720 cm^{-1} is due to stretching vibrations from carbonyl stretch of ubiquitous –COOH functional groups. The peak at 1650 cm^{-1} is due to >C=O, -COOH or amide H while the band at 1443 cm^{-1} is due to CH_3 asymmetric stretch, -CH bend. The peak around 1230 cm^{-1} is attributed to C-O stretching of aryl ethers, -OH deformation of –COOH group, and to C-O stretching. The peak at 1165 cm^{-1} is due to aliphatic OH. The O-H deformations and C-O stretching in polysaccharides and alcoholic groups or Si-O of silicate impurities are representative of ~ 1049 cm^{-1}. In the third or last spectral region small but sharp peaks are often due to aromatic C-H out-of-plane bending (940, 779 cm^{-1}) or mineral peaks (Al-O-Si) between 850-825 cm^{-1} or 797 cm^{-1} (Fe-O-Si) [1,14,33-36].

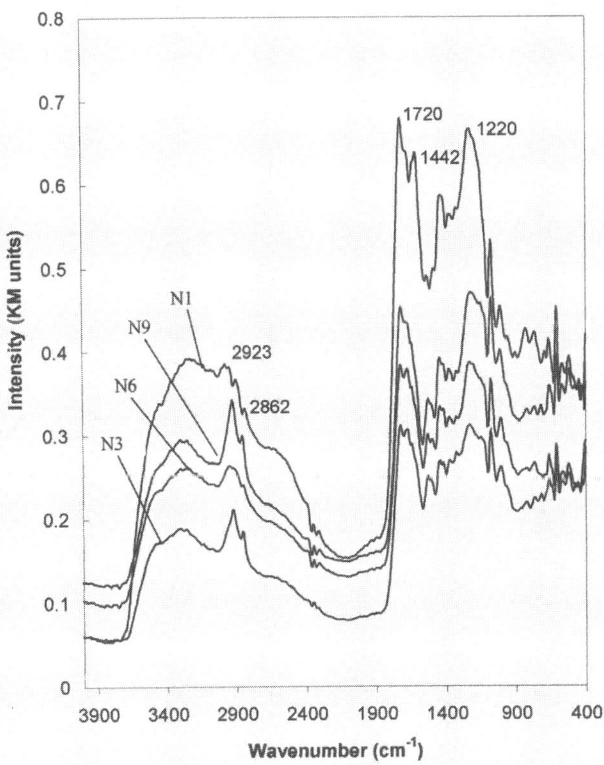

Figure 15.1 DRIFT spectra of soil derived humic acids sequentially (N1, N3, N6 and N9) extracted from the north slope soil of Mt. Toby, Western Massachusetts, USA

As the sequential extraction number increased, the intensity of the aliphatic -CH_2 band at 2923 cm^{-1} and CH_2 symmetric peak at 2862 cm^{-1} increased, and the well-resolved bands emerged for both humic acid series. These data are consistent with previous findings [30].

Table 15.2 shows the aliphatic to aromatic functional group (AL/AR) intensity ratio for each of the samples. From these data it is clear that as the sequential extraction of HA progresses, the number of aliphatic functional groups increases.

The AL/AR ratio of sequentially extracted north slope HA samples increased from 1.50 to 1.92 as the extraction number increased. A similar trend was also found with the south slope humic acids and their AL/AR ratio increased from 1.62 to 1.74 as the sequential extraction of HA progressed. Table 15.3 illustrates the intensity ratio (O/R) of oxygen to recalcitrant functional groups decreases from 0.58 to 0.53 for north slope HA samples while the O/R ratio for south side humic acids remained constant from S1 to S8 at 0.55 but dropped to 0.51 for the S9 HA sample.

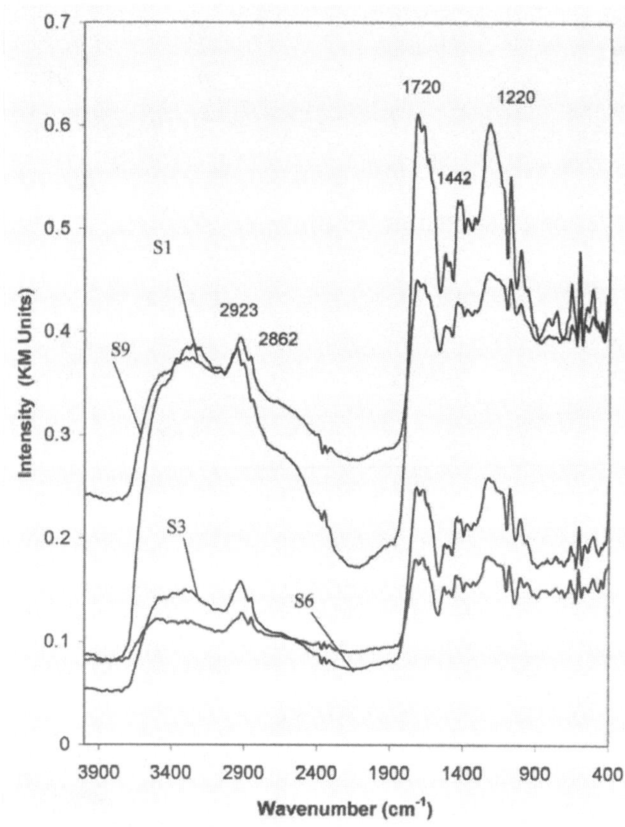

Figure 15.2 DRIFT spectra of soil derived humic acids sequentially (S1, S3, S8 and S9) extracted from the south slope soil of Mt. Toby, Western Massachusetts, USA

Table 15.2 Aliphatic to aromatic functional groups intensity ratio (AL/AR) for the north and south slope soil-derived humic acid fractions

North Slope		South slope	
Extracted HA fraction	AL/AR	Extracted HA fraction	AL/AR
N1	1.50	S1	1.62
N3	1.73	S3	1.73
N6	1.79	S8	1.80
N9	1.92	S9	1.74

$$AL/AR = \frac{(2923 + 2862 + 1442 + 1049 + 1080)\ cm^{-1}}{(3054 + 1652 + 1527)\ cm^{-1}}$$

Table 15.3 Oxygen to recalcitrant functional groups intensity ratio (O/R) for the north and south slope soil derived humic acid fractions.

North Slope		South slope	
Extracted HA fraction	O/R	Extracted HA fraction	O/R
N1	0.58	S1	0.55
N3	0.54	S3	0.55
N6	0.54	S8	0.55
N9	0.53	S9	0.51

$$\frac{O}{R} = \frac{(1720 + 2507 + 1643 + 1165)\ cm^{-1}}{(2923+2862+1442+1049+1080+3054+1612+1527)\ cm^{-1}}$$

15.3.2. Humic Acid Preconcentration by Flow –FFF

Humic acid samples were preconcentrated by a factor ~100 with flow FFF on-channel preconcentration. The concentrated HA fraction was collected in a 0.5 ml final volume. The appearance of HA before and after pre concentration is shown in Figure 15.3. Preparation of higher concentrations of HA samples is limited by solubility in basic media (30 mM TRIS-HNO_3 (pH 7.5), 0.1 M sodium bicarbonate or sodium hydroxide) for dissolution of sequentially extracted HAs. This is most likely due to the reduced polar nature of the sequentially extracted HAs.

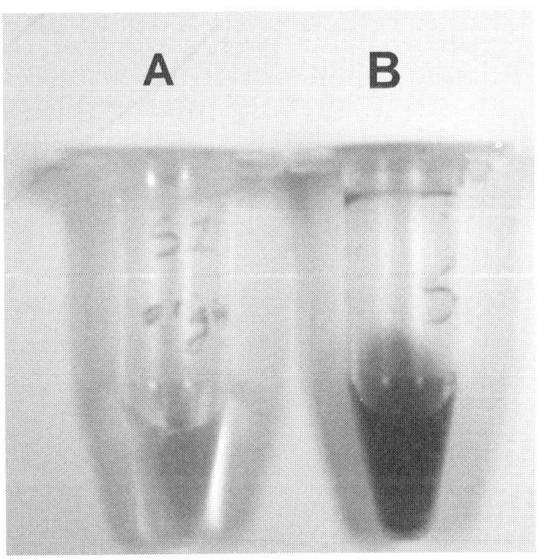

Figure 15.3 Flow field-flow fractionation (flow-FFF) channel pre-concentration of humic acid before CE analysis: (A) before pre-concentration - aliquot of 5 mg humic acid dissolved in 50 mL of 30 mM TRIS/HNO_3 (B) after pre-concentration - preconcentrated HA in 0.5 mL volume

15.3.3. CE-ICP-MS Analyses

A major concern with the speciation of humic acids using CE coupled to element sensitive detection is CE artifacts wherein the metal species chemically interact within the system [37]. Experimental design is crucial to minimize these artifacts. Inclusion of a zwitterionic additive (tricine) in the buffer is used in this study to reduce metal ion interactions with the fused silica capillary.

As illustrated in Figure 15.4, the FlFFF preconcentration step is necessary to detect metal-humic complexes separated by CE-ICP-MS. Preparation of higher concentrations of HA samples is limited by their solubility. Electropherograms for sample S1 without preconcentration are shown in Figure 15.4a for five elements. The elemental peaks of the separated humic acid are barely detectable (i.e., Cu) above the flat baseline. For most elements (Pb, Zn, Mn, Co) no signal is distinguishable above background noise. As shown in Figures 15.4b and 15.4c for a preconcentrated sample, reproducible peaks are observed for elements. Thus, FlFFF preconcentration is essential to detect metal-humic associations by CE-ICP-MS.

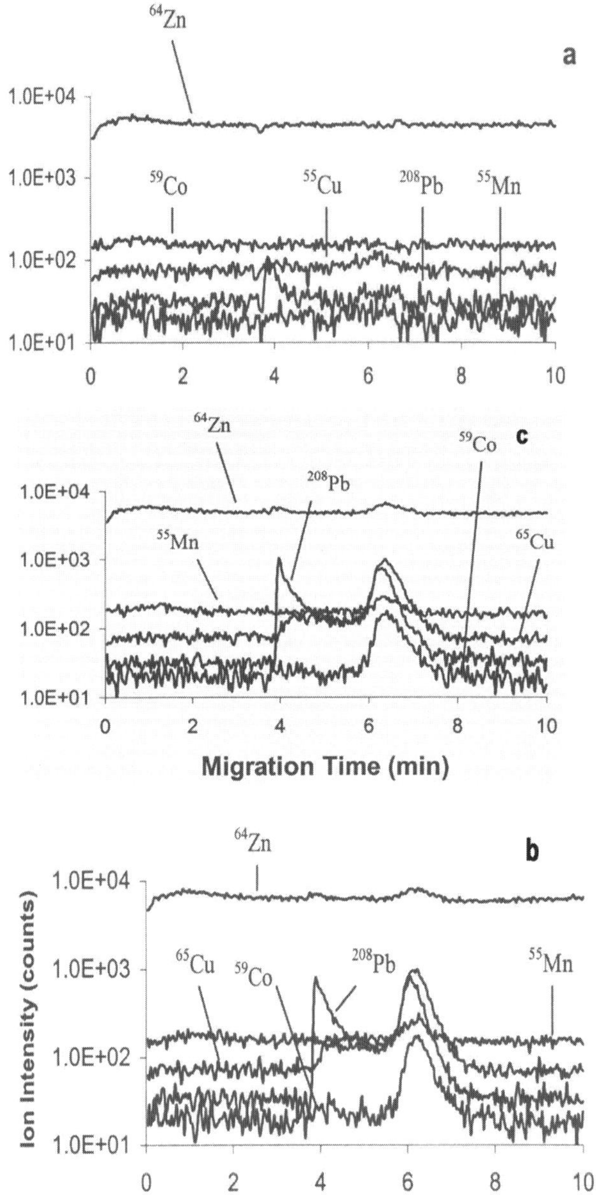

Figure 15.4 Electropherograms of humic acid (S1). 4a: Sample (100 mg L^{-1}) dissolved in 30 mM Tris-HNO_3; 4b and 4c: same sample after flow FFF pre-concentration runs 1 and 2, respectively. Injection volumes 22.8 nL (4a), 11.4 nL (4b, 4c). Note: the *y*-axis is a log scale

The two electropherograms (Figures 15.4b and 15.4c) demonstrate the reproducibility of the separation. Capillary flushing with NaOH between runs and zwitterions addition to the buffer were necessary to obtain these reproducible electropherograms.

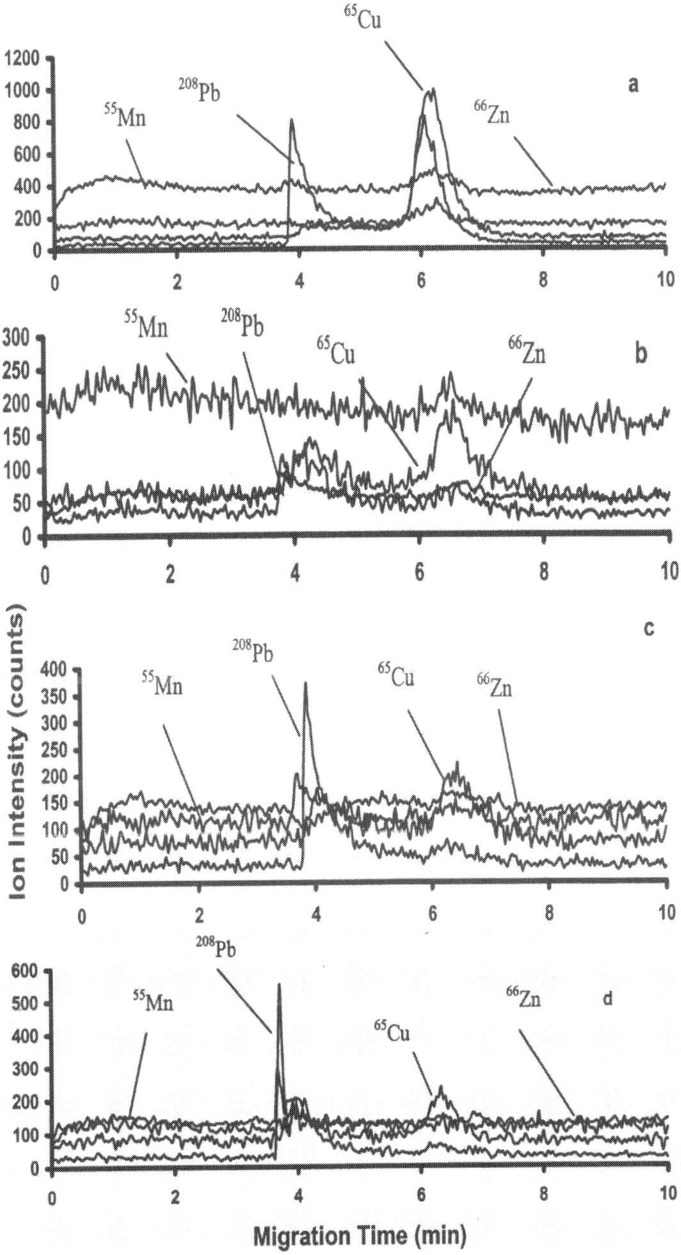

Figure 15.5 Electropherograms of south side humic acids: a) S1, b) S3, c) S8, d) S9. Sample (100 mg L^{-1}) dissolved in 30 mM TRIS-HNO_3 pH 7.5, FlFFF preconcentrated. Injection volume 11.4 nL (hydrodynamic)

Two elemental peaks are apparent in Figures 15.4b and 15.4c for the S1 humic acid. The first peak exhibits a sharp leading edge and appeared at ~ 4 min for Pb and Cu, and the second broad, asymmetric peak at ~ 6 min is observed for all elements. Peaks eluting at ~ 4 min represent the free metal ions. This was confirmed by migration time matching with free ion electropherograms. The peaks appearing at ~ 6 min correspond to metals bound to HA.

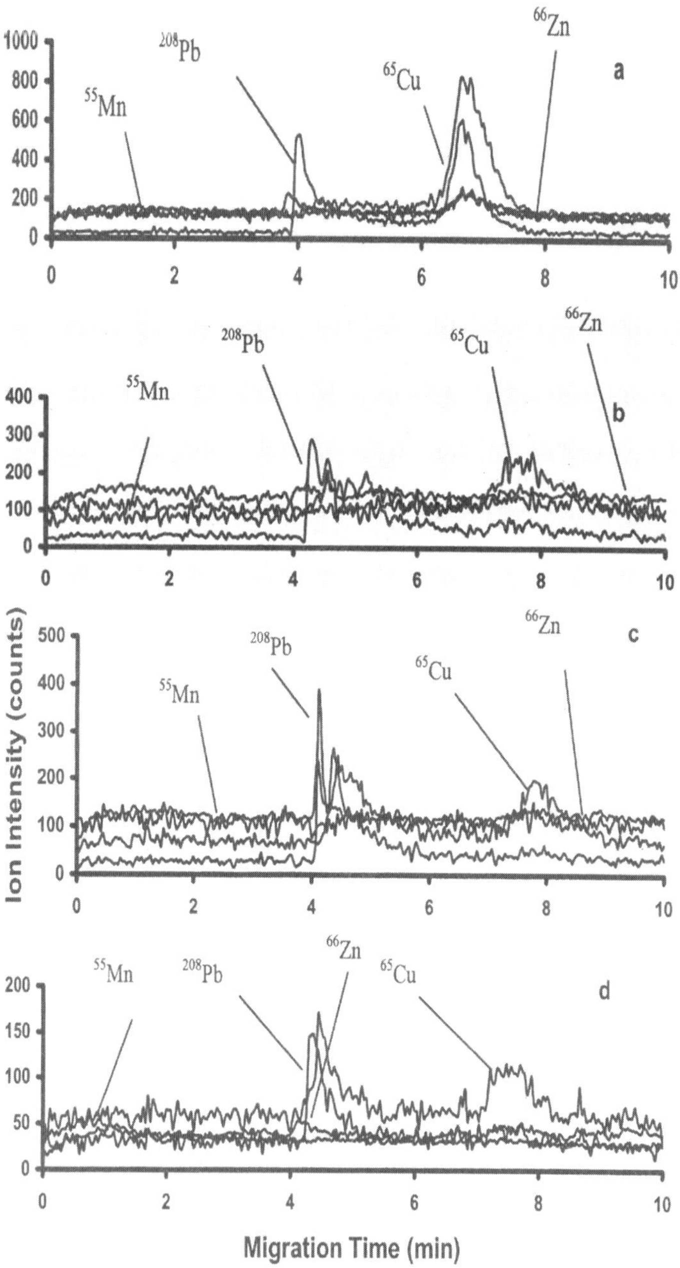

Figure 15.6 Electropherograms of north side humic acids: a) N1, b) N3, c) N6, d) N9. Sample (100 mg L^{-1}) dissolved in 30 mM TRIS-HNO_3 pH 7.5, FlFFF preconcentrated. Injection volume 11.4 nL (hydrodynamic)

Ion electropherograms for the south slope humic acid extracts are shown in Figure 15.5. Two peaks are prominent. One appears at ~ 3.8-4.0 min and is due to unbound trace metals. Humic acid associated trace metals appear as a broad band between 5.5 and 7.0 minutes. The height of the second peak decreased gradually from ~1,000 intensity counts to around ~200 counts for Cu. A similar trend is observed for Pb, Mn and Zn.

The peak height of the trace metal-associated second peak decreased considerably from the S1 humic acid sample to the sequentially extracted third humic acid (S3) sample and finally to a lesser extent from S3 to S8 humic acid samples. No appreciable change is evident for S8 to S9. Similar ion electropherograms of north-slope humic acids are shown in Figure 15.6. Two peaks at ~ 4 to 4.4 min and 6.8 to 7.8 min retention times are observed.

15.4. DISCUSSION

15.4.1. Solubility and Spectroscopic Characterization of Sequentially Extracted Humic Acids

It was observed that the first extracted humic fraction is easily soluble in a basic solution (for example, 0.1 M sodium bicarbonate) in both north and south humic acids but subsequent fractions were increasingly resistant to dissolution in either 0.1 M sodium bicarbonate or sodium hydroxide solution. This further illustrates the increasing hydrophobic nature of the sequentially extracted HA samples.

Infrared Spectra. The degree of aromaticity decreases with sequential extraction of the HAs from the same soil sample. Infrared spectra show that the initially extracted HA fractions were aromatic in nature, while later HA fractions were rich in aliphatic moieties. This trend was observed for both the north and south slope soil-derived humic acid samples. Moreover, the O/R ratio data indicate that the number of oxygen-rich groups (i.e., -COOH, phenolic OH, enolic O) in these humic acids decreases with sequential extraction of humic acids originated from both slopes of Mt. Toby. Since oxygen is an important donor atom in metal binding to HA, the observed decrease in oxygen sites has a major implication on complexing of trace metals by humic acids. Also, the complexing properties of oxygen containing carboxylic groups may participate in metal-HA reactions by ion exchange mechanisms. Therefore, metal binding should be reduced as the HA becomes more aliphatic and rich in less polar functionalities with successive extraction.

15.4.2. Flow FFF Preconcentration

Flow FFF is suitable for on-channel preconcentration [32] owing to its open-channel characteristics. Shear degradation and adsorption are minimized, since no stationary phase or packing material exists in the channel [16,32]. As illustrated in Figure 15.4, the FlFFF preconcentration improved the CE detection of trace metals associated with humic acids dissolved in the nanoliter sample injection volume. This technique can be easily adapted for experimental situations in which the dissolved humic substance concentrations are very dilute (e.g., in the aquatic environment).

15.4.3. CE-ICP-MS Analysis of Sequentially Extracted Humic Acids

Humic substances are well known for their ability to bind and form complexes with metal ions in the environment. They are involved in the transportation of both contaminants and nutritionally important metal ions [1-4,38]. The electropherograms (Figures 15.4b and 15.4c) demonstrate duplicate reproducibility of the technique with optimized CE separation conditions. Metal ion interactions with the fused silica capillary are minimized by adding zwitterionic tricine to the buffer. The tricine increases the ionic strength, thereby decreasing analyte-wall interactions without significantly increasing Joule heating [37]. The slight variation of migration time observed here might be due to both capillary wall interactions and the mobility of the analyte [39]. The mobilities of analyte ions are affected by the compositional and charge changes of humic acids as the oxygen-rich groups (i.e., carboxylic groups -COO^-H^+, phenolic -OH) in humic acids decrease with successive extractions. Temperature control is also important to obtaining reproducible electropherograms [37].

Peaks eluting at ~4 min (See Figures 15.4, 15.5 and 15.6 for HA series samples) are the free metal ions (identified by migration times matching those for free ion electropherograms). Particularly noteworthy is the behavior of Pb, which demonstrated both free ions as well as a complexed form. Lead is kinetically labile and unlikely to form stable complexes on the time scale of a CE separation [20].

The electropherograms (see Figures 15.5 and 15.6) demonstrate the decrease in metal binding capability of progressively extracted humic acids. The structural and compositional differences among the fractions affect their metal binding properties. With continuous extraction the metal binding characteristics of HA fractions decreased (that is, as the O/R ratio decreased). Differences in Cu, Pb, and Zn bound to these fractions are apparent between the north and south side slopes. In addition, the results (not shown here) demonstrate that other trace metals like Al, Ni and Fe are strongly bound to these humic acids.

15.5. CONCLUSIONS

Spectroscopic techniques combined with CE-ICP-MS analysis demonstrated a diversity of trace metal complexing capability of a series of humic acids sequentially extracted from the north and south side soil of Mt. Toby. As the sequential extraction progresses, the successive HA samples exhibit less polar, more aliphatic nature and contain fewer available oxygen-rich functional groups. The flow FFF preconcentration technique improved the CE detection of trace metals complexed to HA. A preconcentration factor of ~100 was achieved, and nanoliter volume samples (*ca* 10 nL) can be used with CE-ICP-MS. Reproducible results were obtained with the addition of tricine to the CE buffer. Further investigations involving the use of neutral coated capillaries and different buffer conditions are underway to lessen interactions between humic acids and the capillary walls. The CE-ICP-MS results demonstrate the presence of tightly bound trace metals (Pb, Cu, Zn, Co, Fe, Al) in all extracted humic acids. The HA fractions showed different metal-HA complexing properties. The extent of trace metals bound to those humic acids declined with declining population of oxygen rich moieties in HA structures in north and south slope samples. Finally, differences were observed between humic acids extracted from south- and north-side slopes of Mt. Toby soils. Further study is needed to relate observed differences to site geology, soil environment and microclimatic conditions.

ACKNOWLEDGEMENTS

Dula Amarasiriwardena gratefully acknowledges the financial and instrument support of the National Science Foundation (BIR 951270), the Kresge Foundation, and a summer Faculty Development Grant. We would also like to thank Mr. Seughun Kang for kindly providing the humic acids. This research was supported in part by ICP Information Newsletter, Inc., Hadley, Massachusetts. Anahita Dua thanks the undergraduate research grant awarded by the Howard Hughes Medical Institute (HHMI, Grant # 71100-503803) to Hampshire College.

REFERENCES

1. Stevenson FJ. *Humus chemistry: Genesis, composition, and reactions*. 2nd Edn., New York: Wiley, 1994.
2. Davies G, Fataftah A, Cherkasskiy A, Ghabbour EA, Radwan A, Jansen SA, Kolla S, Paciolla MD, Sein Jr. LT, Buermann W, Balasubramanian M, Budnick J, Xing B. Tight metal binding by humic acids and its role in biomineralization. *J. Chem. Soc., Dalton Trans.,* 1997; 4047-4060.
3. Davies G, Ghabbour EA, Steelink C. Humic acids: Marvelous products of soil chemistry. *J. Chem. Ed.,* 2001; 78: 1609-1614.
4. Steelink C. Investigating humic acids in soils. *Anal. Chem.,* 2002; 74: 326A-333A.
5. Beckett R, Zhang J, Giddings JC. Determination of molecular weight distributions of fulvic and humic acids using flow field-flow fractionation. *Environ. Sci. Technol.*, 1987; 21: 289-295.
6. Amarasiriwardena D, Siripinyanond A, Barnes RM. Trace elemental distribution in soil and compost-derived humic acid molecular fractions and colloidal organic matter in municipal wastewater by flow field-flow fractionation-inductively coupled plasma mass spectrometry. *J. Anal. At. Spectrom.,* 2001; 16: 978-986.
7. Conte P, Piccolo, A. Conformational arrangement of dissolved humic substances, Influence of solution composition on association of humic molecules, *Environ. Sci. Technol.*, 1999; 33: 1682-1690.
8. Szymczak W, Wolf M, Wittmaack K. Comparison of As-delivered and AFFFF-size fractionated Suwannee River fulvic acid by time-of-flight mass spectrometry, In: Ghabbour EA and Davies G eds. *Humic substances: Nature's most versatile materials.* New York: Taylor and Francis, 2004: 31-38.
9. Garcia-Mina J. Exploring the molecular character and heterogeneity of humic substances via the study of the iron-binding process using an extended polyelectrolyte model, In: Ghabbour EA and Davies G eds. *Humic substances: Nature's most versatile materials.* New York: Taylor and Francis, 2004:195-218.
10. Rottmann L, Heumann KG. Development of an on-line isotope dilution

technique with HPLC/ICP-MS for the accurate determination of elemental species. *Fresenius J. Anal. Chem.*, 1994; 350: 221-227.

11. Chin YP, Aiken G, O'Loughlin E. Molecular weight, polydispersity, and spectroscopic properties of aquatic humic substances. *Environ. Sci. Technol.*, 1994; 28: 1853-1858.
12. Dycus PJM, Healy KD, Stearman GK, Wells MJM. Diffusion coefficients and molecular weight distributions of humic and fulvic acids determined by flow-field-flow fractionation. *Separation Sci. Technol.,* 1995; 30: 1435-1453.
13. Vogl J, Heumann KG. Determination of heavy metal complexes with humic substances by HPLC/ICP-MS coupling using on-line isotope dilution technique. *J. Anal. At. Spectrom.*, 1994; 359: 438-441.
14. Ruiz-Haas P, Amarasiriwardena D, Xing B. Determination of trace metals bound to soil humic acids species by size exclusion chromatography and inductively coupled plasma mass spectrometry. In: Davies G, Ghabbour EA eds. *Humic substances: Structures, properties and uses.* Cambridge, UK: Royal Society of Chemistry, 1998: 147-163.
15. Harvey D. *Modern analytical chemistry*. New York: McGraw-Hill, 1st Edn. 2000: 593-610.
16. Giddings JC. A new separation concept based on a coupling of concentration and flow non-uniformities. *Sep. Science*, 1966; 1: 123-125.
17. Giddings JC. Field-flow fractionation of macromolecules. *J. Chromatogr.*, 1989; 470: 327-335.
18. Siripinyanond A, Barnes RM. Flow-field-flow fractionation-inductively coupled plasma mass spectrometry and metal speciation in proteins: A feasibility study. *J. Anal. At. Spectrom.,* 1999; 14: 1527-1531.
19. Schmitt P, Kettrup A, Freitag D, Garrison AW. Flocculation of humic substances with metal ions as followed by capillary zone electrophoresis. *Fresenius J. Anal. Chem.,* 1996; 354: 915-920.
20. Sonke JE; Salters VJM. Determination of neodymium-fulvic acid binding constants by capillary electrophoresis inductively coupled plasma mass spectrometry (CE-ICP-MS). *J. Anal. At. Spectrom.* 2004; 19: 235-240.
21. Bhandari SA, Amarasiriwardena D, Xing B. Application of high performance size exclusion chromatography (HPSEC) with detection by inductively coupled plasma-mass spectrometry (ICP-MS) for the study of metal complexation properties of soil derived humic acid molecular fractions. In: Ghabbour EA, Davies G eds. *Understanding humic substances: Advanced methods, properties and applications,* Cambridge, UK: Royal Society of Chemistry, 1999: 203-221.
22. Wrobel K, Sadi BBM, Wrobel K, Castillo G, Caruso JA. Effect of metal ions on the molecular weight distributions of humic substances derived from municipal compost: ultrafiltration and size exclusion chromatography with spectrophotometric and inductively coupled plasma-MS detection. *Anal. Chem.,* 2002; 350: 221-227.
23. Anderson T, Shifely L, Amarasiriwardena D, Xing, B, Siripinyanond A, Barnes RM. Characterization of trace metals complexed to humic acids derived from agricultural soils, annelid composts, and sediment by flow field-flow fractionation-inductively coupled plasma mass spectrometry. In: Ghabbour EA, Davies G eds. *Humic substances: Structures, models and functions.* Cambridge, UK: Royal Society of Chemistry, 2001: 165-177.
24. Taylor HE, Garbarino JR, Murphy DM, Beckett R. Inductively coupled plasma mass spectrometry as an elemental detector for field -flow

fractionation particle speciation. *Anal. Chem.*, 1992; 64: 2036-2041.

25. Beckett R. Field-flow fractionation-ICP-MS: A powerful new analytical tool for characterizing macromolecules and particles. *At. Spectrosc.*,1991; 12: 228-232
26. Siripinyanond A, Barnes RM, Amarasiriwardena D. Flow field-flow fractionation-inductively coupled plasma mass spectrometry for sediment bound trace metal characterization. *J. Anal. At. Spectrom.,* 2002; 17:1055-1064.
27. Bell J, Amarasiriwardena D, Siripinyanond A, Xing B, Barnes RM. Investigation of colloidal properties and trace metal complexation characteristics of soil-derived fulvic acids by flow-field-flow fractionation-inductively coupled plasma mass spectrometry (flow FFF-ICP-MS), In: Ghabbour EA, Davies G eds. *Humic substances: Nature's most versatile materials.* New York: Taylor and Francis, 2004: 31-38.
28. Olesik JW, Kinzer JA, Olesik SV. Capillary electrophoresis inductively coupled plasma spectrometry for rapid elemental speciation. *Anal Chem.*, 1995; 67: 1-12.
29. Lu Q, Bird SM, Barnes RM. Interface for capillary electrophoresis and inductively coupled plasma-mass spectrometry. *Anal Chem.*, 1995; 67: 2949-2956.
30. Kang S, Amarasiriwardena D, Veneman P, Xing B. Characterization of ten sequentially extracted humic acids and a humin from a soil in western Massachusetts. *Soil Sci.*, 2003; 168: 880-887.
31. Swift R. Organic matter characterization. In: Sparks DL, Page AL, Helmke PA, Loeppert eds. *Methods of soil analysis, Part 3: Chemical methods*, SSSA and ASA, Madison, WI, 1966:1011-1069.
32. Hassellöv M, Lyvén B, Haraldsson C, Sirinawin W. Determination of continuous size and trace element distribution of colloidal material in natural water by on-line coupling of flow field-flow fractionation with ICP-MS. *Anal. Chem.,* 1999; 71: 3497-3502.
33. Siripinyanond A, Ph.D. Thesis, University of Massachusetts, Amherst, USA, 2002.
34. Baes AU, Bloom PR. Diffuse reflectance and transmission Fourier transform infrared (DRIFT) spectroscopy of humic and fulvic acids. *Soil Sci. Soc. Am. J.*, 1989; 53: 695-700.
35. Niemeyer J, Chen Y, Bollag J-M. Characterization of humic acids, composts, and peat by diffuse reflectance Fourier-transform infrared spectroscopy. *Soil Sci. Soc. Am. J.*, 1992; 56: 135-140.
36. Wander MM, Traina SJ. Organic fractions from organically and conventionally managed soils: II. Characterization of composition. *Soil Sci. Soc. Am. J.*, 1996; 60: 1087-1094.
37. Oda RP, Landers JP. In: Landers JP ed. *Handbook of capillary electrophoresis*, 2nd Edn. Boca Raton, FL: CRC Press, 1997:14, 23, 34-35.
38. Macalady DL, Ranville JF. In: Macalady DL ed. *Perspectives in environmental chemistry*. New York: Oxford University Press, 1998: 94-137.
39. Shihabi ZK. In: Landers JP ed. *Handbook of capillary electrophoresis,* 2nd Edn. Boca Raton, FL: CRC Press, 1997: 473.

Part 4

SOIL AGGREGATION

Chapter 16

SOIL ORGANIC MATTER AND AGGREGATE STABILITY AFFECTED BY TILLAGE

Joseph L. Pikul Jr.,[1] Jane M. F. Johnson,[2] Sara Wright,[3] TheCan Caesar[4] and Michael Ellsbury[1]

[1] USDA-Agricultural Research Service, Northern Grain Insect Research Laboratory, 2923 Medary Ave., Brookings, SD 57006, USA
[2] USDA-ARS, North Central Soil Conservation Research Laboratory, 803 Iowa Ave., Morris, MN 56267, USA
[3] USDA-ARS, Sustainable Agricultural Systems Laboratory, Building 001, 10300 Baltimore Ave., Beltsville, MD 20705, USA
[4] USDA-ARS, Northern Plains Agricultural Research Laboratory, 1500 North Central Ave., Sidney, MT 59270, USA

16.1. INTRODUCTION*

Interaction between crop and soil management practices and soil conditions often are clouded by variability within a system. Further, causal relationships between management and soil quality are difficult to extrapolate from one region to another because of differences in soil type, climate, and management norms. The quantity and quality of soil organic matter (SOM) provides an important diagnostic link between management and sustainability of soil function.

Considerable research has been conducted on relationships among cropping sequence, soil organic matter and various biological and physical soil properties. It is generally accepted that crop production alone has caused a decline in SOM compared to the original grassland levels throughout the Great Plains [1-3]. Tillage has caused soil C losses from 28 to 77%, depending on geographic location (climate) and soil type [4]. Summer fallow, an agricultural practice to conserve

* **Abbreviations: AM,** arbuscular mycorrhizal fungi; **ANOVA**, analysis of variance; **BD**, soil bulk density; **BF,** basidiomycetes fungi; **BFpost**, basidiomycete fungi assay on soil following wet sieving; **BFpre**, basidiomycete fungi assay on soil before wet sieving; **C**, carbon; **C:N**, carbon to nitrogen atomic ratio; **CT**, chisel tillage; **DASD**, dry aggregate size distribution; **EC**, soil electrical conductivity; **ELISA**, enzyme-linked immunosorbent assay; **HA**, humic acid; **IRTG**, immunoreactive total glomalin; **MWD**, mean weight diameter; **N**, nitrogen; **NT**, no tillage; **p-value**, probability test statistic; **POM**, particulate organic matter (fine = 0.5 - 0.053 mm and coarse = 2.0 – 0.5 mm); **SOC**, soil organic carbon; **SOM**, soil organic matter.

water, in the crop sequence such as wheat-fallow crop has been implicated as an additional cause of serious declines in SOM [2,5,6] compared to annual cropping systems.

Soil organic matter influences soil compactibility, friability, soil water-holding capacity, air and water infiltration, nutrient conservation and soil permeability [7]. Boyle et al. [8], in a review of the influence of SOM on soil aggregation and water infiltration, concluded that SOM has a disproportionate effect on soil physical behavior. Hudson [9] reported soils high in SOM have greater available water-holding capacity than soils of similar texture with less SOM. Bauer and Black [10] found that a decline in SOM did not change the available water-holding capacity of moderately coarse-textured soils but increased the available water-holding capacity in medium and fine textured soils because of an increase in soil bulk density (BD). Bruce et al. [11] determined that increased phytomass input to a loamy sand increased aggregate stability and water infiltration. On long-term tillage, residue management, and N-fertility plots Pikul and Zuzel [12] reported that an increase in SOM increased the porosity of surface crusts in a silt loam. In contrast, Mulla et al. [13] were not able to establish a relation between SOM and physical behavior of a Naff silt loam, although surface crusting was observed on the "conventional" but not on the "alternative" farm. The “alternative” farm studied by Mulla et al. [13] used a cropping system that was more diverse than the “conventional” farm. However, tillage was used on both farms. Soane [14] reported that soil compaction was sensitive to small changes in SOM contents and generally decreased with increasing SOM. Adams [15] and Hudson [9] found that a 2% decrease in SOM increased BD by 0.1 Mg m^{-3} or more. Maintenance of SOM seems to be a key to sustaining soil resources and crop productivity [16].

Water stability of soil aggregates depends on the quality of organic materials [17]. Degens [18] provides a review of the function of labile organic bonding and binding agents related to soil aggregation. Soil organic carbon (SOC) accounted for 70 to 90% of the variability in soil aggregate stability of a clay loam soil [19]. Wright and Upadhyaya [20] evaluated thirty-seven soils from four geographic areas in the U. S. for the presence of glomalin, a glycoprotein exudate from arbuscular mycorrhizal fungi. They found a positive correlation of soil aggregate stability with glomalin content. Gale et al. [21] found evidence that in no till, aggregate formation is directly related to root-residue decomposition and POM C dynamics. Johnson et al. [22] found that crude humic acid and aggregate stability of a Langhei clay loam could be increased with the addition of the by-product of corn stover fermentation, which is about 70% lignin. These reports show the dynamic nature of the biological and chemical interactions that likely interplay in aggregate formation. Resolution of this issue will require characterizing field soil physically, chemically and biologically in relation to soil aggregation.

There is poor understanding of the effect of soil and crop management on the composition of SOM in agricultural soils. However, recently there have been important research contributions showing, even in the short-term, that the process of humification under different agricultural systems results in unique chemical constituents of humic materials. Ding et al. [23] investigated the chemical composition of SOM in a Norfolk loamy sand following 20 years of tillage. They showed that the composition of humic acid under conventional tillage (multiple passes with disk, cultivate, and in-row subsoiling) was less aliphatic and more aromatic than humic acid developed under conservation tillage (in-row subsoiling). Bird et al. [24] showed that on a Willows clay, the humic acid and SOM light fraction represent the primary active sink and source of sequestered N, affecting both short-term and long-term soil fertility. On upland, arable calcari-chromic luvisols in Estonia, Reintam et al. [25] found a decrease in the content of C and N and an increase in oxygen content of humic acids developed during three decades of agricultural management. Stearman et al. [26] found that on a Loring silt loam, humic acid under no tillage (treatments with larger amounts of C) had greater aliphatic to aromatic ratios and suggested that this characteristic might be due to earlier stages of decomposition.

Our objectives were to determine the effect of tillage on components of SOM and the stability of soil aggregates in a silty clay loam.

16.2. MATERIALS AND METHODS

16.2.1. Field Site and Sampling

Our study was conducted on the boundary between two farms. Along the boundary, soil forming factors such as weather, slope, slope position, and parent material were the same. One farm chiseled and disked fields each fall (CT), and the other farm used no tillage (NT). On the NT farm, primary tillage was last used in 1992. Under CT, corn (*Zea mays* L.) and soybean {*Glycine max* (L.) Merr.} were row cultivated each year. The farms were located about 24 km SE of Brookings, South Dakota. Crop rotation on each farm was corn-soybean and both farms were in the same crop phase of the rotation each year.

The soil is a Vienna-Brookings silty clay loam complex (Fine-silty, mixed, superactive, frigid Aquic Hapludoll and fine-loamy, mixed superactive, frigid Calcic Hapludoll) on a 0.6% slope. On each farm, four replications were established at two slope positions (near-summit and near-toe-slope positions). Slope positions were about 480 m apart.

To measure soil-profile properties, triplicate soil cores, 32 mm in diameter, were taken from each farm and each replicate at 80 mm increments from the top 240 mm in fall 2001. Soil BD, C, C:N, pH, electrical conductivity (EC), and soil texture were measured for these samples. Soil pH and EC were measured using a 1:1 soil to water mixture. Volumetric soil carbon was calculated as the product of SOC concentration, bulk density, and sample increment.

For determining properties of soil aggregates, about 10 kg of soil was collected from the surface 50 mm in spring and fall 2002. Both farms were in soybean. Soil was collected using a flat scoop from both crop-row and between-row positions in spring 2002 and from two slope positions in fall 2002. After air drying, soil aggregates were separated into six size groups using a rotary sieve [27]. Group 1 was soil <0.4 mm, group 2 was 0.4-0.8 mm, group 3 was 0.8-2 mm, group 4 was 2-6 mm, group 5 was 6-19 mm, and group 6 was >19 mm. Aggregates obtained using the rotary sieve were further processed to measure dry aggregate stability, water stability, SOC, SOM, POM, HA, glomalin and BF.

16.2.2. Measurements

We measured the water stability of dry aggregates to evaluate the treatment effect on soil slaking. Water stability was measured using the sieving procedure described by Kemper and Rosenau [28]. Tests were conducted on dry aggregates from sieve groups 2, 3, 4 and 5; tests on premoistened aggregates were conducted on aggregates from sieve groups 3, 4 and 5. Duplicate measurements were made on all aggregates. Prewetting was accomplished using a humidifier [28]. Soil aggregate stability calculations were corrected for the mass of sand remaining on sieves.

Total C and N were measured by combustion using a LECO† CN 2000 analyzer (Leco Corp., St Joseph, MI). Measurements were made on both soil and HA from aggregate groups 1-6. All samples were ground and passed through a 0.5 mm sieve prior to analysis. Visible pieces of crop residue were removed prior to grinding.

Particulate organic matter was measured by dispersing and sieving using a modification of the method provided by Cambardella and Elliott [29]. Soil in aggregate groups 4, 5 and 6 was crushed and passed through a 2 mm sieve. Soil in all aggregate groups was dispersed in sodium hexametaphosphate for 24 hours, stirred with a malt mixer for five minutes, and transferred to a set of nested sieves having mesh sizes of 0.5 and 0.053 mm. Sieves were rinsed until all material smaller than the mesh size had been washed through. Material on each sieve was transferred to aluminum weigh pans, and the mass of organic material determined by loss on ignition (450°C for 4 hours). Particulate organic matter was expressed as percent of soil organic matter (SOM) in each aggregate group, where SOM was measured by loss on ignition (450°C for 4 hours).

Soil-aggregating basidiomycetes were quantified using an ELISA protocol as described in Caesar-TonThat, et al. [30] Absorbance ($Abs_{450/655}$) was read at dual wavelengths of 450 nm/655 nm using a BioRad 550 microplate reader controlled by a computer with the Plate Reader Manage program (BioRad, Hercules, CA). All incubation steps were performed at room temperature on aggregate material. All samples were processed in triplicate. Two independent tests were made on each aggregate sample. Measurements were made on "whole aggregates" prior to wet sieving (BFpre) and measurements also were made on the soil retained on the sieve following the wet sieve test (BFpost).

Easily extractable glomalin, total glomalin (TG), and immunoreactive glomalin (IRTG) fractions were extracted as described in Wright et al. [31] except that a 1 g of sample was extracted using 8 mL of extraction solution. Bradford and immunoreactive protein assays were performed as described by Wright and Upadhyaya [20] and Wright et al. [32].

Soil was fractionated into humin, HA and fulvic acid (FA) fractions according to Stevenson [33]. Thirty g of soil was extracted for HA isolation. Soils were treated with 0.05 M HCl and washed with reverse-osmosis H_2O to remove carbonates. After washing with HCl, the HA and FA were separated from the humin and inorganic fraction using repeated extractions with 0.5 M NaOH under N_2. The humin fraction is the portion of humus that remains bound to the mineral soil after extraction with dilute alkaline solution [33]. Humin and HA fractions were freeze-dried, and dried fractions were ground and analyzed for total C and N and inorganic C.

Analysis of variance was used to determine differences among soil properties and between aggregate size groups. Linear regression and multiple linear

---

† Mention of trade names is for the benefit of the reader and does not constitute endorsement by the USDA over other products not mentioned.

regressions (best sub-sets) were used to identify those soil properties most important to the prediction of water stability.

16.3. RESULTS AND DISCUSSION

16.3.1. General Site Characteristics

There were no differences in soil texture between farms. The soil was a clay loam, on a 0.6 % slope, having 340 g kg^{-1} clay and 260 g kg^{-1} sand (data not shown). Under NT, bulk density for soil depths of 0-80, 80-160, and 160-240 mm was significantly greater at each depth compared with CT by 9 to 12% (Table 16.1). Soil C distribution was stratified with depth as would be expected according to the tillage method.

Table 16.1 Analysis of variance of soil properties at depths of 0-80, 80-160 and 160-240 mm under no tillage (NT) and chisel tillage (CT) treatments. Soil cores collected after soybean planting in spring 2002

Soil property and tillage system		Soil depth (mm) 0-80	80-160	160-240
Bulk density			g cm^{-3}	
NT		1.18	1.39	1.46
CT		1.05	1.24	1.34
	p-value	0.028	0.002	0.002
Organic carbon			g kg^{-1}	
NT		31.0	23.7	22.9
CT		29.4	26.2	23.3
	p-value	0.05	0.005	ns[a]
Carbon to nitrogen ratio				
NT		10.8	10.5	10.5
CT		10.5	10.2	10.2
p-value		0.028	ns	ns
Soil pH				
NT		6.6	6.4	6.8
CT		6.3	6.2	6.6
p-value		ns	ns	ns
Electrical conductivity			dS m^{-1}	
NT		0.396	0.295	0.308
CT		0.360	0.306	0.316
p-value		ns	ns	ns

[a] ns = not significant

Soil organic carbon profiles (Table 16.1) show that there was a greater concentration of C and a wider C:N in the top 80 mm under NT compared with CT. At the 80 to 160 mm depth, SOC was higher under CT compared with NT. These profiles probably reflect differences in residue placement between tillage treatments. In the case of NT, all residues remained on the surface. Under CT, enriched SOC at the 80 to 160 mm depth, when compared to NT, suggests residue incorporation [34] at this depth. In the top 240 mm, volumetric SOC under NT was 8% greater ($p = 0.05$, data not shown) than under CT. The C:N ratio of the NT was 3% higher in the surface 80 mm compared to the CT, likely reflecting the accumulation of residue at the surface. There were no differences in soil pH or EC between treatments (Table 16.1).

16.3.2. Soil Aggregate Stability

Dry Aggregate Distribution and Stability. The erodible fraction (EF) is defined as the percentage of soil mass with aggregates < 0.84 mm diameter, and this parameter has been related to soil wind erodibility. Merrill et al. [35] have shown that EF was more sensitive to soil management effects than indices describing aggregate size distribution (such as MWD). We found differences in the distribution of dry aggregates between tillage treatments and sampling date. Further, there was a significant interaction between tillage and sample time (Two-way analysis of variance shown in the lower portion of Table 16.2). However, inspection of one-way analysis of variance for time of sampling (Table 16.2, upper portion) shows that, under NT, the distribution of aggregates did not change with sampling time. The significant interaction of tillage and time is largely a consequence of CT.

Under CT, we found a significant increase in the mass of aggregates in size groups 1 through 4 collected in the fall 2002 compared to soil collected in spring 2002 (Table 16.2, one-way ANOVA). In contrast, NT resulted in a dramatic shift in the distribution of aggregate size class such that there was a greater mass of soil in class 6 compared with the smallest aggregate sizes (Table 16.2, two-way analysis of variance). We speculate that the change in aggregate distribution (greater mass of large aggregates under NT) might be explained by the accumulation of root biomass and root exudates under NT consistent with the findings of Gale et al. [21,36]. New aggregation under CT would be disrupted by annual fall tillage.

Soil in groups 1 and 2 are susceptible to wind erosion [35] and both farms were in soybean during 2002. In northern, sub-humid regions of the Great Plains, wind and water erosion are persistent problems. Potential soil losses may be less than other erosion-prone hot spots. For example, in the United States, the southern Great Plains are the soil wind erosion hot spots [37]. However, regardless of the quantity of soil eroded, the consequences are the same. Erosion removes the best quality soil first. Fine soil particles lost to erosion are the richest in organic matter and nutrients. Soil conservation practices that improve soil aggregate stability also help to retard soil loss by maintaining surface conditions resistant to weather vagaries.

There were no differences in DASD between samples collected on row or between row in spring 2002 (data not shown). Further, there were no significant differences in DASD between samples collected up or down slope on respective farms (data not shown).

Multiple passes of aggregates through a rotary sieve provides a sensitive way to test for dry aggregate stability [27]. With a second sieving, aggregates under CT had a greater tendency to abrade into small aggregates (groups 1 and 2) when compared with NT (Table 16.3, change in mass following second sieving). The increase in mass within groups 1 and 2 indicates that aggregates from CT degraded into particle sizes that are susceptible to wind erosion.

Table 16.2 Dry aggregate size distribution (DASD) under no tillage (NT) and chisel tillage (CT) treatments. Dry aggregate size distributions were measured on soil collected from the top 50 mm after soybean planting (spring) and before soybean harvest (fall) in 2002. Group 1 was soil <0.4 mm, group 2 was 0.4-0.8 mm, group 3 was 0.8-2 mm, group 4 was 2-6 mm, group 5 was 6-19 mm, and group 6 was >19 mm

	Aggregate group					
	1	2	3	4	5	6
No tillage	DASD (g kg^{-1})					
spring	27	25	68	169	272	434
fall	24	23	61	145	205	540
p-value	ns[a]	ns	ns	ns	0.006	0.020
Chisel tillage						
spring	106	73	104	132	232	349
fall	152	102	134	160	244	205
p-value	0.001	0.001	0.001	0.001	ns	0.001
Tillage & time	Average DASD (g kg^{-1})					
NT	26	24	65	157	238	487
CT	129	88	119	146	238	277
Spring	67	49	87	150	252	391
Fall	89	62	98	152	224	373
p-value tillage	0.001	0.001	0.001	ns	ns	0.001
p-value time	0.006	0.001	0.030	ns	0.012	ns
p-value interaction	0.003	0.001	0.001	0.009	0.001	0.001

[a] ns = not significant

Wet Aggregate Stability. Water stability of both dry and pre moistened aggregates was greater under NT compared with CT (Table 16.3), and these results corroborate our dry aggregate stability tests. There were no differences in soil forming factors or soil texture between farms, and consequently we hypothesize that the binding agents responsible for increased stability under NT developed as a consequence of no tillage management over a period of about ten years. These two tests of stability (dry and wet) indirectly provide clues as to the nature of the aggregate binding agents. Kemper and Rosenau [28] described a process of soil drying whereby soluble compounds such as silica, carbonates, and organic molecules concentrate and precipitate at particle-to-particle contacts to cement aggregates together. Differences between treatments are more pronounced (NT had about 50 % greater stability than CT) for tests of water stability using dry aggregates (Table 16.3), and this may indicate a greater concentration of water soluble binding agents present under NT.

Water stability of dry aggregates increased ($p < 0.001$) with aggregate size on both the NT and CT farms (ANOVA among sieve sizes not shown). Tisdall and Oades [17] and Tisdall et al. [38] proposed a hierarchical structure of aggregate formation whereby small aggregates bind to form larger aggregates. Further, the binding agents within small aggregates, termed microaggregates by Tisdall and Oades [17], are different from those that bind microaggregates into macro-aggregates. Data in Table 16.3 are an average for samples collected on and off row. There were no significant differences in water stability (dry or premoistened aggregates) due to row position (data not shown).

Table 16.3 Dry stability and water stability of soil aggregates collected from the top 50mm in spring 2002 from the no tillage (NT) and chisel tillage (CT) farms. Change in mass following second sieving is an indicator of dry aggregate stability (negative value indicates loss of soil from that group). Group 1 was soil <0.4 mm, group 2 was 0.4-0.8 mm, group 3 was 0.8-2 mm, group 4 was 2-6 mm, group 5 was 6-19 mm, and group 6 was >19 mm

	Aggregate group					
	1	2	3	4	5	6
Tillage	Change in mass following second sieving (g kg^{-1})					
NT	9.72	6.09	11.4	17.9	9.70	-54.5
CT	26.6	11.2	7.99	6.00	-6.70	-46.5
p-value	0.001	0.001	0.004	0.001	0.002	0.091
	Water stability[a] of dry aggregates (%)					
NT	nm[b]	20.3	27.8	38.0	55.8	nm
CT	nm	9.8	15.3	15.9	28.0	nm
p-value		0.001	0.001	0.001	0.001	
	Water stability[a] of premoistened aggregates (%)					
NT	nm	nm	87.2	91.0	92.4	nm
CT	nm	nm	81.0	78.7	87.7	nm
p-value			0.01	0.003	ns[c]	

[a] average water stability of soil aggregates collected on row and between row. Mean values of water stability for each treatment and size group represent sixteen measurements (8 plots and duplicate determinations on each plot); [b] not measured; [c] not significant.

16.3.3. Soil Aggregate Properties

Aggregate Carbon and Nitrogen. We evaluated spatial variability within our field sites by sampling on and off row and sampling two slope positions. For samples collected in spring 2002, we found no differences in SOC of aggregates collected on or off row (data not shown). On the CT farm, we found no differences in SOC of aggregates due to slope position in fall 2002. Under NT, we found a significant difference in SOC of aggregates between slope positions in fall 2002 (data not shown). Only results for samples collected from upslope positions of both NT and CT are reported hereafter.

Ten years of NT had a significant ($p \leq 0.01$) effect on SOC in all aggregate groups (Table 16.4). Average SOC of aggregates under NT was 32.3 g kg^{-1} compared with 29.6 g kg^{-1} under CT. Further, there was a difference ($p \leq 0.001$) in SOC among soil aggregate groups on both farms (data not shown). Soil organic carbon was not uniformly distributed among soil aggregate groups; the greatest concentration of SOC was in soil aggregate group 3 under both tillage systems (Table 16.4).

The carbon to nitrogen ratio was significantly greater under NT compared with CT for all aggregate groups (Table 16.4). The average C:N ratio of aggregates was 10.7 under NT and 10.2 under CT. There was a significant difference in C:N among soil aggregate groups and treatments, but no correlation among aggregate size and treatments (two-way analysis of variance not shown). Under NT, the widest C:N was for aggregates in group 3, whereas under CT the widest C:N was for aggregates in group 4 (Table 16.4). There were no differences in SOC of aggregates due to time of sampling (spring or fall) except within aggregate group 4. The effect of sample time is likely due to small variations within each field.

Table 16.4 Carbon and the atomic ratio C:N under no tillage (NT) and chisel tillage (CT) treatments. Measurements were made on soil collected from the top 50 mm after soybean planting (spring) and before soybean harvest (fall) in 2002. Group 1 was soil <0.4 mm, group 2 was 0.4-0.8 mm, group 3 was 0.8-2 mm, group 4 was 2-6 mm, group 5 was 6-19 mm, and group 6 was >19 mm

	Aggregate group						
	1	2	3	4	5	6	Ave.
	Total C (g kg^{-1})						
NT	32.4	32.8	34.4	33.2	31.0	30.1	32.3
CT	29.5	29.7	31.0	30.2	28.8	28.4	29.6
p-value tillage	0.01	0.00	0.007	0.001	0.007	0.01	0.001
p-value time	ns[a]	ns	ns	0.003	ns	ns	ns
p-value aggregate							0.001
	C:N						
NT	10.6	10.7	10.8	10.7	10.6	10.6	10.7
CT	10.2	10.1	10.3	10.4	10.1	10.2	10.2
p-value tillage	0.02	0.00	0.001	0.001	0.004	0.00	0.001
p-value time	ns	0.02	ns	ns	ns	ns	ns
p-value aggregate							0.001

[a] not significant

Particulate organic matter. Soil organic matter is composed of material varying in chemical composition and degree of decomposition Particulate organic matter is physically defined as the organic material isolated in the fraction 0.053 to 2.00 mm. It is an intermediate between fresh plant litter and humified SOM, and has been shown to be more sensitive to changes in management than total SOM [39]. In undisturbed soils, POM is derived primarily from roots [21,36]. New microaggregates are thought to form around decomposing pieces of root-derived POM inside macroaggregates [36].

Average SOM of aggregates was greater ($p < 0.001$, Table 16.5) under NT (78.1 mg g^{-1}) compared with CT (75.4 mg g^{-1}). Although this is a small increase (3%), it is notable because measurable changes in SOM occur slowly. Further, there were differences ($p \leq 0.001$) in SOM among soil aggregate groups (Table 16.5), with soil aggregate group 3 under NT and aggregate group 4 under CT having the greatest concentrations. Measurements of SOM corroborate our SOC results and further show that organic materials are not uniformly distributed across all aggregate sizes. For both SOC and SOM, the greatest concentration was found in mid-sized aggregates. However, this was not the case for the distribution of POM.

Table 16.5 Soil organic matter (SOM), fine particulate organic matter (POM), and total POM under no tillage (NT) and chisel tillage (CT) treatments. Aggregate group 1 was soil <0.4 mm, group 2 was 0.4-0.8 mm, group 3 was 0.8-2 mm, group 4 was 2-6 mm, group 5 was 6-19 mm, and group 6 was >19 mm

	Aggregate group						
	1	2	3	4	5	6	Ave.
	SOM (mg g^{-1})						
NT	76.4	74.6	83.8	82.0	76.9	74.6	78.1
CT	76.6	75.1	78.1	79.3	72.0	71.5	75.4
p-value tillage	ns[a]	ns	0.002	ns	0.039	0.060	0.001
p-value							0.001
	Fine POM (0.053 – 0.5 mm) as percent of SOM						
NT	20.1	13.3	13.4	16.2	15.1	12.3	15.1
CT	16.2	9.0	9.1	11.5	10.5	9.6	11.0
p-value tillage	0.022	0.002	0.001	0.001	0.001	0.053	0.001
p-value							0.001
	Total POM (0.053 – 2.0 mm) as percent of SOM						
NT	23.2	23.0	23.4	20.8	17.3	14.1	20.3
CT	17.8	14.0	17	18.8	12.4	10.6	15.1
p-value tillage	0.015	0.001	0.004	ns	0.003	0.043	0.001
p-value							0.001

[a] not significant

Fine POM (0.053 – 0.5 mm) comprised a greater fraction of total POM than coarse POM (0.5 – 2.0 mm) for both NT and CT (Table 16.5). In Table 16.5, POM is expressed as a percentage of SOM and it is important to note that there were no differences in SOM between NT and CT within aggregate groups 1 and 2 (Table 16.5). However, the greatest concentration of POM was measured within the smaller aggregate groups, and the greatest differences in POM (fine and total) between NT and CT were found in the smaller aggregate groups. Average fine and total POM were significantly ($p \leq 0.001$) greater under NT (Table 16.5), and fine POM was significantly greater under NT compared with CT in all aggregate groups. Although there was more fine POM and total POM under NT compared to CT, the distribution of fine and total POM among the aggregate size groups was similar under both tillage systems (for example, both had the high concentration of POM in the smallest aggregate size group, Table 16.5). Greater concentration of POM in the undisturbed soil of NT might be a consequence of an accumulation of less decomposed root materials as described by Gale et al. [21,36].

C:N of Humic Acid. Humic materials are important soil binding agents, and we analyzed C:N of HA to see whether there were gross differences in the composition of HA between the two tillage systems and among aggregate groups. Generally, it is accepted that the C:N ratio of surface soils under NT will be wider than under tillage and our average C:N of soil aggregate groups (Table 16.4) shows that this is the case. Stevenson [33] has suggested a relation between organic matter decomposition and C:N of HA such that conditions that encourage decomposition (mixing and aeration provided by tillage) result in a narrowing of the C:N of HA. Chefetz et al. [40] found evidence obtained by analysis of HA that "coarse size" aggregates contained freshly deposited organic matter and thus should be expected to have a wider C:N ratio. Coarse aggregates had C:N of HA of 12.3 and fine aggregates had C:N of 10.5. The "coarse aggregates" evaluated by Chefetz et al. [40] roughly correspond to the smallest aggregate group of our study.

The average C:N of HA under NT was significantly different compared with HA under CT. Further, there were significant differences among aggregate groups (Table 16.6) and a trend for the smallest aggregate groups to have the widest C:N ratios (11.2 under NT compared with 11.0 under CT). We think that the difference in C:N ratios among aggregate groups provides evidence to suggest that "less humified" materials are found within the smaller aggregates having the wider C:N ratios.

Table 16.6 Carbon to nitrogen ratio (C:N) of humic acid, glomalin, and basidiomycete assay of soil aggregates under no tillage (NT) and chisel tillage (CT) treatments. Basidiomycete assay was conducted on material retained on sieve following wet sieving (BFpost). Aggregate group 1 was soil <0.4 mm, group 2 was 0.4-0.8 mm, group 3 was 0.8-2 mm, group 4 was 2-6 mm, group 5 was 6-19 mm, and group 6 was >19 mm

	Aggregate group						
	1	2	3	4	5	6	Ave.
			C:N of humic acid				
NT	11.2	11.0	10.4	10.4	10.5		10.7
CT	11.0	10.9	10.2	10.3	10.3		10.5
p-value tillage	ns	ns	ns	ns	0.006		0.001
p-value aggregate							0.001
			Immunoreactive glomalin (mg g^{-1})				
NT	0.52	0.55	0.58	0.57	0.54	0.58	0.56
CT	0.44	0.48	0.46	0.48	0.46	0.44	0.46
p-value tillage	ns	ns	ns	ns	ns	ns	0.003
p-value aggregate							ns
			Basidiomycete assay				
NT			0.757	0.519	0.474		0.583
CT			0.532	0.506	0.429		0.489
p-value tillage			0.018	ns	ns		0.003
p-value aggregate							0.001

Soil Fungi. Our study investigated the effect of two common soil fungal groups on soil aggregation. Basidiomycete fungi occupy aerobic sites and are responsible for the degradation of cellulose and lignin in non-living organic matter. Polysaccharide exudates from this group of fungi are important in soil aggregation, but are short lived as binding agents. Arbuscular mycorrhizal (AM) fungi are termed endomycorrhiza and they live in symbiotic association with plants. This fungal group produces an exudate called glomalin, which is a stable, iron-containing glycoprotein [32] important in soil aggregation [31]. Further, Nichols [41] has shown that POM also contains substantial amounts of glomalin, and this finding may help to explain why fine POM is strongly associated with soil aggregation.

The average concentration of IRTG was significantly greater under NT compared with CT, but differences could not be detected when comparing between individual groups (Table 16.6). The concentration of IRTG was the same among aggregate groups for both NT and CT (Table 16.6). The immunoreactive component of glomalin in soil is considered to be recently deposited and has been shown to be highly correlated with exudates of AM fungal hyphae [42].

The average number of soil-aggregating basidiomycetes (BFpost) in soil left following wet sieving was significantly greater under NT compared with CT. Further, there were significant differences among aggregate groups (Table 16.6) and a trend for the smallest aggregate groups to have the greatest absorbance value (absorbance for BFpre not shown). Polysaccharide exudates from basidiomycete fungi are important in soil aggregation. We do not have independent measurements of polysaccharide exudates; however, we assume that a high absorbance value reflects a high concentration of polysaccharide material from BFpost.

Regression Modeling. Linear regression indicated water stability of dry aggregates was correlated with fine POM ($p = 0.003$), coarse POM ($p = 0.032$), C:N of HA ($p = 0.003$) and BFpre ($p = 0.024$), but not with other soil parameters SOM, SOC, total N, soil C:N, concentration of HA, BFpost, IRTG, or IRTG:TG (using a minimum of $p = 0.1$).

Use of multiple regression modeling allows identification of the best predictors from a given dataset. This present data set is unique in that it has characterized SOM biologically (IRTG, BFpre, BFpost), chemically (HA and C:N of HA) and physically (POM and fine POM). Biological constituents represent the role of fungi, POM primarily is intermediate decomposed plant material and HA is a component of stable SOM. A narrow C:N ratio of HA may indicate more advanced decomposition. The best single-component predictor of aggregate stability was fine POM ($r^2 = 0.34$). The best two-component model included fine POM and BFpre ($r^2 = 0.55$). The best three-component model included fine POM, C:N of HA, and BFpre. This model accounted for 63% ($p < 0.001$) of the variability in water stability of dry aggregates (Figure 16.1). There was little improvement in prediction of aggregate stability from including additional parameters.

Our findings support the hypothesis that stable aggregate formation is of a dynamic nature that reflects biological, chemical and physical interactions. A conceptual model by Six et al. [43] of an aggregate 'life cycle' proposed that fine intra-aggregate POM is formed as it becomes encrusted with clay particles and microbial products, and forms within macroaggregates. As macroaggregates degrade, stable microaggregates become the nucleus for the formation of new macroaggregates. Such models are helpful but need to be expanded to include root exudates, microbial products, various POM fractions and humified material. Furthermore, the function of physical, biological or chemical binding agents within different aggregate size groups needs to be defined.

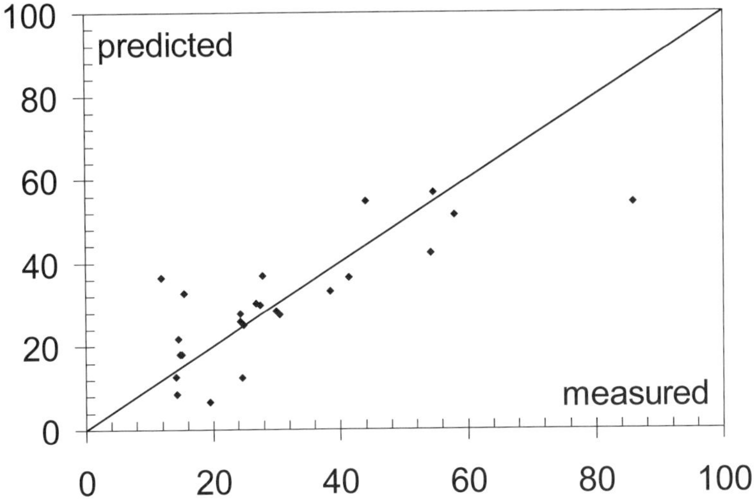

Figure 16.1 Relation of measured water stability and predicted water stability of soil aggregates (data pairs) and a 1:1 line. Predicted values of water stability (WS) were based on predictors of fine POM (FPOM), C:N of humic acid (C:N HA), and basidiomycete assay before sieving (BFpre), providing the following relation: WS = - 388 + 2.69 FPOM + 35.4 C:N HA + 61.4 BFpre

16.4. CONCLUSIONS

Wind and water erosion are persistent problems in northern sub-humid regions of the Great Plains. Soil conservation practices that improve soil aggregate stability also help to retard soil loss by maintaining surface conditions resistant to weather vagaries. Long term field experiments on comparable soil located in close proximity provide a unique and valuable opportunity to compare divergent management strategies. The adjacent farm fields used in this study were under two different tillage management strategies for ten years and were in the same rotation phase. We think that improved soil aggregation and increased surface cover of NT compared to CT will help to keep top soil in place. The multifaceted approach to characterizing SOM as it relates to soil aggregate formation helped to identify how different components of SOM interact to improve soil aggregation. Differences in properties among aggregates show that organic cementing agents (humic materials or microbial exudates) are not uniformly distributed among aggregate groups. Our results show improved soil aggregation as a consequence of no tillage farming when compared to tillage.

ACKNOWLEDGEMENTS

We wish to thank David Harris, Melissa Bossert and Nancy Barbour for skillful analytical work and technical support. We are indebted to South Dakota farmers Dave Diedrich, Larry Diedrich and Larry Thielen for encouragement and the resources of their farms.

REFERENCES

1. Campbell CA, Souster W. Loss of organic matter and potentially mineralizable nitrogen from Saskatchewan soils due to cropping. *Can. J. Soil Sci.*, 1982; 62: 651-656.
2. Monreal CM, Janzen HH. Soil organic-carbon dynamics after 80 years of cropping a Dark Brown Chernozem. *Can. J. Soil Sci.*, 1993; 73:133-136.
3. Allmaras RR, Schomberg HH, Douglas, CL, Dao, TH. Soil organic carbon sequestration potential of adopting conservation tillage in US croplands. *J. Soil Water Cons.*, 2000; 55: 365-373.
4. Paustian K, Collins HP, Paul EA. Management controls on soil carbon, In Paul EA, Paustian K, Elliott ET, Cole, CV. eds. *Soil Organic matter in Temperate Agroecosystems: Long-Term Experiments in North America*. Boca Raton, FL: CRC Press, 1997: 15-49
5. Rasmussen PE, Parton WJ. Long-term effects of residue management in wheat-fallow: I. Inputs, yield, and soil organic matter. *Soil Sci. Soc. Am. J.*, 1994; 58: 523-530.
6. Biederbeck VO, Campbell CA, Zentner RP. Effect of crop rotation and fertilization on some biological properties of a loam in southwestern Saskatchewan. *Can. J. Soil. Sci.*, 1984; 64: 355-367.
7. Carter MR. Soil quality for sustainable land management: Organic matter and aggregation interactions that maintain soil functions. *Agron J.*, 2002; 94: 38-47.
8. Boyle M, Frankenberger Jr. WT, Stolzy LH. The influence of organic matter on soil aggregation and water infiltration. *J. Prod. Agric.*, 1989; 2: 290-299.
9. Hudson BD. Soil organic matter and available water capacity. *J. Soil Water Conserv.*, 1994; 49:189-194.
10. Bauer A, Black AL. Organic carbon concentration effects on available water capacity of three soil textural groups. *Soil Sci. Soc. Am. J.*, 1992; 56: 248-254.
11. Bruce RR, Langdale GW, West LT, Miller WP. Soil surface modification by biomass inputs affecting rainfall infiltration. *Soil Sci. Soc. Am. J.*, 1992; 56: 1614-1620.
12. Pikul Jr. JL, Zuzel JF. Soil crusting and water infiltration affected by long-term tillage and residue management. *Soil Sci. Soc. Am. J.*, 1994; 58: 1524-1530.
13. Mulla DJ, Huyck LM, Reganold JP. Temporal variation in aggregate stability on conventional and alternative farms. *Soil Sci. Soc. Am. J.*, 1992; 56: 1620-1624.
14. Soane BD. The role of organic matter in soil compactibility: A review of some practical aspects. *Soil Tillage Res.*, 1990; 16: 179-201.
15. Adams WA. The effect of organic matter on bulk and true densities of some uncultivated podzolic soils. *J Soil Sci.,* 1973; 24: 10-17.
16. Doran JW, Elliott ET, Paustian K. Soil microbial activity, nitrogen cycling, and long-term changes in organic carbon pools as related to fallow tillage management. *Soil Till. Res.*, 1998; 49: 3-18.
17. Tisdall JM, Oades JM. Organic matter and water stable aggregates. *J. Soil Science*, 1982; 33:141-163.
18. Degens BP. Macroaggregation of soils by biological bonding and binding mechanisms and the factors affecting these: a review. *Aust. J. Soil Res.*, 1997; 35: 431-459.
19. Mbagwu JSC, Bazzoffi P. Properties of soil aggregates as influenced by tillage practices. *Soil Use Management*, 1989; 5: 180-188.

20. Wright SF, Upadhyaya A. A survey of soils for aggregate stability and glomalin, a glycoprotein produced by hyphae of arbuscular mycorrhizal fungi. *Plant Soil*, 1998; 198: 97-107.
21. Gale WJ, Cambardella CA, Bailey TB. Surface residue- and root-derived carbon in stable and unstable aggregates. *Soil Sci. Soc. Am. J.*, 2000; 64: 196-201.
22. Johnson JMF, Reicosky D, Sharratt B, Lindstrom M, Voorhees W, Carpenter-Boggs L. Characterization of soil amended with the by-product of corn stover fermentation. *Soil Sci Soc Am J.*, 2004; 68: 139-147.
23. Ding G, Novak M, Amarasiriwardena D, Hunt PG, Xing B. Soil organic matter characteristics as affected by tillage management. *Soil Sci. Soc. Am. J.*, 2002; 66: 421-429.
24. Bird JA, van Kessel C, Horwath WR. Nitrogen dynamics in humic fractions under alternative straw management in temperate rice. *Soil Sci. Soc. Am. J.*, 2002; 66: 478-488.
25. Reintam L, Kann J, Kailas T, Kährik R. Elemental composition of humic and fulvic acids in the epipedon of some Estonian soils. *Proc. Estonian Acad. Sci. Chem.*, 2000; 49: 131-144.
26. Stearman GK, Lewis RJ, Tortorelli LJ, Tyler DD. Characterization of humic acid from no-tilled and tilled soils using C-13 nuclear magnetic resonance. *Soil Sci. Soc. Am. J.*, 1989; 53: 744-749.
27. Chepil WS. A compact rotary sieve and the importance of dry sieving in physical soil analysis. *Soil Sci. Soc. Am. Proc.*, 1962; 26: 4-6.
28. Kemper WD, Rosenau RC. Aggregate stability and size distribution. In: *Methods of Soil Analysis, Part 1. Physical and Mineralogical Methods*. Madison, WI: Agronomy Monograph No. 9, 2nd ed. ASA, 1986: 425-444.
29. Cambardella CA, Elliott ET. Particulate soil organic-matter changes across a grassland cultivation sequence. *Soil Sci. Soc. Am. J.*, 1992; 56: 777-783.
30. Caesar-TonThat TC, Shelver WL, Thorn RG, Cochran VL. Generation of antibodies for soil aggregating basidiomycete detection as an early indicator of trends in soil quality. *Applied Soil Ecology*, 2001; 18: 99-116.
31. Wright SF, Starr JL, Paltineau IC. Changes in aggregate stability and concentration of glomalin during tillage management transition. *Soil Sci. Soc. Am. J.*, 1999; 63: 1825-1829.
32. Wright SF, Upadhyaya A. Extraction of an abundant and unusual protein from soil and comparison with hyphal protein of arbuscular mycorrhizal fungi. *Soil Sci.*, 1996; 161: 575-586.
33. Stevenson FJ. *Humic chemistry: Genesis, composition, reactions*. 2nd Edn. New York: Wiley, 1994.
34. Allmaras RR, Pikul Jr. JL, Kraft JM, Wilkins DE. A method for measuring incorporated crop residue and associated soil properties. *Soil Sci. Soc. Am. J.*, 1988; 52: 1128-1133.
35. Merrill SD, Black AL, Fryrear DW, Saleh A, Zobeck TM, Halvorson AD, Tanaka DL. Soil wind erosion hazard of spring wheat-fallow as affected by long-term climate and tillage. *Soil Sci. Soc. Am. J.*, 1999; 63: 1768-1777.
36. Gale WJ, Cambardella CA, Bailey TB. Root-derived carbon and the formation and stabilization of aggregates. *Soil Sci. Soc. Am. J.*, 2000; 64: 201-207.
37. Warren A. Erosion by wind, global hot spots of. In: Lal R ed. *Encyclopedia of Soil Science*. New York, NY: Dekker, 2002: 508-511.
38. Tisdall JM, Cockroft B, Uren NC. The stability of soil aggregates as affected

by organic materials, microbial activity and physical disruption. *Aust. J. Soil Res.*, 1978; 16: 9-17.

39. Cambardella CA, Gajda AM, Doran JW, Wienhold BJ, Kettler TA. Estimation of particulate and total organic matter by weight loss-on-ignition, In: Lal R, Kimball JM, Follet RF, Stewart BA eds. *Assessment methods for soil carbon.* Boca Raton, FL: Lewis Publishers, 2001: 349-359.
40. Chefetz B, Tarchitzky J, Deshmukh AP, Hatcher PG, Chen Y. Structural characterization of soil organic matter and humic acids in particle-size fractions of an agricultural soil. *Soil Sci. Soc. Am. J.*, 2002; 66: 129-141.
41. Nichols K. Characterization of glomalin: A glycoprotein produced by arbuscular mycorrhizal fungi. *Ph.D. Dissertation*, 2003. University of Maryland, College Park, Maryland.
42. Wright SF. A fluorescent antibody assay for hyphae and glomalin from arbuscular mycorrhizal fungi. *Plant Soil*, 2000; 226: 171-177.
43. Six J, Elliot ET, Paustian K. Soil macroaggregate turnover and microaggregate formation: A mechanism for C sequestration under no-tillage management. *Soil Bio. Biochem.*, 2000; 32: 2099-2103.

INDEX

D

N

O